Ensemble Machine Learning

Cha Zhang • Yunqian Ma
Editors

Ensemble Machine Learning

Methods and Applications

Editors
Cha Zhang
Microsoft
One Microsoft Road
98052 Redmond
USA

Yunqian Ma
Honeywell
Douglas Drive North 1985
55422 Golden Valley
USA

ISBN 978-1-4899-8817-1 ISBN 978-1-4419-9326-7 (eBook)
DOI 10.1007/978-1-4419-9326-7
Springer New York Dordrecht Heidelberg London

© Springer Science+Business Media, LLC 2012
Softcover reprint of the hardcover 1st edition 2012
All rights reserved. This work may not be translated or copied in whole or in part without the written permission of the publisher (Springer Science+Business Media, LLC, 233 Spring Street, New York, NY 10013, USA), except for brief excerpts in connection with reviews or scholarly analysis. Use in connection with any form of information storage and retrieval, electronic adaptation, computer software, or by similar or dissimilar methodology now known or hereafter developed is forbidden.
The use in this publication of trade names, trademarks, service marks, and similar terms, even if they are not identified as such, is not to be taken as an expression of opinion as to whether or not they are subject to proprietary rights.

Printed on acid-free paper

Springer is part of Springer Science+Business Media (www.springer.com)

Preface

Making decisions based on the input of multiple people or experts has been a common practice in human civilization and serves as the foundation of a democratic society. Over the past few decades, researchers in the computational intelligence and machine learning community have studied schemes that share such a joint decision procedure. These schemes are generally referred to as ensemble learning, which is known to reduce the classifiers' variance and improve the decision system's robustness and accuracy.

However, it was not until recently that researchers were able to fully unleash the power and potential of ensemble learning with new algorithms such as boosting and random forest. Today, ensemble learning has many real-world applications, including object detection and tracking, scene segmentation and analysis, image recognition, information retrieval, bioinformatics, data mining, etc. To give a concrete example, most modern digital cameras are equipped with face detection technology. While the human neural system has evolved for millions of years to recognize human faces efficiently and accurately, detecting faces by computers has long been one of the most challenging problems in computer vision. The problem was largely solved by Viola and Jones, who developed a high-performance face detector based on boosting (more details in Chap. 8). Another example is the random forest-based skeleton tracking algorithm adopted in the Xbox Kinect sensor, which allows people to interact with games freely without game controllers.

Despite the great success of ensemble learning methods recently, we found very few books that were dedicated to this topic, and even fewer that provided insights about how such methods shall be applied in real-world applications. The primary goal of this book is to fill the existing gap in the literature and comprehensively cover the state-of-the-art ensemble learning methods, and provide a set of applications that demonstrate the various usages of ensemble learning methods in the real world. Since ensemble learning is still a research area with rapid developments, we invited well-known experts in the field to make contributions. In particular, this book contains chapters contributed by researchers in both academia and leading industrial research labs. It shall serve the needs of different readers at different levels. For readers who are new to the subject, the book provides an excellent entry point with a high-level introductory view of the topic as well as an in-depth discussion of the key technical details. For researchers in the same area, the book is a handy reference summarizing the up-to-date advances in ensemble learning, their connections, and future directions. For practitioners, the book provides a number of applications for ensemble learning and offers examples of successful, real-world systems.

This book consists of two parts. The first part, from Chaps. 1 to 7, focuses more on the theory aspect of ensemble learning. The second part, from Chaps. 8 to 11, presents a few applications for ensemble learning.

Chapter 1, as an introduction for this book, provides an overview of various methods in ensemble learning. A review of the well-known boosting algorithm is given in Chap. 2. In Chap. 3, the boosting approach is applied for density estimation, regression, and classification, all of which use kernel estimators as weak learners. Chapter 4 describes a "targeted learning" scheme for the estimation of nonpathwise differentiable parameters and considers a loss-based super learner that uses the cross-validated empirical mean of the estimated loss as estimator of risk. Random forest is discussed in detail in Chap. 5. Chapter 6 presents negative correlation-based ensemble learning for improving diversity, which introduces the negatively correlated ensemble learning algorithm and explains that regularization is an important factor to address the overfitting problem for noisy data. Chapter 7 describes a family of algorithms based on mixtures of Nystrom approximations called Ensemble Nystrom algorithms, which yields more accurate low rank approximations than the standard Nystrom method. Ensemble learning applications are presented from Chaps. 8 to 11. Chapter 8 explains how the boosting algorithm can be applied in object detection tasks, where positive examples are rare and the detection speed is critical. Chapter 9 presents various ensemble learning techniques that have been applied to the problem of human activity recognition. Boosting algorithms for medical applications, especially medical image analysis are described in Chap. 10, and random forest for bioinformatics applications is demonstrated in Chap. 11. Overall, this book is intended to provide a solid theoretical background and practical guide of ensemble learning to students and practitioners.

We would like to sincerely thank all the contributors of this book for presenting their research in an easily accessible manner, and for putting such discussion into a historical context. We would like to thank Brett Kurzman of Springer for his strong support to this book.

Redmond, WA — Cha Zhang
Golden Valley, MN — Yunqian Ma

Contents

Chapter 1
Ensemble Learning

Robi Polikar

1.1 Introduction

Over the last couple of decades, multiple classifier systems, also called ensemble systems have enjoyed growing attention within the computational intelligence and machine learning community. This attention has been well deserved, as ensemble systems have proven themselves to be very effective and extremely versatile in a broad spectrum of problem domains and real-world applications. Originally developed to reduce the variance—thereby improving the accuracy—of an automated decision-making system, ensemble systems have since been successfully used to address a variety of machine learning problems, such as feature selection, confidence estimation, missing feature, incremental learning, error correction, class-imbalanced data, learning concept drift from nonstationary distributions, among others. This chapter provides an overview of ensemble systems, their properties, and how they can be applied to such a wide spectrum of applications.

Truth be told, machine learning and computational intelligence researchers have been rather late in discovering the ensemble-based systems, and the benefits offered by such systems in decision making. While there is now a significant body of knowledge and literature on ensemble systems as a result of a couple of decades of intensive research, ensemble-based decision making has in fact been around and part of our daily lives perhaps as long as the civilized communities existed. You see, ensemble-based decision making is nothing new to us; as humans, we use such systems in our daily lives so often that it is perhaps second nature to us. Examples are many: the essence of democracy where a group of people vote to make a decision, whether to choose an elected official or to decide on a new law, is in fact based on ensemble-based decision making. The judicial system in many countries, whether based on a jury of peers or a panel of judges, is also based on

R. Polikar (✉)
Rowan University, Glassboro, NJ 08028, USA
e-mail: polikar@rowan.edu

C. Zhang and Y. Ma (eds.), *Ensemble Machine Learning: Methods and Applications*,
DOI 10.1007/978-1-4419-9326-7_1, © Springer Science+Business Media, LLC 2012

ensemble-based decision making. Perhaps more practically, whenever we are faced with making a decision that has some important consequence, we often seek the opinions of different "experts" to help us make that decision; consulting with several doctors before agreeing to a major medical operation, reading user reviews before purchasing an item, calling references before hiring a potential job applicant, even peer review of this article prior to publication, are all examples of ensemble-based decision making. In the context of this discussion, we will loosely use the terms expert, classifier, hypothesis, and decision interchangeably.

While the original goal for using ensemble systems is in fact similar to the reason we use such mechanisms in our daily lives—that is, to improve our confidence that we are making the right decision, by weighing various opinions, and combining them through some thought process to reach a final decision—there are many other machine-learning specific applications of ensemble systems. These include confidence estimation, feature selection, addressing missing features, incremental learning from sequential data, data fusion of heterogeneous data types, learning non-stationary environments, and addressing imbalanced data problems, among others.

In this chapter, we first provide a background on ensemble systems, including statistical and computational reasons for using them. Next, we discuss the three pillars of the ensemble systems: diversity, training ensemble members, and combining ensemble members. After an overview of commonly used ensemble-based algorithms, we then look at various aforementioned applications of ensemble systems as we try to answer the question "what else can ensemble systems do for you?"

1.1.1 Statistical and Computational Justifications for Ensemble Systems

The premise of using ensemble-based decision systems in our daily lives is fundamentally not different from their use in computational intelligence. We consult with others before making a decision often because of the variability in the past record and accuracy of any of the individual decision makers. If in fact there were such an expert, or perhaps an oracle, whose predictions were always true, we would never need any other decision maker, and there would never be a need for ensemble-based systems. Alas, no such oracle exists; every decision maker has an imperfect past record. In other words, the accuracy of each decision maker's decision has a nonzero variability. Now, note that any classification error is composed of two components that we can control: bias, the accuracy of the classifier; and variance, the precision of the classifier when trained on different training sets. Often, these two components have a trade-off relationship: classifiers with low bias tend to have high variance and vice versa. On the other hand, we also know that averaging has a smoothing (variance-reducing) effect. Hence, the goal of ensemble systems is to create several classifiers with relatively fixed (or similar) bias and then combining their outputs, say by averaging, to reduce the variance.

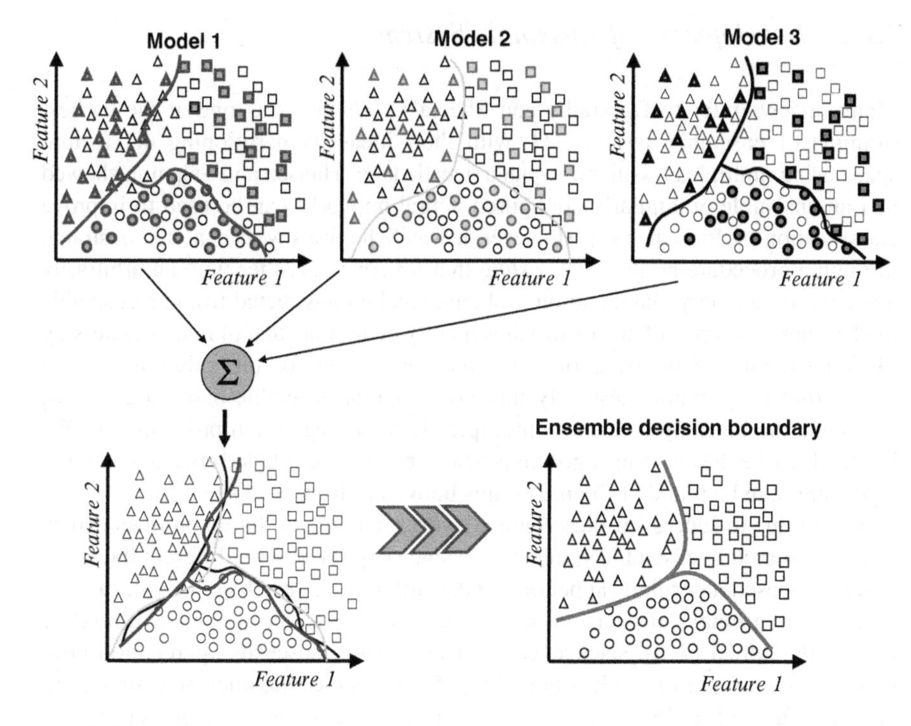

Fig. 1.1 Variability reduction using ensemble systems

The reduction of variability can be thought of as reducing high-frequency (high-variance) noise using a moving average filter, where each sample of the signal is averaged by a neighbor of samples around it. Assuming that noise in each sample is independent, the noise component is averaged out, whereas the information content that is common to all segments of the signal is unaffected by the averaging operation. Increasing classifier accuracy using an ensemble of classifiers works exactly the same way: assuming that classifiers make different errors on each sample, but generally agree on their correct classifications, averaging the classifier outputs reduces the error by averaging out the error components.

It is important to point out two issues here: first, in the context of ensemble systems, there are many ways of combining ensemble members, of which averaging the classifier outputs is only one method. We discuss different combination schemes later in this chapter. Second, combining the classifier outputs does not necessarily lead to a classification performance that is guaranteed to be better than the best classifier in the ensemble. Rather, it reduces our likelihood of choosing a classifier with a poor performance. After all, if we knew a priori which classifier would perform the best, we would only use that classifier and would not need to use an ensemble. A representative illustration of the variance reduction ability of the ensemble of classifiers is shown in Fig. 1.1.

1.1.2 Development of Ensemble Systems

Many reviews refer to Dasarathy and Sheela's 1979 work as one of the earliest example of ensemble systems [1], with their ideas on partitioning the feature space using multiple classifiers. About a decade later, Hansen and Salamon showed that an ensemble of similarly configured neural networks can be used to improve classification performance [2]. However, it was Schapire's work that demonstrated through a procedure he named *boosting* that a strong classifier with an arbitrarily low error on a binary classification problem, can be constructed from an ensemble of classifiers, the error of any of which is merely better than that of random guessing [3]. The theory of boosting provided the foundation for the subsequent suite of *AdaBoost* algorithms, arguably the most popular ensemble-based algorithms, extending the boosting concept to multiple class and regression problems [4]. We briefly describe the boosting algorithms below, but a more detailed coverage of these algorithms can be found in Chap. 2 of this book, and Kuncheva's text [5].

In part due to success of these seminal works, and in part based on independent efforts, research in ensemble systems have since exploded, with different flavors of ensemble-based algorithms appearing under different names: bagging [6], random forests (an ensemble of decision trees), composite classifier systems [1], mixture of experts (MoE) [7, 8], stacked generalization [9], consensus aggregation [10], combination of multiple classifiers [11–15], dynamic classifier selection [15], classifier fusion [16–18], committee of neural networks [19], classifier ensembles [19, 20], among many others. These algorithms, and in general all ensemble-based systems, typically differ from each other based on the selection of training data for individual classifiers, the specific procedure used for generating ensemble members, and/or the combination rule for obtaining the ensemble decision. As we will see, these are the three pillars of any ensemble system.

In most cases, ensemble members are used in one of two general settings: classifier selection and classifier fusion [5, 15, 21]. In *classifier selection*, each classifier is trained as a local expert in some local neighborhood of the entire feature space. Given a new instance, the classifier trained with data closest to the vicinity of this instance, in some distance metric sense, is then chosen to make the final decision, or given the highest weight in contributing to the final decision [7, 15, 22, 23]. In *classifier fusion* all classifiers are trained over the entire feature space, and then combined to obtain a composite classifier with lower variance (and hence lower error). Bagging [6], random forests [24], arc-x4 [25], and boosting/AdaBoost [3, 4] are examples of this approach. Combining the individual classifiers can be based on the labels only, or based on class-specific continuous valued outputs [18, 26, 27], for which classifier outputs are first normalized to the [0, 1] interval to be interpreted as the support given by the classifier to each class [18, 28]. Such interpretation leads to algebraic combination rules (simple or weighted majority voting, maximum/minimum/sum/product, or other combinations class-specific outputs) [12, 27, 29], the Dempster–Shafer-based classifier fusion [13, 30], or decision templates [18, 21, 26, 31]. Many of these combination rules are discussed below in more detail.

A sample of the immense literature on classifier combination can be found in Kuncheva's book [5] (and references therein), an excellent text devoted to theory and implementation of ensemble-based classifiers.

1.2 Building an Ensemble System

Three strategies need to be chosen for building an effective ensemble system. We have previously referred to these as the three pillars of ensemble systems: (1) data sampling/selection; (2) training member classifiers; and (3) combining classifiers.

1.2.1 Data Sampling and Selection: Diversity

Making different errors on any given sample is of paramount importance in ensemble-based systems. After all, if all ensemble members provide the same output, there is nothing to be gained from their combination. Therefore, we need diversity in the decisions of ensemble members, particularly when they are making an error. The importance of diversity for ensemble systems is well established [32, 33]. Ideally, classifier outputs should be independent or preferably negatively correlated [34, 35].

Diversity in ensembles can be achieved through several strategies, although using different subsets of the training data is the most common approach, also illustrated in Fig. 1.1. Different sampling strategies lead to different ensemble algorithms. For example, using bootstrapped replicas of the training data leads to bagging, whereas sampling from a distribution that favors previously misclassified samples is the core of boosting algorithms. On the other hand, one can also use different subsets of the available features to train each classifier, which leads to *random subspace methods* [36]. Other less common approaches also include using different parameters of the base classifier (such as training an ensemble of multilayer perceptrons, each with a different number of hidden layer nodes), or even using different base classifiers as the ensemble members. Definitions of different types of diversity measures can be found in [5, 37, 38]. We should also note that while the importance of diversity, and lack of diversity leading to inferior ensemble performance has been wellestablished, an explicit relationship between diversity and ensemble accuracy has not been identified [38, 39].

1.2.2 Training Member Classifiers

At the core of any ensemble-based system is the strategy used to train individual ensemble members. Numerous competing algorithms have been developed for training ensemble classifiers; however, bagging (and related algorithms arc-x4 and random forests), boosting (and its many variations), stack generalization and hierarchical MoE remain as the most commonly employed approaches. These approaches are discussed in more detail below, in Sect. 1.3.

1.2.3 Combining Ensemble Members

The last step in any ensemble-based system is the mechanism used to combine the individual classifiers. The strategy used in this step depends, in part, on the type of classifiers used as ensemble members. For example, some classifiers, such as support vector machines, provide only discrete-valued label outputs. The most commonly used combination rules for such classifiers is (simple or weighted) majority voting followed at a distant second by the Borda count. Other classifiers, such as multilayer perceptron or (naïve) Bayes classifier, provide continuous valued class-specific outputs, which are interpreted as the support given by the classifier to each class. A wider array of options is available for such classifiers, such as arithmetic (sum, product, mean, etc.) combiners or more sophisticated decision templates, in addition to voting-based approaches. Many of these combiners can be used immediately after the training is complete, whereas more complex combination algorithms may require an additional training step (as used in stacked generalization or hierarchical MoE). We now briefly discuss some of these approaches.

1.2.3.1 Combining Class Labels

Let us first assume that only the class labels are available from the classifier outputs, and define the decision of the tth classifier as $d_{t,c} \in \{0,1\}, t = 1, \ldots, T$ and $c = 1, \ldots, C$, where T is the number of classifiers and C is the number of classes. If t^{th} classifier (or hypothesis) h_t chooses class ω_c, then $d_{t,c} = 1$, and 0, otherwise. Note that the continuous valued outputs can easily be converted to label outputs (by assigning $d_{t,c} = 1$ for the class with the highest output), but not vice versa. Therefore, the combination rules described in this section can also be used by classifiers providing specific class supports.

Majority Voting

Majority voting has three flavors, depending on whether the ensemble decision is the class (1) on which all classifiers agree (*unanimous voting*); (2) predicted by at least one more than half the number of classifiers (*simple majority*); or (3) that receives the highest number of votes, whether or not the sum of those votes exceeds 50% (*plurality voting*). When not specified otherwise, majority voting usually refers to plurality voting, which can be mathematically defined as follows: choose class ω_{c^*}, if

$$\sum_{t=1}^{T} d_{t,c^*} = \max_c \sum_{t=1}^{T} d_{t,c} \tag{1.1}$$

If the classifier outputs are independent, then it can be shown that majority voting is the optimal combination rule. To see this, consider an odd number of T classifiers, with each classifier having a probability of correct classification p. Then, the ensemble makes the correct decision if at least $\lfloor T/2 \rfloor + 1$ of these classifiers choose the correct label. Here, the floor function $\lfloor \blacksquare \rfloor$ returns the largest integer less than or equal to its argument. The accuracy of the ensemble is governed by the binomial distribution; the probability of having $k \geq T/2 + 1$ out of T classifiers returning the correct class. Since each classifier has a success rate of p, the probability of ensemble success is then

$$p_{\text{ens}} = \sum_{k=\frac{T}{2}+1}^{T} \binom{T}{k} p^k (1-p)^{T-k} \tag{1.2}$$

Note that P_{ens} approaches 1 as $T \to \infty$, if $p > 0.5$; and it approaches 0 if $p < 0.5$. This result is also known as the Condorcet Jury theorem (1786), as it formalizes the probability of a plurality-based jury decision to be the correct one. Equation (1.2) makes a powerful statement: if the probability of a member classifier giving the correct answer is higher than ½, which really is the least we can expect from a classifier on a binary class problem, then the probability of success approaches 1 very quickly. If we have a multiclass problem, the same concept holds as long as each classifier has a probability of success better than random guessing (i.e., $p > 1/4$ for a four class problem). An extensive and excellent analysis of the majority voting approach can be found in [5].

Weighted Majority Voting

If we have reason to believe that some of the classifiers are more likely to be correct than others, weighting the decisions of those classifiers more heavily can further improve the overall performance compared to that of plurality voting. Let us assume that we have a mechanism for predicting the (future) approximate generalization performance of each classifier. We can then assign a weight W_t to classifier h_t in proportion of its estimated generalization performance. The ensemble, combined according to weighted majority voting then chooses class c^*, if

$$\sum\nolimits_{t=1}^{T} w_t d_{t,c^*} = \max_c \sum\nolimits_{t=1}^{T} w_t d_{t,c} \tag{1.3}$$

that is, if the total weighted vote received by class ω_{c*} is higher than the total vote received by any other class. In general, voting weights are normalized such that they add up to 1.

So, how do we assign the weights? If we knew, a priori, which classifiers would work better, we would only use those classifiers. In the absence of such information, a plausible and commonly used strategy is to use the performance of a classifier on a separate validation (or even training) dataset, as an estimate of that classifier's generalization performance. As we will see in the later sections, AdaBoost follows such an approach. A detailed discussion on weighted majority voting can also be found in [40].

Borda Count

Voting approaches typically use a winner-take-all strategy, i.e., only the class that is chosen by each classifier receives a vote, ignoring any support that nonwinning classes may receive. Borda count uses a different approach, feasible if we can rank order the classifier outputs, that is, if we know the class with the most support (the winning class), as well as the class with the second most support, etc. Of course, if the classifiers provide continuous outputs, the classes can easily be rank ordered with respect to the support they receive from the classifier.

In Borda count, devised in 1770 by Jean Charles de Borda, each classifier (decision maker) rank orders the classes. If there are C candidates, the winning class receives C-1 votes, the class with the second highest support receives C-2 votes, and the class with the ith highest support receives C-i votes. The class with the lowest support receives no votes. The votes are then added up, and the class with the most votes is chosen as the ensemble decision.

1.2.3.2 Combining Continuous Outputs

If a classifier provides continuous output for each class (such as multilayer perceptron or radial basis function networks, naïve Bayes, relevance vector machines, etc.), such outputs—upon proper normalization (such as softmax normalization in (1.4) [41])—can be interpreted as the degree of support given to that class, and under certain conditions can also be interpreted as an estimate of the posterior probability for that class. Representing the actual classifier output corresponding to class ω_c for instance $\boldsymbol{x}$ as $g_c(\boldsymbol{x})$, and the normalized values as $\tilde{g}_c(\boldsymbol{x})$, approximated posterior probabilities $P(\omega_c|\boldsymbol{x})$ can be obtained as

$$P(\omega_c|\boldsymbol{x}) \approx \tilde{g}_c(\boldsymbol{x}) = \frac{e^{g_c(x)}}{\sum_{i=1}^{C} e^{g_i(x)}} \Rightarrow \sum_{i=1}^{C} \tilde{g}_i(\boldsymbol{x}) = 1 \tag{1.4}$$

Fig. 1.2 Decision profile for a given instance $\boldsymbol{x}$

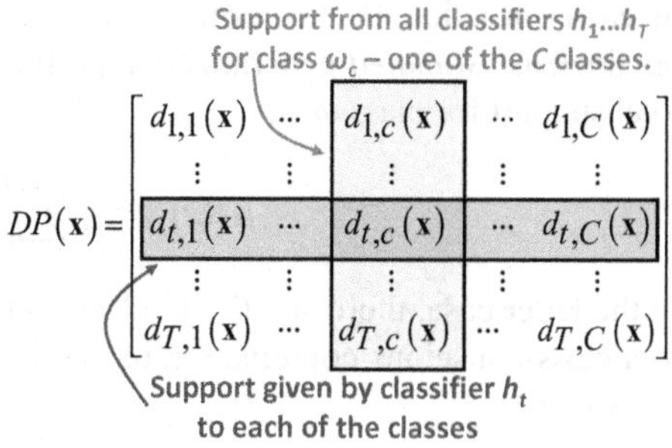

$$DP(\mathbf{x}) = \begin{bmatrix} d_{1,1}(\mathbf{x}) & \cdots & d_{1,c}(\mathbf{x}) & \cdots & d_{1,C}(\mathbf{x}) \\ \vdots & \vdots & \vdots & \vdots & \vdots \\ d_{t,1}(\mathbf{x}) & \cdots & d_{t,c}(\mathbf{x}) & \cdots & d_{t,C}(\mathbf{x}) \\ \vdots & \vdots & \vdots & \vdots & \vdots \\ d_{T,1}(\mathbf{x}) & \cdots & d_{T,c}(\mathbf{x}) & \cdots & d_{T,C}(\mathbf{x}) \end{bmatrix}$$

In order to consolidate different combination rules, we use Kuncheva's *decision profile* matrix $DP(\boldsymbol{x})$ [18], whose elements $d_{t,c} \in [0, 1]$ represent the support given by the tth classifier to class ω_c. Specifically, as illustrated in Fig. 1.2, the rows of $DP(\boldsymbol{x})$ represent the support given by individual classifiers to each of the classes, whereas the columns represent the support received by a particular class c from all classifiers.

Algebraic Combiners

In algebraic combiners, the total support for each class is obtained as a simple algebraic function of the supports received by individual classifiers. Following the notation used in [18], let us represent the total support received by class ω_c, the cth column of the decision profile $DP(\boldsymbol{x})$, as

$$\mu_c(\boldsymbol{x}) = F\,[d_{1,c}(\boldsymbol{x}), \ldots, d_{T,C}(\boldsymbol{x})] \tag{1.5}$$

where $F[\blacksquare]$ is one of the following combination functions.

Mean Rule: The support for class ω_c is the average of all classifiers' c^{th} outputs,

$$\mu_c(\boldsymbol{x}) = \frac{1}{T}\sum\nolimits_{t=1}^{T} d_{t,c}(\boldsymbol{x}) \tag{1.6}$$

hence the function $F[\cdot]$ is the averaging function. Note that the mean rule results in the identical final classification as the sum rule, which only differs from the mean rule by the $1/T$ normalization factor. In either case, the final decision is the class ω_c for which the total support $\mu_c(\boldsymbol{x})$ is the highest.

Weighted Average: The weighted average rule combines the mean and the weighted majority voting rules, where the weights are applied not to class labels, but to the actual continuous outputs. The weights can be obtained during the ensemble generation as part of the regular training, as in AdaBoost, or a separate training can be used to obtain the weights, such as in a MoE. Usually, each classifier h_t receives a weight, although it is also possible to assign a weight to each class output of each classifier. In the former case, we have T weights, $w_1, \ldots, w_T$, usually obtained as *estimated* generalization performances based on training data, with the total support for class ω_c as

$$\mu_c(\boldsymbol{x}) = \frac{1}{T}\sum\nolimits_{t=1}^{T} w_t d_{t,c}(\boldsymbol{x}) \tag{1.7}$$

In the latter case, there are $T * C$ class and classifier-specific weights, which leads to a class-conscious combination of classifier outputs [18]. Total support for class ω_c is then

$$\mu_c(\boldsymbol{x}) = \frac{1}{T}\sum\nolimits_{t=1}^{T} w_{t,c} d_{t,c}(\boldsymbol{x}) \tag{1.8}$$

where $w_{t,c}$ is the weight of the tth classifier for classifying class ω_c instances.

Trimmed mean: Sometimes classifiers may erroneously give unusually low or high support to a particular class such that the correct decisions of other classifiers are not enough to undo the damage done by this unusual vote. This problem can be avoided by discarding the decisions of those classifiers with the highest and lowest support before calculating the mean. This is called trimmed mean. For a $R\%$ trimmed mean, $R\%$ of the support from each end is removed, with the mean calculated on the remaining supports, avoiding the extreme values of support. Note that 50% trimmed mean is equivalent to the median rule discussed below.

Minimum/Maximum/Median Rule: These functions simply take the minimum, maximum, or the median among the classifiers' individual outputs.

$$\begin{aligned} \mu_c(\boldsymbol{x}) &= \min\nolimits_{t=1,\ldots,T}\{d_{t,c}(\boldsymbol{x})\} \\ \mu_c(\boldsymbol{x}) &= \max\nolimits_{t=1,\ldots,T}\{d_{t,c}(\boldsymbol{x})\} \\ \mu_c(\boldsymbol{x}) &= \text{median}_{t=1,\ldots,T}\{d_{t,c}(\boldsymbol{x})\} \end{aligned} \tag{1.9}$$

where the ensemble decision is chosen as the class for which total support is largest. Note that the *minimum rule* chooses the class for which the *minimum support* among the classifiers is highest.

Product Rule: The product rule chooses the class whose product of supports from each classifier is the highest. Due to the nulling nature of multiplying with zero, this rule decimates any class that receives at least one zero (or very small) support.

$$\mu_c(\boldsymbol{x}) = \frac{1}{T}\prod\nolimits_{t=1}^{T} d_{t,c}(\boldsymbol{x}) \tag{1.10}$$

Generalized Mean: All of the aforementioned rules are in fact special cases of the generalized mean,

$$\mu_c(\boldsymbol{x}) = \left(\frac{1}{T}\sum\nolimits_{t=1}^{T} (d_{t,c}(\boldsymbol{x}))^{\alpha}\right)^{1/\alpha} \tag{1.11}$$

where different choices of α lead to different combination rules. For example, $\alpha \rightarrow -\infty$, leads to minimum rule, and $\alpha \rightarrow 0$, leads to

$$\mu_c(x) = \left(\prod_{t=1}^{T} (d_{t,c}(x))\right)^{1/T} \tag{1.12}$$

which is the geometric mean, a modified version of the product rule. For $\alpha \rightarrow 1$, we get the mean rule, and $\alpha \rightarrow \infty$ leads to the maximum rule.

Decision Template: Consider computing the average decision profile observed for each class throughout training. Kuncheva defines this average decision profile as the *decision template* of that class [18]. We can then compare the decision profile of a given instance to the decision templates (i.e., average decision profiles) of each class, choosing the class whose decision template is closest to the decision profile of the current instance, in some similarity measure. The decision template for class ω_c is then computed as

$$DT_c = 1/N_c \sum_{X_c \in \omega_c} DP\,(X_c) \tag{1.13}$$

as the average decision profile obtained from $\boldsymbol{X}_c$, the set of training instances (of cardinality N_c) whose true class is ω_c. Given an unlabeled test instance $\boldsymbol{x}$, we first construct its decision profile $DP(\boldsymbol{x})$ from the ensemble outputs and calculate the similarity S between $DP(\boldsymbol{x})$ and the decision template DT_c for each class ω_c as the degree of support given to class ω_c.

$$\mu_c(\boldsymbol{x}) = S(DP(x), DT_c), c = 1, \ldots, C \tag{1.14}$$

where the similarity measure S is usually a squared Euclidean distance,

$$\mu_c(\boldsymbol{x}) = 1 - \frac{1}{T \times C} \sum_{t=1}^{T} \sum_{i=1}^{C} (DT_c(t,i) - d_{t,i}(\boldsymbol{x}))^2 \tag{1.15}$$

and where $DT_c(t,i)$ is the decision template support given by the tth classifier to class ω_i, i.e., the support given by the tth classifier to class ω_i, averaged over all class ω_c training instances. We expect this support to be high when $i = c$, and low otherwise. The second term $d_{t,i}(x)$ is the support given by the tth classifier to class ω_i for the given instance $\boldsymbol{x}$. The class with the highest total support is then chosen as the ensemble decision.

1.3 Popular Ensemble-Based Algorithms

A rich collection of ensemble-based classifiers have been developed over the last several years. However, many of these are some variation of the select few well-established algorithms whose capabilities have also been extensively tested and widely reported. In this section, we present an overview of some of the most prominent ensemble algorithms.

Algorithm 1 Bagging

Inputs: Training data S; supervised learning algorithm, **BaseClassifier**, integer T specifying ensemble size; percent R to create bootstrapped training data.
Do $t = 1, \ldots, T$

1. Take a bootstrapped replica S_t by randomly drawing $R\%$ of S.
2. Call **BaseClassifier** with S_t and receive the hypothesis (classifier) h_t.
3. Add h_t to the ensemble, $\mathcal{E} \leftarrow \mathcal{E} \cup \ h_t$.

End
Ensemble Combination: Simple Majority Voting—Given unlabeled instance $\boldsymbol{x}$

1. Evaluate the ensemble $\mathcal{E} = \{h_1, \ldots, h_T\}$ on $\boldsymbol{x}$.
2. Let $v_{t,c} = 1$ if h_t chooses class ω_c, and 0, otherwise.
3. Obtain total vote received by each class

$$V_c = \sum\nolimits_{t=1}^{T} v_{t,c}, \ c = 1, ..., C \tag{1.16}$$

Output: Class with the highest V_c.

1.3.1 Bagging

Breiman's bagging (short for Bootstrap Aggregation) algorithm is one of the earliest and simplest, yet effective, ensemble-based algorithms. Given a training dataset S of cardinality N, bagging simply trains T independent classifiers, each trained by sampling, with replacement, N instances (or some percentage of N) from S. The diversity in the ensemble is ensured by the variations within the bootstrapped replicas on which each classifier is trained, as well as by using a relatively *weak classifier* whose decision boundaries measurably vary with respect to relatively small perturbations in the training data. Linear classifiers, such as decision stumps, linear SVM, and single layer perceptrons are good candidates for this purpose. The classifiers so trained are then combined via simple majority voting. The pseudocode for bagging is provided in Algorithm 1.

Bagging is best suited for problems with relatively small available training datasets. A variation of bagging, called *Pasting Small Votes* [42], designed for problems with large training datasets, follows a similar approach, but partitioning the large dataset into smaller segments. Individual classifiers are trained with these segments, called *bites*, before combining them via majority voting.

Another creative version of bagging is the *Random Forest* algorithm, essentially an ensemble of decision trees trained with a bagging mechanism [24]. In addition to choosing instances, however, a random forest can also incorporate random subset selection of features as described in Ho's random subspace models [36].

1.3.2 Boosting and AdaBoost

Boosting, introduced in Schapire's seminal work *strength of weak learning* [3], is an iterative approach for generating a strong classifier, one that is capable of achieving arbitrarily low training error, from an ensemble of weak classifiers, each of which can barely do better than random guessing. While boosting also combines an ensemble of weak classifiers using simple majority voting, it differs from bagging in one crucial way. In bagging, instances selected to train individual classifiers are bootstrapped replicas of the training data, which means that each instance has equal chance of being in each training dataset. In boosting, however, the training dataset for each subsequent classifier increasingly focuses on instances misclassified by previously generated classifiers.

Boosting, designed for binary class problems, creates sets of three weak classifiers at a time: the first classifier (or hypothesis) h_1 is trained on a random subset of the available training data, similar to bagging. The second classifier, h_2, is trained on a different subset of the original dataset, precisely half of which is correctly identified by h_1, and the other half is misclassified. Such a training subset is said to be the "most informative," given the decision of h_1. The third classifier h_3 is then trained with instances on which h_1 and h_2 disagree. These three classifiers are then combined through a three-way majority vote. Schapire proved that the training error of this three-classifier ensemble is bounded above by $g(\varepsilon) < 3\varepsilon^2 - 2\varepsilon^3$, where ε is the error of any of the three classifiers, provided that each classifier has an error rate $\varepsilon < 0.5$, the least we can expect from a classifier on a binary classification problem.

AdaBoost (short for *Adaptive Boosting*) [4], and its several variations later extended the original boosting algorithm to multiple classes (AdaBoost.M1, AdaBost.M2), as well as to regression problems (AdaBoost.R). Here we describe the AdaBoost.M1, the most popular version of the AdaBoost algorithms.

AdaBoost has two fundamental differences from boosting: (1) instances are drawn into the subsequent datasets from an iteratively updated *sample distribution* of the training data; and (2) the classifiers are combined through weighted majority voting, where *voting weights* are based on classifiers' training errors, which themselves are weighted according to the sample distribution. The sample distribution ensures that harder samples, i.e., instances misclassified by the previous classifier are more likely to be included in the training data of the next classifier.

The pseudocode of the AdaBoost.M1 is provided in Algorithm 2. The sample distribution, $D_t(i)$ essentially assigns a weight to each training instance $\boldsymbol{x}_i$, $i = 1, \ldots, N$, from which training data subsets S_t are drawn for each consecutive classifier (hypothesis) h_t. The distribution is initialized to be uniform; hence, all instances have equal probability to be drawn into the first training dataset. The training error ε_t of classifier h_t is then computed as the sum of these distribution weights of the instances misclassified by h_t ((1.17), where $[\![\blacksquare]\!]$ is 1 if its argument is true and 0 otherwise). AdaBoost.M1 requires that this error be less than $1/2$, which is then normalized to obtain β_t, such that $0 < \beta_t < 1$ for $0 < \varepsilon_t < 1/2$.

Algorithm 2 AdaBoost.M1

Inputs: Training data = $\{x_i, y_i\}, i = 1, \ldots, N$ $y_i \in \{\omega_1, \ldots, \omega_C\}$, supervised learner **BaseClassifier**; ensemble size T.
Initialize $D_1(i) = 1/N$.
Do for $t = 1, 2, \ldots, T$:

1. Draw training subset S_t from the distribution D_t.
2. Train **BaseClassifier** on S_t, receive hypothesis $h_t: X \to Y$
3. Calculate the error of h_t:

$$\varepsilon_t = \sum_i I[\![h_t(x_i \neq y_i)]\!] D_t(x_i) \tag{1.17}$$

   If $\varepsilon_t > 1/2$ **abort.**
4. Set

$$\beta_t = \varepsilon_t/(1 - \varepsilon_t) \tag{1.18}$$

5. Update sampling distribution

$$D_{t+1}(i) = \frac{D_t(i)}{Z_t} \cdot \begin{cases} \beta_t, \text{ if } h_t(x_i) = y_i \\ 1, \quad \text{otherwise} \end{cases} \tag{1.19}$$

   where $Z_t = \sum_i D_t(i)$ is a normalization constant to ensure that D_{t+1} is a proper distribution function.

End
Weighted Majority Voting: Given unlabeled instance z,
obtain total vote received by each class

$$V_c = \sum_{t:h_t(z)=\omega_c} \log\left(\frac{1}{\beta_t}\right), \quad c = 1, \ldots, C \tag{1.20}$$

Output: Class with the highest V_c.

The heart of AdaBoost.M1 is the distribution update rule shown in (1.19): the distribution weights of the instances correctly classified by the current hypothesis h_t are reduced by a factor of β_t, whereas the weights of the misclassified instances are left unchanged. When the updated weights are renormalized by Z_t to ensure that D_{t+1} is a proper distribution, the weights of the misclassified instances are effectively increased. Hence, with each new classifier added to the ensemble, AdaBoost focuses on increasingly difficult instances. At each iteration t, (1.19) raises the weights of misclassified instances such that they add up to $1/2$, and lowers those of correctly classified ones, such that they too add up to $1/2$. Since the base model learning algorithm **BaseClassifier** is required to have an error less than $1/2$, it is guaranteed to correctly classify at least one previously misclassified training example. When it is unable to do so, AdaBoost aborts; otherwise, it continues until T classifiers are generated, which are then combined using the weighted majority voting.

Note that the reciprocals of the normalized errors of individual classifiers are used as voting weights in *weighted majority voting* in AdaBoost.M1; hence, classifiers that have shown good performance during training (low β_t) are rewarded with higher voting weights. Since the performance of a classifier on its own training data can be very close to zero, β_t can be quite large, causing numerical instabilities. Such instabilities are avoided by the use of the logarithm in the voting weights (1.20).

Much of the popularity of AdaBoost.M1 is not only due to its intuitive and extremely effective structure but also due to Freund and Schapire's elegant proof that shows the training error of AdaBoost.M1 as bounded above

$$E_{\text{ensemble}} < 2^T \prod_{t=1}^{T} \sqrt{\varepsilon_t(1-\varepsilon_t)} \tag{1.21}$$

Since $\varepsilon_t < 1/2$, E_{ensemble}, the error of the ensemble, is guaranteed to decrease as the ensemble grows. It is interesting, however, to note that AdaBoost.M1 still requires the classifiers to have a (weighted) error that is less than $1/2$ even on nonbinary class problems. Achieving this threshold becomes increasingly difficult as the number of classes increase. Freund and Schapire recognized that there is information even in the classifiers' nonselected class outputs. For example, in handwritten character recognition problem, the characters "1" and "7" look alike, and the classifier may give a high support to both of these classes, and low support to all others. AdaBoost.M2 takes advantage of the supports given to nonchosen classes and defines a pseudo-loss, and unlike the error in AdaBoost.M1, is no longer required to be less than $1/2$. Yet AdaBoost.M2 has a very similar upper bound for training error as AdaBoost.M1. AdaBoost.R is another variation—designed for function approximation problems—that essentially replaces classification error with regression error [4].

1.3.3 Stacked Generalization

The algorithms described so far use nontrainable combiners, where the combination weights are established once the member classifiers are trained. Such a combination rule does not allow determining which member classifier has learned which partition of the feature space. Using trainable combiners, it is possible to determine which classifiers are likely to be successful in which part of the feature space and combine them accordingly. Specifically, the ensemble members can be combined using a separate classifier, trained on the outputs of the ensemble members, which leads to the stacked generalization model.

Wolpert's stacked generalization [9], illustrated in Fig. 1.3, first creates T Tier-1 classifiers, $C_1, \ldots, C_T$, based on a cross-validation partition of the training data. To do so, the entire training dataset is divided into B blocks, and each Tier-1 classifier is first trained on (a different set of) $B-1$ blocks of the training data. Each classifier is then evaluated on the Bth (pseudo-test) block, not seen during training. The outputs of these classifiers on their pseudo-training blocks constitute the training data for

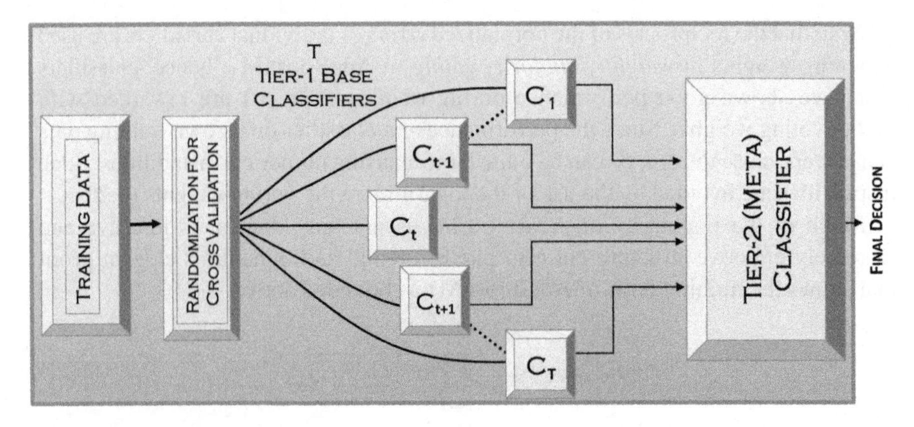

Fig. 1.3 Stacked generalization

the Tier-2 (meta) classifier, which effectively serves as the combination rule for the Tier-1 classifiers. Note that the meta-classifier is not trained on the original feature space, but rather on the decision space of Tier-1 classifiers.

Once the meta-classifier is trained, all Tier-1 classifiers (each of which has been trained B times on overlapping subsets of the original training data) are discarded, and each is retrained on the combined entire training data. The stacked generalization model is then ready to evaluate previously unseen field data.

1.3.4 Mixture of Experts

Mixture of experts is a similar algorithm, also using a trainable combiner. MoE, also trains an ensemble of (Tier-1) classifiers using a suitable sampling technique. Classifiers are then combined through a weighted combination rule, where the weights are determined through a gating network [7], which itself is typically trained using expectation-maximization (EM) algorithm [8,43] on the original training data. Hence, the weights determined by the gating network are dynamically assigned based on the given input, as the MoE effectively learns which portion of the feature space is learned by each ensemble member. Figure 1.4 illustrates the structure of the MoE algorithm.

Mixture-of-experts can also be seen as a classifier selection algorithm, where individual classifiers are trained to become experts in some portion of the feature space. In this setting, individual classifiers are indeed trained to become experts, and hence are usually not weak classifiers. The combination rule then selects the most appropriate classifier, or classifiers weighted with respect to their expertise, for each given instance. The pooling/combining system may then choose a single classifier with the highest weight, or calculate a weighted sum of the classifier outputs for each class, and pick the class that receives the highest weighted sum.

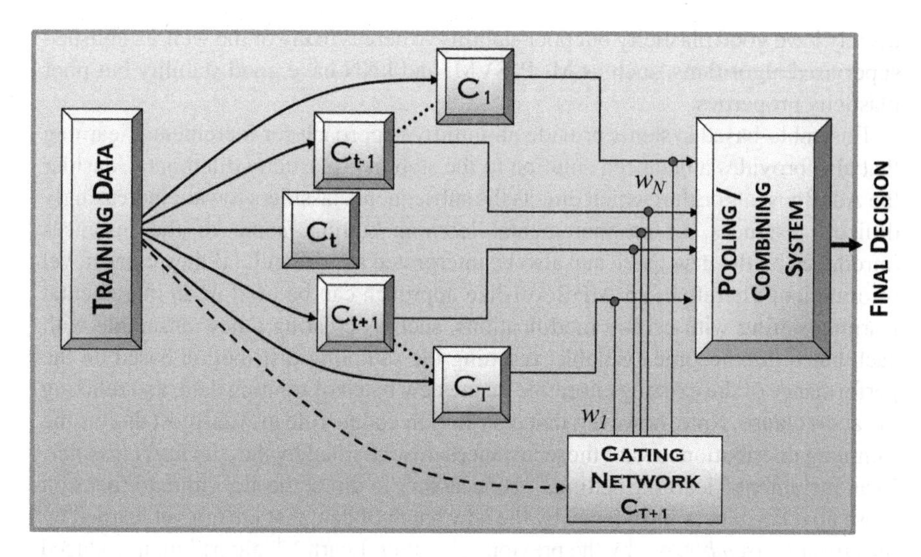

Fig. 1.4 Mixture of experts model

1.4 What Else Can Ensemble Systems Do for You?

While ensemble systems were originally developed to reduce the variability in classifier decision and thereby increase generalization performance, there are many additional problem domains where ensemble systems have proven to be extremely effective. In this section, we discuss some of these emerging applications of ensemble systems along with a family of algorithms, called Learn^{++}, which are designed for these applications.

1.4.1 Incremental Learning

In many real-world applications, particularly those that generate large volumes of data, such data often become available in batches over a period of time. These applications need a mechanism to incorporate the additional data into the knowledge base in an incremental manner, preferably without needing access to the previous data. Formally speaking, incremental learning refers to sequentially updating a hypothesis using current data and previous hypotheses—but not previous data—such that the current hypothesis describes all data that have been acquired thus far. Incremental learning is associated with the well-known stability–plasticity dilemma, where stability refers to the algorithm's ability to retain existing knowledge and plasticity refers to the algorithm's ability to acquire new data. Improving one usually comes at the expense of the other. For example, online data streaming algorithms usually have good plasticity but poor stability, whereas many of the well-established supervised algorithms, such as MLP, SVM, and kNN have good stability but poor plasticity properties.

Ensemble-based systems provide an intuitive approach for incremental learning that also provides a balanced solution to the stability–plasticity dilemma. Consider the AdaBoost algorithm which directs the subsequent classifiers toward increasingly difficult instances. In an incremental learning setting, some of the instances introduced by the new batch can also be interpreted as "difficult" if they carry novel information. Therefore, an AdaBoost-like approach can be used in an incremental learning setting with certain modifications, such as creating a new ensemble with each batch that become available; resetting the sampling distribution based on the performance of the existing ensemble on the new batch of training data, and relaxing the abort clause. Note, however, that distribution update rule in AdaBoost directs the sampling distribution toward those instances misclassified by the *previous classifier*. In an incremental learning setting, it is necessary to direct the algorithm to focus on those novel instances introduced by the new batch of data that are not yet learned by the *current ensemble*, not by the previous classifier. Learn^{++} algorithm, introduced in [44, 45], incorporate these ideas.

The incremental learning problem becomes particularly challenging if the new data also introduce new classes. This is because classifiers previously trained on earlier batches of data inevitably misclassify instances of the new class on which they were not trained. Only the new classifiers are able to recognize the new class(es). Therefore, any decision by the *new* classifiers correctly choosing the new class is outvoted by the earlier classifiers, until there are enough new classifiers to counteract the total vote of those original classifiers. Hence, a relatively large number of new classifiers that recognize the new class are needed, so that their total weight can overwrite the incorrect votes of the original classifiers.

The Learn^{++}.NC (for *N*ew *C*lasses), described in Algorithm 3, addresses these issues [46] by assigning dynamic weights to ensemble members, based on its prediction of which classifiers are likely to perform well on which classes. Learn^{++}.NC cross-references the predictions of each classifier—with those of others—with respect to classes on which they were trained. Looking at the decisions of other classifiers, each classifier decides whether its decision is in line with the predictions of others, and the classes on which it was trained. If not, the classifier reduces its vote, or possibly refrains from voting altogether. As an example, consider an ensemble of classifiers, E_1, trained with instances from two classes ω_1, and ω_2; and a second ensemble, E_2, trained on instances from classes ω_1, ω_2, and a new class, ω_3. An instance from the new class ω_3 is shown to all classifiers. Since E_1 classifiers do not recognize class ω_3, they incorrectly choose ω_1 or ω_2, whereas E_2 classifiers correctly recognize ω_3. Learn^{++}.NC keeps track of which classifiers are trained on which classes. In this example, knowing that E_2 classifiers have seen ω_3 instances, and that E_1 classifiers have not, it is reasonable to believe that E_2 classifiers are correct, particularly if they overwhelmingly choose ω_3 for that instance. To the extent E_2 classifiers are confident of their decision, the voting weights of E_1 classifiers can therefore be reduced. Then, E_2 no longer needs a large number of classifiers: in fact, if E_2 classifiers agree with each other on their correct decision, then very few classifiers are adequate to remove any bias induced by E_1. This voting process, described in Algorithm 4, is called *dynamically weighted consult-and-vote* (DW-CAV) [46].

Algorithm 3 Learn^{++}.NC

Input: For each dataset $k = 1, \ldots, K$, training data $S_k = \{\mathbf{x}_i, y_i\}$, $i = 1, \ldots, N_k$ $y_i \in \Omega = \{\omega_1, \ldots, \omega_C\}$, supervised learner **BaseClassifier**; ensemble size T_k.
Do for $k = 1, \ldots, K$.

1. Initialize instance weights $w_1^k(i) = 1/N_k$.
2. **If** $k \neq 1$, **Set** $t = 0$ and **Go to** Step 5 to adjust initialization weights.

Do for $t = 1, \ldots, T_k$.

1. Set

$$D_t^k = \mathbf{w}_t^k \Big/ \sum_{i=1}^{N_k} w_t^k(i) \tag{1.22}$$

so that D_t^k is a distribution.
2. Train **BaseClassifier** on $TR_t^k \subset S_k$ drawn from D_t^k, receive h_t^k.
3. Calculate error

$$\varepsilon_t^k = \sum_i I\left[\!\left[h_t^k(\mathbf{x}_i \neq y_i)\right]\!\right] D_t^k(\mathbf{x}_i) \tag{1.23}$$

4. If $\varepsilon_t^k > \frac{1}{2}$ discard h_t^k and go to Step 2. Otherwise, normalize ε_t^k:
Normalize ε_t^k :

$$\beta_t^k = \varepsilon_t^k / \left(1 - \varepsilon_t^k\right) \tag{1.24}$$

5. Let CL_t^k be the set of class labels used in training h_t^k for dataset S_k.
6. Call **DW-CAV** to obtain the composite hypothesis H_t^k.
7. Compute the error of the composite hypothesis

$$E_t^k = \sum_i I\left[\!\left[H_t^k(\mathbf{x}_i \neq y_i)\right]\!\right] D_t^k(\mathbf{x}_i) \tag{1.25}$$

8. Normalize E_t^k: $B_t^k = E_t^k / \left(1 - E_t^k\right)$, and update the weights:

$$W_{t+1}^k(i) = w_t^k(i) \cdot \begin{cases} B_t^k, & H_t^k(\mathbf{x}_i = y_i) \\ 1, & \text{otherwise} \end{cases} \tag{1.26}$$

End
End
Call DW-CAV to obtain the final hypothesis, H_{final}.

Algorithm 4 DW-CAV (Dynamically Weighed—Consult and Vote).

Inputs: Instance $\boldsymbol{x}_i$ to be classified; all classifiers h_t^k generated thus far; normalized error values, β_t^k; class labels, CL_t^k used in training h_t^k.
Initialize classifier voting weights $W_t^k = \log\left(1/\beta_t^k\right)$.
Calculate for each $\omega_c \in \{\omega_1, \ldots, \omega_C\}$.

1. Normalization factor:

$$Z_c = \sum\nolimits_k \sum\nolimits_{t:c\in CL_t^k} W_t^k \tag{1.27}$$

2. Class-specific confidence:

$$P_c(i) = \frac{\sum_k \sum_{t:h_k^t(x_i)=\omega_c} W_t^k}{Z_c} \tag{1.28}$$

3. If $P_k(i) = P_l(i), k \neq l$ such that $\mathcal{E}_k \cap \mathcal{E}_l = \emptyset\, P_k(i) = P_l(i) = 0$

where $\mathcal{E}_k$ is the set of classifiers that have seen class ω_k.
Update voting weights for instance x_i

$$W_t^k(i) = W_t^k \cdot \prod\nolimits_{c:\omega_c \notin CL_t^k} (1 - P_c(i)) \tag{1.29}$$

Compute final (current composite) hypothesis

$$H_{\text{final}}(x_i) = \arg\max_{\omega\in\Omega} \sum\nolimits_k \sum\nolimits_{t:h_k^t(x_i)=\omega_c} W_t^k(i) \tag{1.30}$$

Specifically, Learn^{++}.NC updates its sampling distribution based on the composite hypothesis H ((1.25)), which is the ensemble decision of all classifiers generated thus far. The composite hypothesis H_t^k for the first t classifiers from the kth batch is computed by the weighted majority voting of all classifiers using the weights W_t^k, which themselves are weighted based on each classifiers class-specific confidence P_c ((1.27) and (1.28)).

The class-specific confidence $P_c(i)$ for instance $\boldsymbol{x}_i$ is the ratio of total weight of all classifiers that choose class ω_c (for instance $\boldsymbol{x}_i$), to the total weight of all classifiers that have seen class ω_c. Hence, $P_c(i)$ represents the collective confidence of classifiers trained on class ω_c in choosing class ω_c for instance $\boldsymbol{x}_i$. A high value of $P_c(i)$, close to 1, indicates that classifiers trained to recognize class ω_c have in fact overwhelmingly picked class ω_c, and hence those that were not trained on ω_c should not vote (or reduce their voting weight) for that instance.

Extensive experiments with Learn^{++}.NC showed that the algorithm can very quickly learn new classes when they are present, and in fact is also able to remember a class, when it is no longer present in future data batches [46].

1.4.2 Data Fusion

A common problem in many large-scale data analysis and automated decision making applications is to combine information from different data sources that often provide heterogeneous data. Diagnosing a disease from several blood or behavioral tests, imaging results, and time series data (such as EEG or ECG) is such an application. Detecting the health of a system or predicting weather patterns based on data from a variety of sensors, or the health of a company based on several sources of financial indicators are other examples of data fusion. In most data fusion applications, the data are heterogeneous, that is, they are of different format, dimensionality, or structure: some are scalar variables (such as blood pressure, temperature, humidity, speed), some are time series data (such as electrocardiogram, stock prices over a period of time, etc.), some are images (such as MRI or PET images, 3D visualizations, etc.).

Ensemble systems provide a naturally suited solution for such problems: individual classifiers (or even an ensemble of classifiers) can be trained on each data source and then combined through a suitable combiner. The stacked generalization or MoEs structures are particularly well suited for data fusion applications. In both cases, each classifier (or even a model of ensemble of classifiers) can be trained on a separate data source. Then, a subsequent meta-classifier or a gating network can be trained to learn which models or experts have better prediction accuracy, or which ones have learned which feature space. Figure 1.5 illustrates this structure.

A comprehensive review of using ensemble-based systems for data fusion, as well as detailed description of Learn^{++} implementation for data fusion—shown to be quite successful on a variety of data fusions problems—can be found in [47]. Other ensemble-based fusion approaches include combining classifiers using Dempster–Shafer-based combination [48–50], ARTMAP [51], genetic algorithms [52], and other combinations of boosting/voting methods [53–55]. Using diversity metrics for ensemble-based data fusion is discussed in [56].

1.4.3 Feature Selection and Classifying with Missing Data

While most ensemble-based systems create individual classifiers by altering the training data instances—but keeping all features for a given instance—individual features can also be altered by using all of the training data available. In such a setting, individual classifiers are trained with different subsets of the entire feature set. Algorithms that use different feature subsets are commonly referred to as random subspace methods, a term coined by Ho [36]. While Ho used this approach for creating random forests, the approach can also be used for feature selection as well as diversity enhancement.

Another interesting application of RSM-related methods is to use the ensemble approach to classify data that have missing features. Most classification algorithms have matrix multiplications that require the entire feature vector to be available.

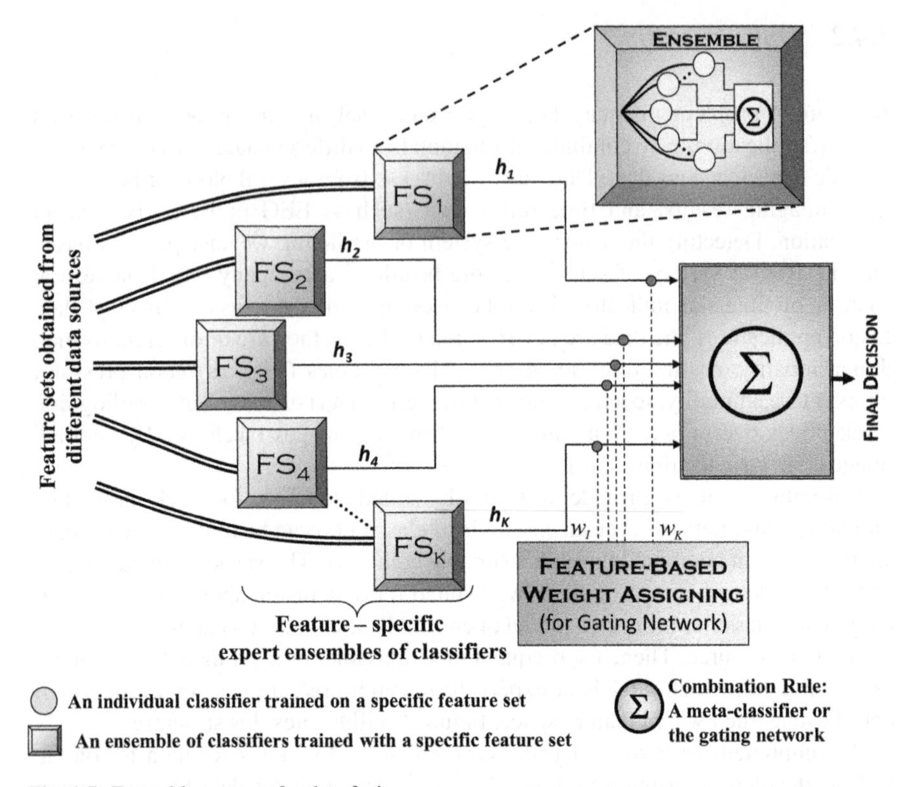

Fig. 1.5 Ensemble systems for data fusion

However, missing data is quite common in real-world applications: bad sensors, failed pixels, unanswered questions in surveys, malfunctioning equipment, medical tests that cannot be administered under certain conditions, etc. are all common scenarios in practice that can result in missing attributes. Feature values that are beyond the expected dynamic range of the data due to extreme noise, signal saturation, data corruption, etc. can also be treated as missing data.

Typical solutions to missing features include imputation algorithms where the value of the missing variable is estimated based on other observed values of that variable. Imputation-based algorithms (such as expectation maximization, mean imputation, k-nearest neighbor imputation, etc.), are popular because they are theoretically justified and tractable; however, they are also prone to significant estimation errors particularly for large dimensional and/or noisy datasets.

An ensemble-based solution to this problem was offered in Learn^{++}.MF [57] (MF for *M*issing *F*eatures), which generates a large number of classifiers, each of which is trained using only random subsets of the available features. The instance sampling distribution in other versions of Learn^{++} algorithms is replaced with a

Fig. 1.6 (**a**) Training classifiers with random subsets of the features; (**b**) classifying an instance missing feature f_2. Only shaded classifiers can be used

a

C_1	C_2	C_3	C_4	C_5	C_6	C_7	C_8	C_9	C_{10}
f1	X	f1	X	X	f1	X	f1	f1	X
X	f2	X	X	f2	X	f2	f2	X	X
X	X	f3	X	f3	X	f3	X	f3	f3
X	f4	X	f4	f4	f4	X	X	f4	f4
f5	X	f5	f5	X	f5	f5	X	X	f5
f6	f6	X	f6	X	X	X	f6	X	X

b

C_1	C_2	C_3	C_4	C_5	C_6	C_7	C_8	C_9	C_{10}
f1	X	f1	X	X	f1	X	f1	f1	X
X	f2	X	X	f2	X	f2	f2	X	X
X	X	f3	X	f3	X	f3	X	f3	f3
X	f4	X	f4	f4	f4	X	X	f4	f4
f5	X	f5	f5	X	f5	f5	X	X	f5
f6	f6	X	f6	X	X	X	f6	X	X

feature sampling distribution, which favors those features that have not been well represented in the previous classifiers' feature sets. Then, a data instance with missing features is classified using the majority voting of only those classifiers whose feature sets did not include the missing attributes. This is conceptually illustrated in Fig. 1.6a, which shows 10 classifiers, each trained on three of the six features available in the dataset. Features that are not used during training are indicated with an "X." Then, at the time of testing, let us assume that feature number 2, f_2, is missing. This means that those classifiers whose training feature sets included f_2, that is, classifiers C_2, C_5, C_7, and C_8, cannot be used in classifying this instance. However, the remaining classifiers, shaded in Fig. 1.6b, did not use f_2 during their training, therefore those classifiers can still be used.

Learn^{++}.MF is listed in Algorithm 5 below. Perhaps the most important parameter of the algorithm is *nof*, the number of features, out of a total of f, to be used to train each classifier. Choosing a smaller *nof* allows a larger number of missing features to be accommodated by the algorithm. However, choosing a larger *nof* usually improves individual classifier performances. The primary assumption made by Learn^{++}.MF is that the dataset includes a redundant set of features, and the problem is at least partially solvable using a subset of the features, whose identities are unknown to us. Of course, if we knew the identities of those features, we would only use those features in the first place.

A theoretical analysis of this algorithm, including probability of finding at least one useable classifier in the absence of m missing features, when each classifier is trained using *nof* of a total of f features, as well as the number of classifiers needed to guarantee at least one useable classifier are provided in [57].

Algorithm 5 Learn^{++}.MF

Inputs: Sentinel value *sen*, **BaseClassifier**; the number of classifiers, T.
Training dataset $S = \{x_i, y_i\}, i = 1, \ldots, N$, with N instances of f features from c classes, number of features used to train each classifier, *nof*;
Initialize feature distribution $D_1(j) = 1/f, \forall j, j = 1, \ldots, f$;
Do for $t = 1, \ldots, T$.

1. Normalize D_t to make it a proper distribution.
2. Draw *nof* features from D_t to form selected features: $F_{selection}(t)$.
3. Call BaseClassifier to train classifier C_t using only those features in $F_{selection}(t)$.
4. Add C_t to the ensemble $\mathcal{E}$
5. Obtain *Perf*(t) the classification performance on S. If *Perf*$(t) < 1/c$, discard C_t and go to Step 2.
6. Update feature distribution

$$D_{t+1}\left(F_{\text{selection}}(t)\right) = (nof/f) \cdot D_t\left(F_{\text{selection}}(t)\right) \tag{1.31}$$

End

Using trained ensemble
Given test/field data z,

1. Determine missing features $M(z)$ =arg$(z(j)$ ==*sen*$)$, $\forall j$
2. Obtain ensemble decision as the class with the most votes among the outputs of classifiers C_t^* trained on the nonmissing features:

$$\varepsilon(z) = \arg\max_y \sum_{t:C_t^*(z)=y} [\![M(z) \cap F_{\text{selection}}(t) \neq \emptyset]\!] \tag{1.32}$$

1.4.4 *Learning from Nonstationary Environments: Concept Drift*

Much of computational intelligence literature is devoted to algorithms that can learn from data that are assumed to be drawn from a fixed but unknown distribution. For a great many applications, however, this assumption is simply not true. For example, predicting future weather patterns from current and past climate data, predicting future stock returns from current and past financial data, identifying e-mail spam from current and past e-mail content, determining which online adds a user will respond based on the user's past web surfing record, predicting future energy demand and prices based on current and past data are all examples of applications where the nature and characteristics of the data—and the underlying phenomena that generate such data—may change over time. Therefore, a learning model trained at a fixed point in time—and a decision boundary generated by such a model—may not reflect the current state of nature due to a change in the underlying environment. Such an environment is referred to as a nonstationary environment, and the problem of learning in such an environment is often referred to as learning *concept drift*. More specifically, given the Bayes posterior probability of class ω that a given instance x belongs, $P(\omega|x) = P(x|\omega)P(\omega)/P(x)$, concept drift can be formally defined as any scenario where the posterior probability changes over time, i.e., $P^{t+1}(\omega|x) \neq P^{t}(\omega|x)$.

To be sure, this is a very challenging problem in machine learning because the underlying change may be gradual or rapid, cyclical or noncyclical, systematic or random, with fixed or variable rate of drift, and with local or global activity in the feature space that spans the data. Furthermore, concept drift can also be perceived, rather than real, as a result of insufficient, unknown, or unobservable features in a dataset, a phenomenon known as *hidden context* [58]. In such a case, an underlying phenomenon provides a true and static description of the environment over time, which, unfortunately, is hidden from the learner's view. Having the benefit of knowing this hidden context would make the problem to have a fixed (and hence stationary) distribution.

Concept drift problems are usually associated with incremental learning or learning from a stream of data, where new data become available over time. Combining several authors' suggestions for desired properties of a concept drift algorithms, Elwell and Polikar provided the following guidelines for addressing concept drift problems: (1) any given instance of data—whether provided online or in batches—can only be used once for training (one-pass incremental learning); (2) knowledge should be labeled with respect to its relevance to the current environment, and be dynamically updated as new data continuously arrive; (3) the learner should have a mechanism to reconcile when existing and newly acquired knowledge conflict with each other; (4) the learner should be able—not only to temporarily forget information that is no longer relevant to the current environment but also to recall prior knowledge if the drift/change in the environment follow a cyclical nature; and (5) knowledge should be incrementally and periodically stored so that it can be recalled to produce the best hypothesis for an unknown (unlabeled) data instance at any time during the learning process [59].

Earliest examples of concept drift algorithms use a single classifier to learn from the latest batch of data available, using some form of windowing to control the batch size. Successful examples of this *instance selection* approach include STAGGER [60] and FLORA [58] algorithms, which use a sliding window to choose a block of (new) instances to train a new classifier. The window size can be dynamically updated using a "window adjustment heuristic," based on how fast the environment is changing. Instances that fall outside of the window are then assumed irrelevant and hence the information carried by them are irrecoverably forgotten. Other examples of this window-based approach include [61–63], which use different drift detection mechanisms or base classifiers. Such approaches are often either not truly incremental as they may access prior data, or cannot handle cyclic environments. Some approaches include a novelty (anomaly) detection to determine the precise moment when changes occur, typically by using statistical measures, such as control charts based CUSUM [64, 65], confidence interval on error [66, 67], or other statistical approaches [68]. A new classifier trained on new data since the last detection of change then replaces the earlier classifier(s).

The ensemble-based algorithms provide an alternate approach to concept drift problems. These algorithms generally belong to one of three categories [69]: (1) update the combination rules or voting weights of a fixed ensemble, such as [70, 71]; an approach loosely based on Littlestone's Winnow [72] and Freund and Schapire's Hedge (a precursor of AdaBoost) [4]; (2) update the parameters of existing ensemble members using an online learner [66, 73]; and/or (3) add new members to build an ensemble with each incoming dataset. Most algorithms fall into this last category, where the oldest (e.g., Streaming Ensemble Algorithm (SEA) [74] or Recursive Ensemble Approach (REA) [75]) or the least contributing ensemble members are replaced with new ones (as in Dynamic Integration [76], or Dynamic Weighted Majority (DWM) [77]). While many ensemble approaches use some form of voting, there is some disagreement on whether the voting should be weighted, e.g., giving higher weight to a classifier if its training data were in the same region as the testing example [76], or unweighted, as in [78, 79], where the authors argue that weights based on previous data, whose distribution may have changed, are uninformative for future datasets. Other efforts that combine ensemble systems with drift detection include Bifet's adaptive sliding window (ADWIN) [80, 81], also available within the WEKA-like software suite, Massive Online Analysis (MOA) at [82].

More recently, a new addition to Learn^{++} suite of algorithms, Learn^{++}.NSE, has been introduced as a general framework to learning concept drift that does not make any restriction on the nature of the drift. Learn^{++}.NSE (for NonStationary Environments) inherits the dynamic distribution-guided ensemble structure and incremental learning abilities of all Learn^{++} algorithms (hence strictly follows the one-pass rule). Learn^{++}.NSE trains a new classifier for each batch of data it receives, and combines the classifiers using a dynamically weighted majority voting. The novelty of the approach is in determining the voting weights, based on each classifier's time-adjusted accuracy on current and past environments, allowing the algorithm to recognize, and act accordingly, to changes in underlying data distributions, including possible reoccurrence of an earlier distribution [59].

The Learn^{++}.NSE algorithm is listed in Algorithm 6, which receives the training dataset $D^t = \{x_i^t \in X; y_i^t \in Y\}$, $i = 1, \ldots, m^t$, at time t. Hence x_i^t is the ith instance of the dataset, drawn from an unknown distribution $P^t(\boldsymbol{x}, y)$, which is the currently available representation of a possibly drifting distribution at time t. At time $t + 1$, a new batch of data is drawn from $P^{t+1}(\boldsymbol{x}, y)$. Between any two consecutive batches, the environment may experience a change whose rate is not known, nor assumed to be constant. Previously seen data are not available to the algorithm, allowing Learn^{++}.NSE to operate in a truly incremental fashion.

Algorithm 6 Learn^{++}.NSE

Input: For each dataset D^t $t = 1,2,\ldots.$
Training data $\{x^t(i) \in X; y^t(i) \in Y = \{1,\ldots, c\}\}$, $i = 1, \ldots, m^t$; Supervised learning algorithm **BaseClassifier**; Sigmoid parameters a (slope) and b (infliction point).
Do for $t = 1, 2, \ldots.$
If $t = 1$, **Initialize** $D^1(i) = w^t(i) = 1/m^1, \forall i$, Go to step 3. **Endif**

1. Compute error of the existing ensemble on new data

$$E_t = \sum_{i=1}^{m^t} 1/m^t \cdot [\![H^{t-1}(x^t(i)) \neq y^t(i)]\!] \tag{1.33}$$

2. Update and normalize instance weights

$$w_i^t = \frac{1}{m^t} \cdot \begin{cases} E^t, & H^{t-1}(x^t(i)) = y^t(i) \\ 1, & \text{otherwise} \end{cases} \tag{1.34}$$

Set

$$D^t = \mathbf{w}^t \Big/ \sum_{i=1}^{m^t} w^t(i) \Rightarrow D^t \tag{1.35}$$

is a distribution.

3. Call **BaseClassifier** with D^t, obtain $h^t: X \rightarrow Y$.
4. Evaluate all existing classifiers on new data D^t

$$\varepsilon_k^t = \sum_{i=1}^{m^t} D^t(i) [\![h_k(x^t(i)) \neq y^t(i)]\!] \quad \text{for } k = 1, \ldots, t \tag{1.36}$$

If $\varepsilon_{k=t}^t > 1/2$, generate a new h_t. If $\varepsilon_{k<t}^t > 1/2$, set $\varepsilon_k^t = 1/2$,

$$\beta_k^t = \varepsilon_k^t/(1 - \varepsilon_k^t), \text{ for } k = 1, \ldots, t \rightarrow 0 \leq \beta_k^t \leq 1 \tag{1.37}$$

5. Sigmoid-based time averaging of normalized errors of h_k: For $a, b \in \mathbf{R}$

$$\omega_k^t = 1\Big/\left(1 + e^{-a(t-k-b)}\right), \omega_k^t = \omega_k^t \Big/ \sum_{j=0}^{t-k} \omega_k^{t-j} \tag{1.38}$$

$$\bar{\beta}_k^t = \sum_{j=0}^{t-k} \omega_k^{t-j} \beta_k^{t-j}, \quad \text{for } k = 1, \ldots, t \tag{1.39}$$

6. Calculate classifier voting weights

$$W_k^t = \log\left(1/\bar{\beta}_k^t\right), \quad \text{for} \quad k = 1, \ldots, t \tag{1.40}$$

7. Compute the composite hypothesis (the ensemble decision) as

$$H^t(x^t(i)) = \arg\max_c \sum\nolimits_k W_k^t \cdot [\![h_k\left(x^t(i)\right) = c]\!] \tag{1.41}$$

End Do.
Return the final hypothesis as the current composite hypothesis.

The algorithm is initialized with a single classifier on the first batch of data. With the arrival of each subsequent batch of data, the current ensemble, H^{t-1}—the composite hypothesis of all individual hypotheses previously generated, is first evaluated on the new data (Step 1 in Algorithm 6). In Step 2, the algorithm identifies those examples of the new environment that are not recognized by the existing ensemble, H^{t-1}, and updates the *penalty distribution* D^t. This distribution is used not for instance selection, but rather to assign penalties to classifiers on their ability to identify previously seen or unseen instances. A new classifier h^t, is then trained on the current training data in Step 3. In Step 4, each classifier generated thus far is evaluated on the training data weighted with respect to the penalty distribution. Note that since classifiers are generated at different times, each classifier receives a different number of evaluations: at time t, h^t receives its first evaluation, whereas h^1 is evaluated for t^{th} time. We use $\varepsilon_k^t, k = 1, \ldots, t$ to denote the error of h_k—the classifier generated at time step k—on dataset D^t. Higher weight is given to classifiers that correctly identify previously unknown instances, while classifiers that misclassify previously known data are penalized. Note that if the newest classifier has a weighted error greater than $1/2$, i.e., if $\varepsilon_{k=t}^t \geq 1/2$, this classifier is discarded and replaced with a new classifier. Older classifiers, with error $\varepsilon_{k<t}^t \geq 1/2$, however, are retained but have their error saturated at $1/2$ (which later corresponds to zero vote on that environment). The errors are then normalized, creating β_k^t that fall in the [0, 1] range.

In Step 5, classifier error is further weighted (using a sigmoid function) with respect to time so that recent competence (error rate) is considered more heavily. Such a sigmoid-based weighted averaging also serves to smooth out potential large swings in classifiers errors that may be due to noisy data rather than actual drift. Final voting weights are determined in Step 6 as log-normalized reciprocals of the weighted errors: if a classifier performs poorly on the current environment, it receives little or no weight, and is effectively—but only temporarily—removed from the ensemble. The classifier is not discarded; however, it is recalled through assignment of higher voting weights if it performs well on future environments. Learn^{++}.NSE forgets only temporarily, which is particularly useful in cyclical environments. The final decision is obtained in Step 7 as the weighted majority voting of the current ensemble members.

Learn^{++}.NSE has been evaluated and benchmarked against other algorithms, on a broad spectrum of real-world as well as carefully designed synthetic datasets—including gradual and rapid drift, variable rate of drift, cyclical environments, as well as environments that introduce or remove concepts. These experiments and their results are reported in [59], which shows that the algorithm can serve as a general framework for learning concept drift regardless of the environment that characterizes the drift.

1.4.5 Confidence Estimation

In addition to the various machine learning problems described above, ensemble systems can also be used to address other challenges that are difficult or impossible using a single classifier-based systems.

One such application is to determine the confidence of the (ensemble-based) classifier in its own decision. The idea is extremely intuitive as it directly follows the use of ensemble systems in our daily lives. Consider reading user reviews of a particular product, or consulting the opinions of several physicians on the risks of a particular medical procedure. If all—or at least most—users agree in their opinion that the product reviewed is very good, we would have higher confidence in our decision to purchase that item. Similarly, if all physicians agree on the effectiveness of a particular medical operation, then we would feel more comfortable with that procedure. On the other hand, if some of the reviews are highly complementary, whereas others are highly critical that casts doubt in our decision to purchase that item. Of course, in order for our confidence in the "ensemble of reviewers" to be valid, we must believe that the reviewers are independent of each other, and indeed independently review the items. If certain reviewers were writing reviews based on other reviewers' reviews they read, the confidence based on the ensemble becomes meaningless.

This idea can be naturally extended to classifiers. If considerable majority of the classifiers in an ensemble agree on their decisions, than we can interpret that outcome as ensemble having higher confidence in its decision, as opposed to only a mere majority of classifiers choosing a particular class. In fact, under certain conditions, the consistency of the classifier outputs can also be used to estimate the true posterior probability of each class [28]. Of course, similar to the examples given above, the classifier decisions must be independent for this confidence—and the posterior probabilities—to be meaningful.

1.5 Summary

Ensemble-based systems provide intuitive, simple, elegant, and powerful solutions to a variety of machine learning problems. Originally developed to improve classification accuracy by reducing the variance in classifier outputs, ensemble-based systems have since proven to be very effective in a number of problem domains that are difficult to address using a single model-based system.

A typical ensemble-based system consists of three components: a mechanism to choose instances (or features), which adds to the diversity of the ensemble; a mechanism for training component classifiers of the ensemble; and a mechanism to combine the classifiers. The selection of instances can either be done completely at random, as in bagging, or by following a strategy implemented through a dynamically updated distribution, as in boosting family of algorithms. In general, most ensemble-based systems are independent of the type of base classifier used to create the ensemble, a significant advantage that allows using a specific type of classifier that may be known to be best suited for a given application. In that sense, ensemble-based systems are also known as *algorithm-free-algorithms*.

Finally, a number of different strategies can be used to combine the classifiers, though sum rule, simple majority voting and weighted majority voting are the most commonly used ones due to certain theoretical guarantees they provide.

We also discussed a number of problem domains on which ensemble systems can be used effectively. These include incremental learning from additional data, feature selection, addressing missing features, data fusion, and learning from nonstationary data distributions. Each of these areas has several algorithms developed to address the relevant specific issue, which are summarized in this chapter. We also described a suite of algorithms, collectively known as Learn^{++} family of algorithms that is capable of addressing all of these problems with proper modifications to the base approach: all Learn^{++} algorithms are incremental algorithms that use an ensemble of classifiers trained on the current data only, then combined through majority voting. The individual members of Learn^{++} differ from each other according to the particular distribution update rule along with a creative weight assignment that is specific to the problem.

References

1. B. V. Dasarathy and B. V. Sheela, "Composite classifier system design: concepts and methodology," *Proceedings of the IEEE*, vol. 67, no. 5, pp. 708–713, 1979
2. L. K. Hansen and P. Salamon, "Neural network ensembles," *IEEE Transactions on Pattern Analysis and Machine Intelligence*, vol. 12, no. 10, pp. 993–1001, 1990
3. R. E. Schapire, "The strength of weak learnability," *Machine Learning*, vol. 5, no. 2, pp. 197–227, June 1990
4. Y. Freund and R. E. Schapire, "Decision-theoretic generalization of on-line learning and an application to boosting," *Journal of Computer and System Sciences*, vol. 55, no. 1, pp. 119–139, 1997
5. L. I. Kuncheva, *Combining pattern classifiers, methods and algorithms*. New York, NY: Wiley Interscience, 2005
6. L. Breiman, "Bagging predictors," *Machine Learning*, vol. 24, no. 2, pp. 123–140, 1996
7. R. A. Jacobs, M. I. Jordan, S. J. Nowlan, and G. E. Hinton, "Adaptive mixtures of local experts," *Neural Computation*, vol. 3, no. 1, pp. 79–87, 1991
8. M. J. Jordan and R. A. Jacobs, "Hierarchical mixtures of experts and the EM algorithm," *Neural Computation*, vol. 6, no. 2, pp. 181–214, 1994
9. D. H. Wolpert, "Stacked generalization," *Neural Networks*, vol. 5, no. 2, pp. 241–259, 1992

10. J. A. Benediktsson and P. H. Swain, "Consensus theoretic classification methods," *IEEE Transactions on Systems, Man and Cybernetics*, vol. 22, no. 4, pp. 688–704, 1992
11. L. Xu, A. Krzyzak, and C. Y. Suen, "Methods of combining multiple classifiers and their applications to handwriting recognition," *IEEE Transactions on Systems, Man and Cybernetics*, vol. 22, no. 3, pp. 418–435, 1992
12. T. K. Ho, J. J. Hull, and S. N. Srihari, "Decision combination in multiple classifier systems," *IEEE Transactions on Pattern Analysis and Machine Intelligence*, vol. 16, no. 1, pp. 66–75, 1994
13. G. Rogova, "Combining the results of several neural network classifiers," *Neural Networks*, vol. 7, no. 5, pp. 777–781, 1994
14. L. Lam and C. Y. Suen, "Optimal combinations of pattern classifiers," *Pattern Recognition Letters*, vol. 16, no. 9, pp. 945–954, 1995
15. K. Woods, W. P. J. Kegelmeyer, and K. Bowyer, "Combination of multiple classifiers using local accuracy estimates," *IEEE Transactions on Pattern Analysis and Machine Intelligence*, vol. 19, no. 4, pp. 405–410, 1997
16. I. Bloch, "Information combination operators for data fusion: A comparative review with classification," *IEEE Transactions on Systems, Man, and Cybernetics Part A:Systems and Humans*, vol. 26, no. 1, pp. 52–67, 1996
17. S. B. Cho and J. H. Kim, "Combining multiple neural networks by fuzzy integral for robust classification," *IEEE Transactions on Systems, Man and Cybernetics*, vol. 25, no. 2, pp. 380–384, 1995
18. L. I. Kuncheva, J. C. Bezdek, and R. P. W. Duin, "Decision templates for multiple classifier fusion: an experimental comparison," *Pattern Recognition*, vol. 34, no. 2, pp. 299–314, 2001
19. H. Drucker, C. Cortes, L. D. Jackel, Y. LeCun, and V. Vapnik, "Boosting and other ensemble methods," *Neural Computation*, vol. 6, no. 6, pp. 1289–1301, 1994
20. L. I. Kuncheva, "Classifier ensembles for changing environments," *5th International Workshop on Multiple Classifier Systems* in Lecture Notes in Computer Science, eds. F. Roli, J. Kittler, and T. Windeatt, vol. 3077, pp. 1–15, Cagliari, Italy, 2004
21. L. I. Kuncheva, "Switching between selection and fusion in combining classifiers: An experiment," *IEEE Transactions on Systems, Man, and Cybernetics, Part B: Cybernetics*, vol. 32, no. 2, pp. 146–156, 2002
22. E. Alpaydin and M. I. Jordan, "Local linear perceptrons for classification," *IEEE Transactions on Neural Networks*, vol. 7, no. 3, pp. 788–792, 1996
23. G. Giacinto and F. Roli, "Approach to the automatic design of multiple classifier systems," *Pattern Recognition Letters*, vol. 22, no. 1, pp. 25–33, 2001
24. L. Breiman, "Random forests," *Machine Learning*, vol. 45, no. 1, pp. 5–32, 2001
25. L. Breiman, "Arcing classifiers," *Annals of Statistics*, vol. 26, no. 3, pp. 801–849, 1998
26. F. M. Alkoot and J. Kittler, "Experimental evaluation of expert fusion strategies," *Pattern Recognition Letters*, vol. 20, no. 11–13, pp. 1361–1369, Nov. 1999
27. J. Kittler, M. Hatef, R. P. W. Duin, and J. Mates, "On combining classifiers," *IEEE Transactions on Pattern Analysis and Machine Intelligence*, vol. 20, no. 3, pp. 226–239, 1998
28. M. Muhlbaier, A. Topalis, and R. Polikar, "Ensemble confidence estimates posterior probability," *6th Int. Workshop on Multiple Classifier Systems*, Lecture Notes on Computer Science, eds. N. C. Oza, R. Polikar, J. Kittler, and F. Roli, Eds., vol. 3541, pp. 326–335, Monterey, CA, 2005
29. L. I. Kuncheva, "A theoretical study on six classifier fusion strategies," *IEEE Transactions on Pattern Analysis and Machine Intelligence*, vol. 24, no. 2, pp. 281–286, 2002
30. Y. Lu, "Knowledge integration in a multiple classifier system," *Applied Intelligence*, vol. 6, no. 2, pp. 75–86, 1996
31. D. M. J. Tax, M. van Breukelen, R. P. W. Duin, and J. Kittler, "Combining multiple classifiers by averaging or by multiplying?" *Pattern Recognition*, vol. 33, no. 9, pp. 1475–1485, 2000
32. G. Brown, "Diversity in neural network ensembles." PhD, University of Birmingham, UK, 2004

33. G. Brown, J. Wyatt, R. Harris, and X. Yao, "Diversity creation methods: a survey and categorisation," *Information Fusion*, vol. 6, no. 1, pp. 5–20, 2005
34. A. Chandra and X. Yao, "Evolving hybrid ensembles of learning machines for better generalisation," *Neurocomputing*, vol. 69, no. 7–9, pp. 686–700, Mar. 2006
35. Y. Liu and X. Yao, "Ensemble learning via negative correlation," *Neural Networks*, vol. 12, no. 10, pp. 1399–1404, 1999
36. T. K. Ho, "Random subspace method for constructing decision forests," *IEEE Transactions on Pattern Analysis and Machine Intelligence*, vol. 20, no. 8, pp. 832–844, 1998
37. R. E. Banfield, L. O. Hall, K. W. Bowyer, and W. P. Kegelmeyer, "Ensemble diversity measures and their application to thinning," *Information Fusion*, vol. 6, no. 1, pp. 49–62, 2005
38. L. I. Kuncheva and C. J. Whitaker, "Measures of diversity in classifier ensembles and their relationship with the ensemble accuracy," *Machine Learning*, vol. 51, no. 2, pp. 181–207, 2003
39. L. I. Kuncheva, That elusive diversity in classifier ensembles," *Pattern Recognition and Image Analysis, Lecture Notes in Computer Science*, vol. 2652, 2003, pp. 1126–1138
40. N. Littlestone and M. Warmuth, "Weighted majority algorithm," *Information and Computation*, vol. 108, pp. 212–261, 1994
41. R. O. Duda, P. E. Hart, and D. Stork, "Algorithm independent techniques," in *Pattern classification,* 2 edn New York: Wiley, 2001, pp. 453–516
42. L. Breiman, "Pasting small votes for classification in large databases and on-line," *Machine Learning*, vol. 36, no. 1–2, pp. 85–103, 1999
43. M. I. Jordan and L. Xu, "Convergence results for the EM approach to mixtures of experts architectures," *Neural Networks*, vol. 8, no. 9, pp. 1409–1431, 1995
44. R. Polikar, L. Udpa, S. S. Udpa, and V. Honavar, "Learn++: An incremental learning algorithm for supervised neural networks," *IEEE Transactions on Systems, Man and Cybernetics Part C: Applications and Reviews*, vol. 31, no. 4, pp. 497–508, 2001
45. H. S. Mohammed, J. Leander, M. Marbach, Polikar, and R. Polikar, "Can AdaBoost.M1 learn incrementally? A comparison to Learn++ under different combination rules," *International Conference on Artificial Neural Networks (ICANN2006)* in Lecture Notes in Computer Science, vol. 4131, pp. 254–263, Springer, 2006
46. M. D. Muhlbaier, A. Topalis, and R. Polikar, "Learn++.NC: combining ensemble of classifiers with dynamically weighted consult-and-vote for efficient incremental learning of new classes," *IEEE Transactions on Neural Networks*, vol. 20, no. 1, pp. 152–168, 2009
47. D. Parikh and R. Polikar, "An ensemble-based incremental learning approach to data fusion," *IEEE Transactions on Systems, Man, and Cybernetics, Part B: Cybernetics*, vol. 37, no. 2, pp. 437–450, 2007
48. H. Altincay and M. Demirekler, "Speaker identification by combining multiple classifiers using Dempster-Shafer theory of evidence," *Speech Communication*, vol. 41, no. 4, pp. 531–547, 2003
49. Y. Bi, D. Bell, H. Wang, G. Guo, and K. Greer, "Combining multiple classifiers using dempster's rule of combination for text categorization," *First International Conference, MDAI 2004, Aug 2–4 2004* in Lecture Notes in Artificial Intelligence, vol. 3131, Barcelona, Spain, pp. 127–138, 2004
50. T. Denoeux, "Neural network classifier based on Dempster-Shafer theory," *IEEE Transactions on Systems, Man, and Cybernetics Part A:Systems and Humans*, vol. 30, no. 2, pp. 131–150, 2000
51. G. A. Carpenter, S. Martens, and O. J. Ogas, "Self-organizing information fusion and hierarchical knowledge discovery: a new framework using ARTMAP neural networks," *Neural Networks*, vol. 18, no. 3, pp. 287–295, 2005
52. B. F. Buxton, W. B. Langdon, and S. J. Barrett, "Data fusion by intelligent classifier combination," *Measurement and Control*, vol. 34, no. 8, pp. 229–234, 2001
53. G. J. Briem, J. A. Benediktsson, and J. R. Sveinsson, "Use of multiple classifiers in classification of data from multiple data sources," *2001 International Geoscience and Remote Sensing Symposium (IGARSS 2001)*, vol. 2, Sydney, NSW: Institute of Electrical and Electronics Engineers Inc., pp. 882–884, 2001

54. W. Fan, M. Gordon, and P. Pathak, "On linear mixture of expert approaches to information retrieval," *Decision Support Systems*, vol. 42, no. 2, pp. 975–987, 2005
55. S. Jianbo, W. Jun, and X. Yugeng, "Incremental learning with balanced update on receptive fields for multi-sensor data fusion," *IEEE Transactions on Systems, Man and Cybernetics (B)*, vol. 34, no. 1, pp. 659–665, 2004
56. D. Leonard, D. Lillis, L. Zhang, F. Toolan, R. Collier, and J. Dunnion, "Applying machine learning diversity metrics to data fusion in information retrieval," in *Advances in Information Retrieval,* Lecture Notes in Computer Science, vol. 6611, P. Clough, C. Foley, C. Gurrin, G. Jones, W. Kraaij, H. Lee, and V. Mudoch, eds. Springer, Berlin/Heidelberg, 2011, pp. 695–698
57. R. Polikar, J. DePasquale, H. Syed Mohammed, G. Brown, and L. I. Kuncheva, "Learn++.MF: A random subspace approach for the missing feature problem," *Pattern Recognition*, vol. 43, no. 11, pp. 3817–3832, 2010
58. G. Widmer and M. Kubat, "Learning in the presence of concept drift and hidden contexts," *Machine Learning*, vol. 23, no. 1, pp. 69–101, 1996
59. R. Elwell and R. Polikar, "Incremental learning of concept drift in nonstationary environments," *IEEE Transactions on Neural Networks*, doi: 10.1109/TNN.2011.2160459, vol. 22, no. 10, pp. 1517–1531, October 2011
60. J. C. Schlimmer and R. H. Granger, "Incremental learning from noisy data," *Machine Learning*, vol. 1, no. 3, pp. 317–354, Sept. 1986
61. R. Klinkenberg, "Learning drifting concepts: example selection vs. example weighting," *Intelligent Data Analysis, Special Issue on Incremental Learning Systems Capable of Dealing with Concept Drift*, vol. 8, no. 3, pp. 281–300, 2004
62. M. Nunez, R. Fidalgo, and R. Morales, "Learning in environments with unknown dynamics: towards more robust concept learners," *Journal of Machine Learning Research*, vol. 8, pp. 2595–2628, 2007
63. P. Wang, H. Wang, X. Wu, W. Wang, and B. Shi, "A low-granularity classifier for data streams with concept drifts and biased class distribution," *IEEE Transactions on Knowledge and Data Engineering,* vol. 19, no. 9, pp. 1202–1213, 2007
64. C. Alippi and M. Roveri, "Just-in-time adaptive classifiers; part I: detecting nonstationary changes," *IEEE Transactions on Neural Networks*, vol. 19, no. 7, pp. 1145–1153, 2008
65. C. Alippi and M. Roveri, "Just-in-time adaptive classifiers; part II: designing the classifier," *IEEE Transactions on Neural Networks*, vol. 19, no. 12, pp. 2053–2064, 2008
66. J. Gama, P. Medas, G. Castillo, and P. Rodrigues, "Learning with drift detection," *Advances in Artificial Intelligence—SBIA 2004* in Lecture Notes in Computer Science, vol. 3171, pp. 286–295, 2004
67. L. Cohen, G. Avrahami-Bakish, M. Last, A. Kandel, and O. Kipersztok, "Real-time data mining of non-stationary data streams from sensor networks," *Information Fusion*, vol. 9, no. 3, pp. 344–353, 2008
68. M. Markou and S. Singh, "Novelty detection: a review—part 2: neural network based approaches," *Signal Processing*, vol. 83, no. 12, pp. 2499–2521, 2003
69. L. I. Kuncheva, "Classifier ensembles for changing environments," *Multiple Classifier Systems (MCS 2004) in Lecture Notes in Computer Science*, vol. 3077, pp. 1–15, 2004
70. A. Blum, "Empirical support for winnow and weighted-majority algorithms: results on a calendar scheduling domain," *Machine Learning*, vol. 26, no. 1, pp. 5–23, 1997
71. Z. Xingquan, W. Xindong, and Y. Ying, "Dynamic classifier selection for effective mining from noisy data streams," *Fourth IEEE International Conference on Data Mining (ICDM '04)*, pp. 305–312, 2004
72. N. Littlestone, "Learning quickly when irrelevant attributes abound: A new linear-threshold algorithm," *Machine Learning*, vol. 2, no. 4, pp. 285–318, Apr. 1988
73. N. Oza, "Online ensemble learning." Ph.D. Dissertation, University of California, Berkeley, 2001
74. W. N. Street and Y. Kim, "A streaming ensemble algorithm (SEA) for large-scale classification," *Seventh ACM SIGKDD International Conference on Knowledge Discovery & Data Mining (KDD-01)*, pp. 377–382, 2001

75. S. Chen and H. He, "Towards incremental learning of nonstationary imbalanced data stream: a multiple selectively recursive approach," *Evolving Systems*, vol. in press 2011
76. A. Tsymbal, M. Pechenizkiy, P. Cunningham, and S. Puuronen, "Dynamic integration of classifiers for handling concept drift," *Information Fusion*, vol. 9, no. 1, pp. 56–68, Jan. 2008
77. J. Z. Kolter and M. A. Maloof, "Dynamic weighted majority: an ensemble method for drifting concepts," *Journal of Machine Learning Research*, vol. 8, pp. 2755–2790, 2007
78. J. Gao, W. Fan, and J. Han, "On appropriate assumptions to mine data streams: analysis and practice," *International Conference on Data Mining*, pp. 143–152, 2007
79. J. Gao, B. Ding, F. Wei, H. Jiawei, and P. S. Yu, "Classifying data streams with skewed class distributions and concept drifts," *IEEE Internet Computing*, vol. 12, no. 6, pp. 37–49, 2008
80. A. Bifet, "Adaptive learning and mining for data streams and frequent patterns." Ph.D. Dissertation, Universitat Politècnica de Catalunya, 2009
81. A. Bifet, E. Frank, G. Holmes, and B. Pfahringer, "Accurate ensembles for data streams: Combining restricted Hoeffding trees using stacking," *2nd Asian Conference on Machine Learning* in Journal of Machine Learning Research, vol. 13, Tokyo, 2010
82. A. Bifet, MOA: Massive Online Analysis, Available at: http://moa.cs.waikato.ac.nz/. Lastaccessed:7/22/2011

Chapter 2
Boosting Algorithms: A Review of Methods, Theory, and Applications

Artur J. Ferreira and Mário A.T. Figueiredo

2.1 Introduction

Boosting is a class of machine learning methods based on the idea that a combination of simple classifiers (obtained by a *weak learner*) can perform better than any of the simple classifiers alone. A *weak learner* (WL) is a learning algorithm capable of producing classifiers with probability of error strictly (but only slightly) less than that of random guessing (0.5, in the binary case). On the other hand, a *strong learner* (SL) is able (given enough training data) to yield classifiers with arbitrarily small error probability.

An ensemble (or committee) of classifiers is a classifier build upon some combination of WLs. The strategy of boosting, and ensembles of classifiers, is to learn many *weak* classifiers and combine them in some way, instead of trying to learn a single *strong* classifier. This idea of building ensembles of classifiers has gained interest in the last decade [67]; the rationale is that it may be easier to train several simple classifiers and combine them into a more complex classifier than to learn a single complex classifier. For instance, instead of training a large *neural network* (NN), we may train several simpler NNs and combine their individual outputs in order to produce the final output (as illustrated in Fig. 2.1).

Letting $H_m\colon \mathscr{X} \rightarrow \{-1, +1\}$ be the mth weak binary classifier (for $m = 1, \ldots, M$), and $\mathbf{x} \in \mathscr{X}$ some input pattern to be classified, there are many ways to combine the outputs $H_1(\mathbf{x}), \ldots, H_M(\mathbf{x})$ into a single class prediction [67].

A.J. Ferreira (✉)
Instituto de Telecomunicações, and Instituto Superior de Engenharia de Lisboa – Polytechnic Institute of Lisbon, ADEETC – Gabinete 16, Rua Conselheiro Emidio Navarro, 1959-007 Lisboa, Portugal
e-mail: arturj@isel.pt

M.A.T. Figueiredo
Instituto de Telecomunicações, and Instituto Superior Técnico – Technical University of Lisbon, Torre Norte, Piso 10, Av. Rovisco Pais, 1049-001 Lisboa, Portugal
e-mail: mario.figueiredo@lx.it.pt

C. Zhang and Y. Ma (eds.), *Ensemble Machine Learning: Methods and Applications*,
DOI 10.1007/978-1-4419-9326-7_2, © Springer Science+Business Media, LLC 2012

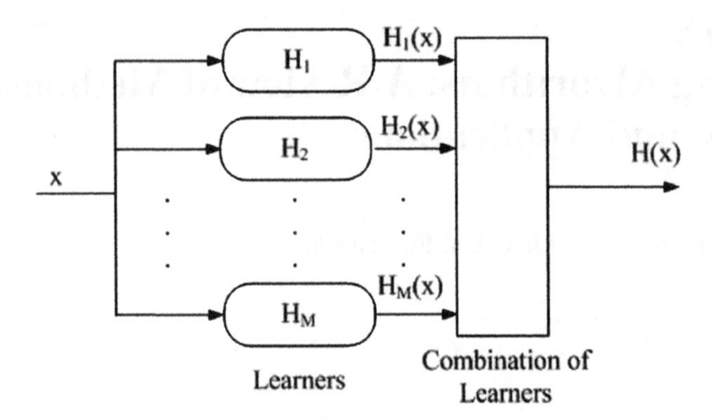

Fig. 2.1 The concept of ensemble of classifiers. The outputs of the weak learners $H_m(\mathbf{x})$ with $m \in \{1, \ldots, M\}$ are combined to produce the output of the ensemble of classifiers given by $H(\mathbf{x})$

For example, assuming that the classifiers err independently of each other, a majority vote combination should yield a lower probability of error than any of the individual classifiers. Considering a weighted linear combination of the outputs of the weak classifiers, the ensemble prediction function $H : \mathscr{X} \rightarrow \{-1, +1\}$ is given by

$$H(\mathbf{x}) = \operatorname{sign}\left(\sum_{m=1}^{M} \alpha_m \; H_m(\mathbf{x})\right), \tag{2.1}$$

where $\alpha_1, \ldots, \alpha_M$ is a set of weights (a simple majority vote results if all the weights are equal).

Among the many different ways in which ensembles of classifiers can be learned and combined [67], boosting techniques exhibit, in addition to good practical performance, several theoretical and algorithmic features that makes them particularly attractive [58, 82, 98]. Essentially, boosting consists of repeatedly using the base weak learning algorithm, on differently weighted versions of the training data, yielding a sequence of weak classifiers that are combined as in (2.1). The weighting of each instance in the training data, at each round of the algorithm, depends on the accuracy of the previous classifiers, thus allowing the algorithm to focus its *attention* on those samples that are still incorrectly classified. The several variants of boosting algorithms differ in their choice of base learners and criterion for updating the weights of the training samples. AdaBoost (which stands for *adaptive boosting*) is arguably the best-known boosting algorithm, and was responsible for sparking the explosion of interest in this class of algorithms that happened after the publication of the seminal works of Freund and Schapire [47–50],

2.1.1 Chapter Outline

The remaining sections of this chapter are organized as follows. Section 2.2 addresses the foundations and origins of boosting algorithms, as a class of methods to improve the accuracy of learning algorithms, by building ensembles of classifiers; the connection of boosting with other machine learning techniques, such as bootstrap and bagging is mentioned. Section 2.3 describes AdaBoost and discusses some of its theoretical properties, regarding training error (TE), generalization error (GE), and the problem of overfitting. In Section 2.4, we describe variants of AdaBoost and their properties, including extensions for multiclass problems, while boosting algorithms for semi-supervised learning (SSL) are discussed in Section 2.5. Section 2.6 discusses several successful applications of batch and online boosting algorithms and presents an experimental evaluation of some boosting algorithms, compared to other machine learning techniques on standard benchmark datasets. Section 2.7 provides a summary and a discussion on boosting algorithms. Finally, Section 2.8 ends the chapter with some bibliographic and historical remarks.

2.2 The Origins of Boosting and Adaptive Boosting

2.2.1 Bootstrapping and Bagging

Bootstrapping [37, 38] is a general purpose sample-based statistical method in which several (nondisjoint) training sets are obtained by drawing randomly, *with* replacement, from a single base dataset. In a dataset with N samples, each instance is selected with probability $1/N$; consequently, after N draws (with large N), the probability that a given instance was not selected is

$$\left(1 - \frac{1}{N}\right)^N \approx \exp(-1) \approx 0.368; \tag{2.2}$$

the validity of this approximation is illustrated in Fig. 2.2, showing that it is quite accurate even with only a moderately large N. This implies that each sample contains roughly 63.2% of the instances.

Classically, bootstrapping is used to infer some statistic $T(P)$ about a (say infinitely large) population P, from N samples thereof: $Z = \{z_1, \ldots, z_N\}$. The idea is to obtain B sets $Z_b^* \subseteq Z$, for $b = 1, \ldots, B$, each containing N random samples (with replacement) from Z, from which B estimates of $T(P)$ are obtained. These estimates are then averaged into a final estimate; it is also possible to obtain variance estimates or confidence intervals. The procedure is formally described in Algorithm 1.

Bagging (which stands for *bootstrap aggregation* [11]) is a technique which uses bootstrap sampling to reduce the variance and/or improve the accuracy of some predictor (it may be used in classification and regression). Consider a size-N dataset $Z = \{z_1, z_2, \ldots, z_N\}$, where now $z_i = (\mathbf{x}_i, y_i)$, where y_i is a class label, in classification problems, or a real number, in regression problems. The rationale of bagging is to learn a set of B predictors (each from a bootstrap sample $Z_b^* \subseteq Z$,

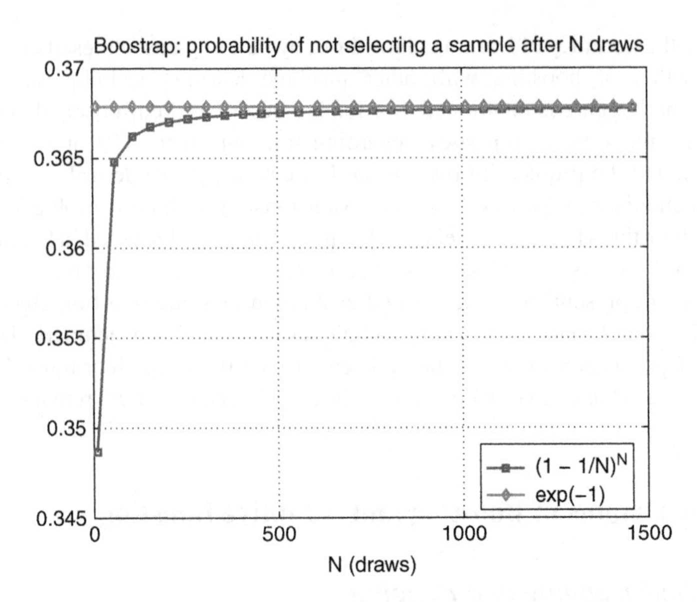

Fig. 2.2 The Bootstrap procedure: probability of not selecting a training sample after N draws and its approximation $\exp(-1)$

Algorithm 1 Bootstrap Procedure

Input: Size-N sample $Z = \{z_1, z_2, \ldots, z_N\}$ of a (potentially infinite) population P.
B, number of bootstrap samples.
Output: Estimate $\widehat{T}(P)$ of the population statistic.

1: **for** $b = 1$ to B **do**
2: Draw, with replacement, N samples from Z, obtaining the bth bootstrap sample Z_b^*.
3: Compute, for each sample Z_b^*, the estimate of the statistic $\widehat{T}(Z_b^*)$.
4: **end for**
5: Compute the bootstrap estimate $\widehat{T}(P)$ as the average of $\widehat{T}(Z_1^*),\ldots,\widehat{T}(Z_B^*)$.
6: Compute the accuracy of the estimate, using, e.g., the variance of $\widehat{T}(Z_1^*),\ldots,\widehat{T}(Z_B^*)$.

for $b = 1, \ldots, B$) and then produce a final predictor by combining (by averaging, in regression, or majority voting, in classification) this set of predictors. The combination of multiple predictors decreases the expected error because it reduces the variance component of the bias–variance decomposition [58]. The reduction on this variance component is proportional to the number of classifiers applied in the ensemble. The bagging procedure, for binary classification, is described in Algorithm 2.

As compared to the process of learning a classifier in a conventional way, that is, from the full training set, bagging has two main advantages:

- increases classifier stability and accuracy;
- reduces classifier variance, in terms of the bias–variance decomposition [58].

Algorithm 2 Bagging Procedure for Classification

Input: Dataset $Z = \{z_1, z_2, \ldots, z_N\}$, with $z_i = (\mathbf{x}_i, y_i)$, where $\mathbf{x}_i \in \mathcal{X}$ and $y_i \in \{-1, +1\}$.
B, number of bootstrap samples.
Output: $H : \mathcal{X} \rightarrow \{-1, +1\}$, the final classifier.

1: **for** $b = 1$ to B **do**
2: Draw, with replacement, N samples from Z, obtaining the bth bootstrap sample Z_b^*.
3: From each bootstrap sample Z_b^*, learn classifier H_b.
4: **end for**
5: Produce the final classifier by a majority vote of $H_1, \ldots, H_B$, that is, $H(\mathbf{x}) = \text{sign}\left(\sum_{b=1}^{B} H_b(\mathbf{x})\right)$.

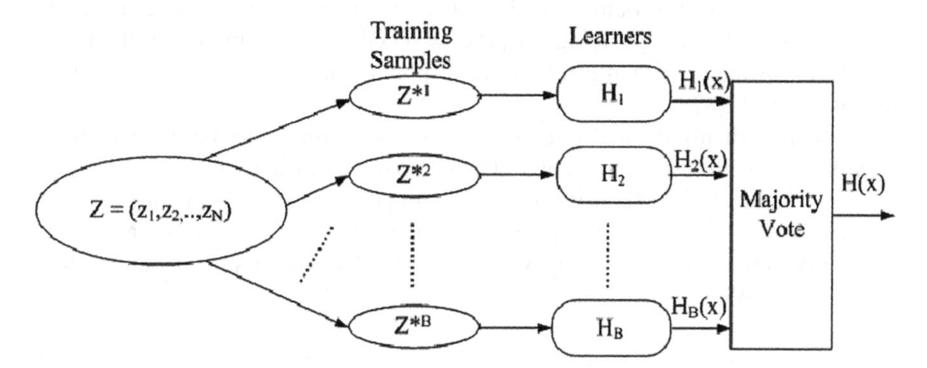

Fig. 2.3 The bagging approach to classification. Using bootstrap, we produce several training samples; each of these samples is fed into a weak learner. The final classification decision is produced by a majority vote on the weak learners output

The use of the bagging technique improves the classification results whenever the base classifiers are unstable, this being the main reason why the bagging approach works well for classification. Figure 2.3 depicts the bagging approach for classification.

For further reading on bagging, see [12, 13, 95, 139, 140]. In [139], the authors argue that for very weak learners (e.g., decision stumps, which are tree classifier with only one inner node), the base classifiers built from bootstrap samples are strongly correlated. As a consequence, a simple bagged classifier with these very weak learners has very little improvement compared to a single classifier trained from the same data. To overcome this problem, they propose the *local lazy learning bagging* (LLLB) approach, where base learners are trained from a small subset surrounding each test instance. The experimental results on real-world datasets show that the LLLB method significantly outperforms standard bagging.

2.2.2 Weak and Strong Learners

Weak and *strong* learning are fundamental concepts at the heart of boosting algorithms, so we briefly review their formal definitions. These concepts are rooted in the theory of PAC (probably approximately correct) learning [114], where they are defined as follows. Consider an hypothesis, i.e., a classification rule $f : \mathscr{X} \rightarrow \{-1, +1\}$, such that $f \in \mathscr{F}$, where $\mathscr{F}$ is some class of functions from $\mathscr{X}$ to $\{-1, +1\}$. Consider also a set of examples of that hypothesis, i.e., a set of pairs $\{(\mathbf{x}_i, y_i), i = 1, \ldots, N\}$ such that $y_i = f(\mathbf{x}_i)$ and the $\mathbf{x}_i$ are samples of some distribution P. A *strong* learner is capable of, given enough data, producing an arbitrarily good classifier with high probability, that is, for every P, $f \in \mathscr{F}$, $\varepsilon \geq 0$, and $\delta \leq 1/2$, it outputs, with probability no less than $1 - \delta$, a classifier $h : \mathscr{X} \rightarrow \{-1, +1\}$ satisfying $\mathbb{P}_P [h(\mathbf{x}) \neq f(\mathbf{x})] \leq \varepsilon$. Furthermore, the time complexity of the algorithm can be at most polynomial in $1/\varepsilon$, $1/\delta$, N, and the dimension of $\mathscr{X}$ [82].

A WL is formally defined in a similar way as a strong one, but with weaker quantification with respect to ε and δ. Given a particular (rather than "for every") pair $\varepsilon_0 \geq 0$, and $\delta_0 \leq 1/2$, a WL outputs, with probability no less than $1 - \delta_0$, a classifier $h : \mathscr{X} \rightarrow \{-1, +1\}$ satisfying $\mathbb{P}_P [h(\mathbf{x}) \neq f(\mathbf{x})] \leq \varepsilon_0$. Underlying the idea of boosting is the fact, proved by Schapire [95] that it is possible to obtain a SL by combining WLs.

2.2.3 Boosting Algorithms

The first boosting procedure was proposed by Schapire in [95], where the key result is that weak and strong learnability are equivalent, in the sense that strong learning can be performed by combining WLs. The boosting procedure proposed in [95] is described in detail in Algorithm 3.

Algorithm 3 Boosting Procedure for Classification

Input: Dataset $Z = \{z_1, z_2, \ldots, z_N\}$, with $z_i = (\mathbf{x_i}, y_i)$, where $\mathbf{x}_i \in \mathscr{X}$ and $y_i \in \{-1, +1\}$.
Output: A classifier $H : \mathscr{X} \rightarrow \{-1, +1\}$.

1: Randomly select, without replacement, $L_1 < N$ samples from Z to obtain Z_1^*.
2: Run the WL on Z_1^*, yielding classifier H_1.
3: Select $L_2 < N$ samples from Z, with half of the samples misclassified by H_1, to obtain Z_2^*.
4: Run the WL on Z_2^*, yielding classifier H_2.
5: Select all samples from Z on which H_1 and H_2 disagree, producing Z_3^*.
6: Run the WL on Z_3^*, yielding classifier H_3.
7: Produce the final classifier as a majority vote: $H(\mathbf{x}) = \text{sign}\left(\sum_{b=1}^{3} H_b(\mathbf{x})\right)$.

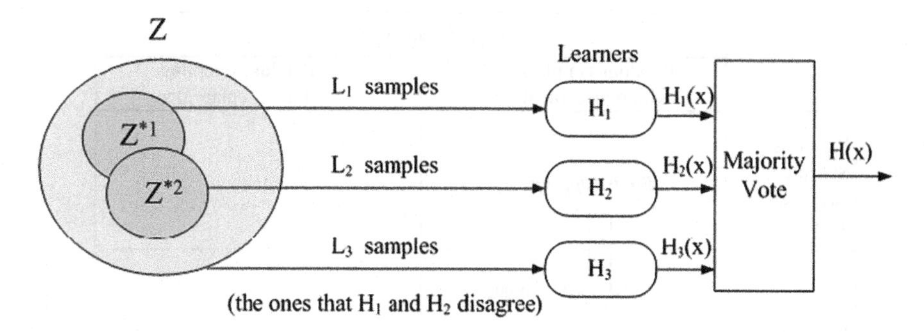

Fig. 2.4 A graphical idea of the first boosting approach proposed in [95]. Notice that each learner can be itself learned by the boosting algorithm in a recursive fashion

As can be seen in Algorithm 3, the training set is randomly divided without replacement into three partitions, Z_1^*, Z_2^*, and Z_3^*. For a given instance, if the first two classifiers (H_1 and H_2) agree on the class label, this is the final decision for that instance. The set of instances on which they disagree defines the partition Z_3^*, which is used to learn H_3. Schapire has shown that this learning method is strong, in the sense defined above. Moreover, the error can be further reduced by using this approach recursively, that is, each learner can itself be obtained by a boosting procedure. Figure 2.4 illustrates the boosting approach.

After this proposal by Schapire, Freund [44] proposed a new boosting algorithm based on, and improving, the ideas presented in [95]. That algorithm improves the accuracy of algorithms for learning binary classifiers, by combining a large number of classifiers, each of which is obtained by running the given learning method on a different set of examples. As in [95], Freund's new proposals also suffered from several drawbacks, namely the need for a very large training set, due to the fact that this set is divided into subsets.

2.2.4 *Relationship Between Boosting, Bagging, and Bootstrapping*

Figure 2.5 shows the connection between bootstrapping, bagging, and boosting, focusing on what they produce and how they handle the training data. The figure emphasizes the fact that these three techniques are all built upon random sampling, being that bootstrapping and bagging perform sampling with replacement while boosting does not. Bagging and boosting have in common the fact that both provide final classifiers that are majority votes of the individual classifiers.

In [29], a comparison of the effectiveness of randomization, bagging, and boosting for improving the performance of the decision-tree algorithm C4.5 [88] is presented. The experimental results show that for cases with little or no classification

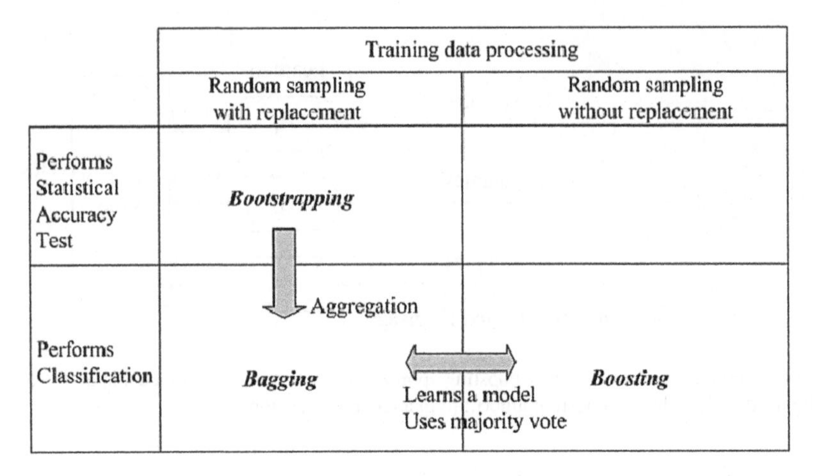

Training data processing
Random sampling with replacement
Random sampling without replacement
Performs Statistical Accuracy Test
Bootstrapping
Aggregation
Performs Classification
Bagging
Learns a model
Uses majority vote
Boosting

Fig. 2.5 Bootstrapping, bagging, and boosting: what they yield and how they handle the training data

noise, randomization is competitive with (and perhaps slightly superior to) bagging but not as accurate as boosting. For situations with substantial classification noise, bagging is much better than boosting, and sometimes better than randomization.

2.3 The AdaBoost Algorithm

After their initial separate work on boosting algorithms, Freund and Schapire proposed the *adaptive boosting* (AdaBoost) algorithm [47], [48], [50]. The key idea behind AdaBoost is to use *weighted* versions of the *same* training data instead of randomly subsamples thereof. The same training set is repeatedly used and, for this reason, it does not need to be very large, unlike earlier boosting methods.

The AdaBoost algorithm is now a well known and deeply studied method to build ensembles of classifiers with very good performance [58]. The algorithm learns a set of classifiers, using a WL, in order to produce the final classifier of the form (2.1). The weak classifiers[1] are obtained sequentially, using reweighted versions of the training data, with the weights depending on the accuracy of the previous classifiers. The training set is always the same at each iteration, with each training instance weighted according to its (mis)classification by the previous classifiers. This allows the WL at each iteration to focus on patterns that were not well classified by the previous weak classifiers. It is important to chose WLs to obtain the base classifiers, allowing them to learn without decreasing significantly the weight of the previously correctly classified instances. If the base learner is too strong, it may achieve high accuracy, leaving only outliers and noisy instances with significant weight to be

[1]We refer to a classifier learned by a WL as a weak classifier.

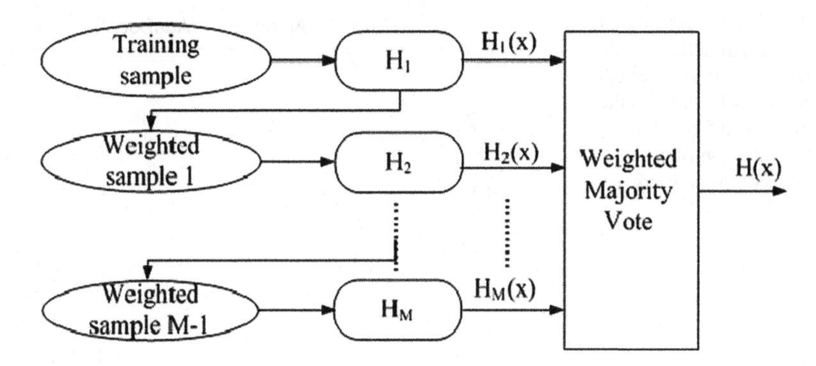

Fig. 2.6 Graphical idea of the adaptive boosting algorithm (adapted from [58]). Each weak learner is trained on a different weighted version of the training data sample. There is no sampling of the training data and the weight of each instance for the following round depends on the performance of the previous learner

Algorithm 4 (Discrete) AdaBoost algorithm for binary classification

Input: Dataset $Z = \{z_1, z_2, \ldots, z_N\}$, with $z_i = (\mathbf{x_i}, y_i)$, where $\mathbf{x}_i \in \mathcal{X}$ and $y_i \in \{-1, +1\}$.
M, the maximum number of classifiers.

Output: A classifier $H : \mathcal{X} \to \{-1, +1\}$.

1: Initialize the weights $w_i^{(1)} = 1/N, i \in \{1, \ldots, N\}$, and set $m = 1$.
2: **while** $m \leq M$ **do**
3: Run weak learner on Z, using weights $w_i^{(m)}$, yielding classifier $H_m : \mathcal{X} \to \{-1, +1\}$.
4: Compute $\text{err}_m = \sum_{i=1}^{N} w_i^{(m)}\ h\left(-y_i\ H_m(\mathbf{x_i})\right)$, the weighted error of H_m.
5: Compute $\alpha_m = \frac{1}{2} \log\left(\frac{1-\text{err}_m}{\text{err}_m}\right)$. {/* Weight of weak classifier. */}
6: For each sample $i = 1, \ldots, N$, update the weight $v_i^{(m)} = w_i^{(m)} \exp(-\alpha_m\ y_i\ H_m(\mathbf{x_i}))$.
7: Renormalize the weights: compute $S_m = \sum_{j=1}^{N} v_j$ and, for $i = 1, \ldots, N$, $w_i^{(m+1)} = v_i^{(m)} \big/ S_m$.
8: Increment the iteration counter: $m \leftarrow m + 1$
9: **end while**
10: Final classifier: $H(\mathbf{x}) = \text{sign}\left(\sum_{j=1}^{M} \alpha_j\ H_j(\mathbf{x})\right)$.

learned in the following rounds. Figure 2.6 depicts the structure of AdaBoost, which is described in detail in Algorithm 4.

The function $h : \mathbb{R} \to \{0, 1\}$ used in line 4 of the algorithm is the Heaviside function, defined as $h(x) = 1$, if $x \geq 0$, and $h(x) = 0$, if $x < 0$. Consequently, since both y_i and $H_m(\mathbf{x_i})$ take values in $\{-1, +1\}$, we have that $h\left(-y_i\ H_m(\mathbf{x_i})\right) = 1$, if $y_i \neq H_m(\mathbf{x_i})$, and $h\left(-y_i\ H_m(\mathbf{x_i})\right) = 0$, if $y_i = H_m(\mathbf{x_i})$, and err_m is the weighted error rate of the mth classifier.

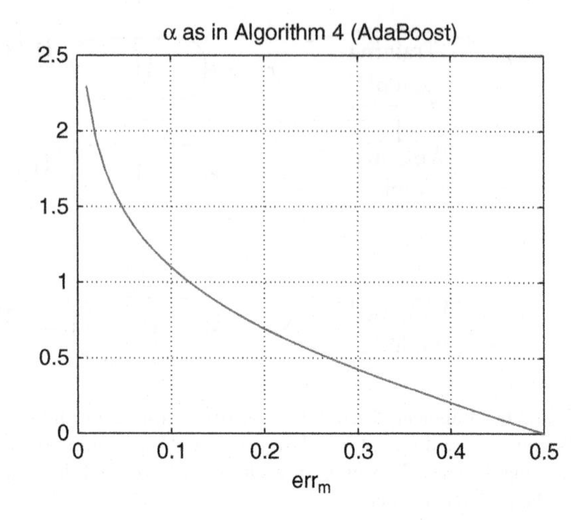

Fig. 2.7 The computation of α_m ($\geq$0) as a function of the weighted classification error for each weak learner. As the error tends to 0.5 (random guessing) the contribution (importance) of the weak learner decreases

Line 3 requires some explanation: what does it mean to run a weak learning algorithm on a weighted version of the training set? It means that the goal of the WL is to obtain a classifier, say H_m, belonging to a given family of classifiers $\mathcal{H}$, that satisfies

$$\sum_{i=1}^{N} w_i \, h\,(-y_i \; H_m(\mathbf{x_i})) \leq \frac{1}{2} - \varepsilon, \tag{2.3}$$

for some small positive ε. Notice that only if $w_i = 1/N$, for $i = 1, \ldots, N$ (e.g., at the first iteration of AdaBoost) does the left hand side of (2.3) coincides with the classical error rate on the training set. The existence of such weak classifiers is an important ingredient of boosting, and the interested reader is referred to [82] for more details. Notice that the *weakness* of the classifier is usually controlled by letting $\mathcal{H}$ contain only simple classifiers; for example, when $\mathcal{X} = \mathbb{R}^d$, the family $\mathcal{H}$ may contain only linear rules of the form $H(\mathbf{x}) = \text{sign}(\mathbf{u}^T\mathbf{x} + r)$, where $\mathbf{u} \in \mathbb{R}^d$ and $r \in \mathbb{R}$, which is sometimes known as a perceptron, or rules based on a single component of the input, i.e., of the form $H(\mathbf{x}) = \text{sign}(u \; x_j + t)$, where $u \in \{-1, +1\}$ and $t \in \mathbb{R}$, which is called a decision stump.

Notice that the AdaBoost algorithm can actually handle weak classifiers with weighted error rate larger than 1/2; of course, by simply inverting the output of such a classifier, we obtain a classifier with weighted error rate less than 1/2. Such an inversion is automatically performed by AdaBoost, because if $\text{err}_m > 1/2$, the corresponding weight α_m is negative, as is clear from its expression in line 5 of Algorithm 4. Figure 2.7 shows how α_m evolves as a function of the weighted classification error err_m for each weak learner.

Table 2.1 shows the connection between the boosting algorithm (Algorithm 3) and AdaBoost (Algorithm 4). We compare these algorithms in terms of how the training data is processed, the number of classifiers and how the final decision is produced.

Table 2.1 Summary of the main differences between Algorithms 3 (Boosting) and 4 (AdaBoost), regarding how training data is used, the number of samples, the number of classifiers, and the decision mechanism

	Boosting (Algorithm 3)	*AdaBoost (Algorithm 4)*
Data usage	Random sampling, no replacement	Weighting (no sampling)
Number of samples	Three	One
Number of classifiers	Three	Up to M
Decision	Majority vote	Weighted majority vote

A key issue when using the weights for the instances, is that the following learner is provided with more information about the importance of each instance and how the previous learners were (or not) able to deal with that instance. This does not happen in bagging nor boosting.

Notice that a straightforward consequence of the instance weighting scheme is that, after the M AdaBoost rounds, the misclassified patterns assigned with higher weights are "hard" patterns to learn; these patterns are probably outliers. This is a kind of side effect of AdaBoost, which can be used for outlier detection on a given training set.

The AdaBoost algorithm has also been extended for regression tasks. In [3], the prediction error is compared against a threshold to mark it as an error or not and then the AdaBoost version for classification is used. In [31], the probabilities kept by the algorithm are modified based on the magnitude of the error; instances with large error on the previous learners have a higher probability of being chosen to train the following base learner. The median or weighted average is then applied to combine the predictions of the different base learners.

2.3.1 Some Theoretical Properties

We now review several properties of AdaBoost that were shown by Freund and Schapire [50, 100], namely the exponential decay of the TE rate.

The first result shows that the TE of the classifier obtained after M boosting rounds is upper bounded by the product of the normalizing constants of the weights of all the rounds, that is,

$$\mathrm{TE} = \frac{1}{N}\sum_{i=1}^{N} h(-y_i\, H(\mathbf{x}_i)) \leq \prod_{j=1}^{M} S_j, \tag{2.4}$$

where S_j is the normalizing constant used in line 7 at iteration j (the proof of this result can be found in Appendix A).

The second result shows how the TE depends on the weighted error rates of the weak classifiers (denoted err_m). Assume that $\mathrm{err}_m = 1/2 - \gamma_m$, with $\gamma_m > \gamma > 0$, for all $m = 1, \ldots, M$. Then

$$\mathrm{TE} \leq \exp\left(-2\, M\, \gamma^2\right), \tag{2.5}$$

that is, the TE decreases exponentially with M, and does so at a rate that depends on γ (the proof of this result can be found in Appendix A).

The *expected test error* commonly addressed as the generalization error (*GE*) also has an upper bound, as demonstrated in [50]. The GE of the final classifier is upper bounded, with high probability by

$$\mathrm{TE} + \tilde{O}\left(\sqrt{\frac{M\,d}{N}}\right), \tag{2.6}$$

where d is the Vapnik–Chervonenkis (VC) [9, 117] dimension of the set of base classifiers. This result shows that there is a trade-off controlled by the "richness" or "complexity" of the base (weak) classifiers; "stronger" base classifiers allow the TE to be lower, but correspond to a larger CV dimension; on the other hand, simpler classifiers have a lower CV dimension, but require more boosting rounds to decrease the TE.

It has been found empirically that the GE usually does not increase as the size of ensemble becomes very large; moreover, it is often observed that the GE continues to decrease even after the TE has reached zero (see Fig. 2.12). In [99], it is shown that this behavior is related to the distribution of margins of the training examples with respect to the generated voting classification rule. The margin of an example is defined as the difference between the number of correct votes and the maximum number of votes received by any incorrect label.

2.3.2 Different Views of AdaBoost

It has been argued that one explanation for the success of AdaBoost is its ability to increase the *margin* between positive and negative examples [99]. This view provides a connection between margin-based discriminative learning (as in *support vector machines*—SVM [102]) and boosting.

The adaptive boosting techniques can be considered as a greedy optimization method for minimizing the exponential loss function

$$\frac{1}{N}\sum_{i=1}^{N}\exp\left(-y_i f(\mathbf{x_i})\right) = \sum_{i=1}^{N}\exp\left(-y_i \sum_{m=1}^{M}\alpha_m H_m(\mathbf{x_i})\right), \tag{2.7}$$

by learning H_m and choosing the most adequate value of α_m at each round. Detailed analysis of boosting and different views of how this learning procedure behaves can be found in [35, 50, 58, 80, 81].

In [21], a unified view of boosting and logistic regression [58] is described. These learning problems are cast in terms of optimization of Bregman distances, due to their high similarity under this framework. For both problems, new sequential and parallel algorithms are proposed and their potential advantages over existing methods are shown. A general proof of convergence for AdaBoost is also presented.

Some connections of AdaBoost with game-theory, linear programming, logistic regression, and estimation of probabilities and outliers are discussed in [97, 100].

An evaluation of bagging and boosting using both NNs and decision trees as learners is carried out in [77]. The experimental results show two important conclusions. The first is that, even though bagging almost always produces a better classifier than any of its individual component classifiers and is relatively impervious to overfitting, it does not generalize any better than a baseline NN ensemble method. The second is that, although boosting is a powerful technique that can usually produce better ensembles than bagging, it is more susceptible to noise and overfitting.

In [42], AdaBoost is evaluated on synthetic and real data using two types of WLs: generative classifiers and radial basis function classifiers. The AdaBoost algorithm with these WLs shows good convergence properties. On benchmark data, boosting of these WLs attains results close to the Real AdaBoost algorithm (with decision trees) and SVM, constituting a low computational complexity competitive choice.

2.4 Variants of AdaBoost

In this section, we review several variants of AdaBoost (although we do not claim to have an exhaustive list), both for binary and multiclass supervised learning problems. Many of these variants have proven to be successful in different types of learning scenarios.

The proposal of the AdaBoost algorithm stimulated a significant amount of research on this type of learning technique, exploiting its theoretical properties and experimental performance. From this research, several variants of AdaBoost have emerged, some targeted at specific problems, such as, for example, face detection and text categorization (TC). Those variants follow the overall structure of AdaBoost (learn a weak classifier, compute the amount of error, update the weights of the training patterns and repeat the process), but introduce changes on several aspects, such as the weight update expression and classifier management.

2.4.1 Detailed Analysis of Some Variants

After AdaBoost was introduced, several modified versions (variants) have been proposed, developed, and compared with AdaBoost. This section addresses some of these variants (shown in the timeline of Fig. 2.8) for supervised learning of binary classifiers.

The following subsections describe in detail some of these variants for binary classification. These variants were selected to be presented in more detail, because they are either the first variants to appear after AdaBoost was proposed or they bring quite different new ideas into the adaptive boosting scheme. These variants have in common the fact that all of them were proved to be successful in real-world machine learning problems.

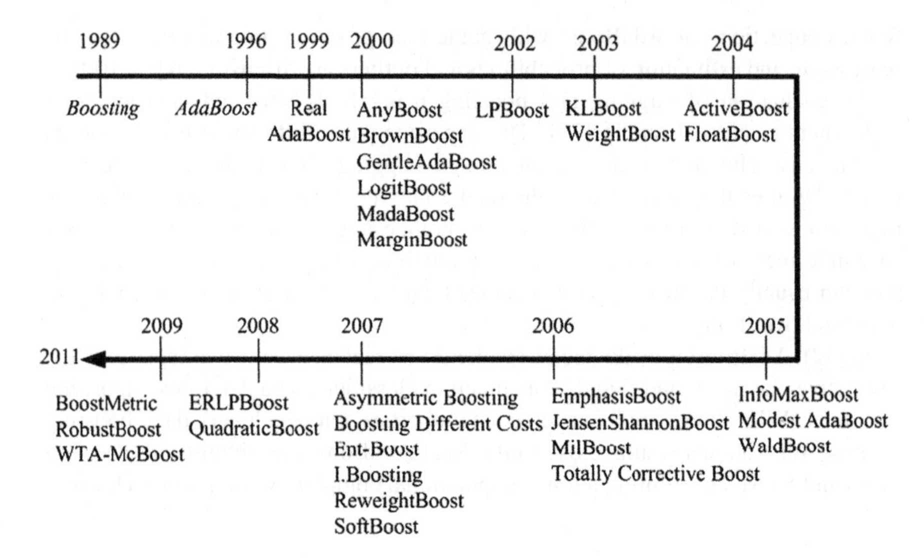

Fig. 2.8 A (possibly incomplete) timeline of AdaBoost variants for supervised learning of binary classifiers, as of 2011

Algorithm 5 Real AdaBoost

Input: Dataset $Z = \{z_1, z_2, \ldots, z_N\}$, with $z_i = (\mathbf{x}_i, y_i)$, where $\mathbf{x}_i \in \mathcal{X}$ and $y_i \in \{-1, +1\}$.
M, the maximum number of classifiers.

Output: A classifier $H : \mathcal{X} \rightarrow \{-1, +1\}$.

1: Initialize the weights $w_i = 1/N, i \in \{1, \ldots, N\}$.
2: **for** $m = 1$ to M **do**
3: Fit the class probability estimate $p_m(\mathbf{x}) = \hat{P}_w(y = 1|\mathbf{x})$, using w_i.
4: Set $\mathrm{H}_m = \frac{1}{2} \log((1 - p_m(\mathbf{x}))p_m(\mathbf{x})) \in \mathcal{R}$.
5: Update the weights: $w_i \leftarrow w_i \exp(-y_i H_m(\mathbf{x}_i))$
6: Renormalize to weights.
7: **end for**
8: Final classifier: $H(\mathbf{x}) = \text{sign}\left(\sum_{j=1}^{M} \alpha_j \; H_j(\mathbf{x})\right)$.

2.4.1.1 Real AdaBoost

The first variant we consider is Real AdaBoost [52, 100], where the term *real* refers to the fact that the algorithm uses real-valued “classifiers” (i.e., before thresholding). This real value can be seen as the probability, or degree of confidence, that a given input pattern belongs to a class, considering the current weight distribution for the training set. The Real AdaBoost algorithm is presented as Algorithm 5.

Comparing Real AdaBoost with AdaBoost, we see that the major differences are in lines 3 and 4. In the Real AdaBoost algorithm, these steps consist of computing

Algorithm 6 Logit Boost

Input: Dataset $Z = \{z_1, z_2, \ldots, z_N\}$, with $z_i = (\mathbf{x_i}, y_i)$, where $\mathbf{x}_i \in \mathscr{X}$ and $y_i \in \{-1, +1\}$.
M, the maximum number of classifiers.

Output: A classifier $H : \mathscr{X} \rightarrow \{-1, +1\}$.

1: Initialize the weights $w_i = 1/N, i \in \{1, \ldots, N\}$.
2: **for** $m = 1$ to M and while $H_m \neq 0$ **do**
3: Compute the working response $z_i = \dfrac{y_i^* - p(\mathbf{x_i})}{p(\mathbf{x_i})(1 - p(\mathbf{x_i}))}$ and weights $w_i = p(\mathbf{x_i})(1 - p(\mathbf{x_i}))$.
4: Fit $H_m(\mathbf{x})$ by a weighted least-squares of z_i to $\mathbf{x_i}$, with weights w_i.
5: Set $H(\mathbf{x}) = H(\mathbf{x}) + \frac{1}{2} H_m(\mathbf{x})$ and $p(\mathbf{x}) = \dfrac{\exp(H(\mathbf{x}))}{\exp(H(\mathbf{x})) + \exp(-H(\mathbf{x}))}$.
6: **end for**
7: Final classifier: $H(\mathbf{x}) = \text{sign}\left(\sum_{j=1}^{M} \alpha_j \; H_j(\mathbf{x})\right)$.

and using estimates of the probabilities that each training pattern belongs to a class, under the current weight distribution. Standard AdaBoost classifies the input patterns and computes the weighted error rate.

2.4.1.2 Logit Boost

The Logit Boost variant consists of using adaptive Newton steps to fit an additive logistic model [51, 52]. Instead of minimizing the exponential loss, Logit Boost minimizes the logistic loss (negative conditional log-likelihood). Algorithm 6 details the Logit Boost algorithm.

2.4.1.3 Gentle AdaBoost

The Gentle AdaBoost [52] algorithm improves over Real AdaBoost by using Newton steps, providing a more reliable and stable ensemble, since it puts less emphasis on outliers. Instead of fitting a class probability estimate, Gentle AdaBoost (described in Algorithm 7) uses weighted least-squares regression [58] at each iteration. The main difference between Gentle and Real AdaBoost is on the use of the estimates of the weighted class probabilities in order to perform the update. The algorithm is *gentle* because it is considered to be both conservative and more stable as compared to Real AdaBoost. Gentle AdaBoost does not require the computation of log ratios which can be numerically unstable (since they involve quotients, maybe with the denominator approaching zero). Experimental results on benchmark data show that the conservative Gentle AdaBoost has similar performance to Real AdaBoost and Logit Boost, and in many cases outperforms these other two variants.

Algorithm 7 Gentle AdaBoost

Input: Dataset $Z = \{z_1, z_2, \ldots, z_N\}$, with $z_i = (\mathbf{x_i}, y_i)$, where $\mathbf{x}_i \in \mathscr{X}$ and $y_i \in \{-1, +1\}$.
M, the maximum number of classifiers.

Output: A classifier $H : \mathscr{X} \rightarrow \{-1, +1\}$.

1: Initialize the weights $w_i = 1/N, i \in \{1, \ldots, N\}$.
2: **for** $m = 1$ to M **do**
3: Train $H_m(\mathbf{x})$ by weighted least-squares of y_i to $\mathbf{x_i}$, with weights w_i.
4: Update $H(\mathbf{x}) \leftarrow H(\mathbf{x}) + H_m(\mathbf{x})$.
5: Update $w_i \leftarrow w_i \exp(-y_i H_m(\mathbf{x_i}))$ and renormalize to $\sum_i w_i = 1$.
6: **end for**
7: Final classifier: $H(\mathbf{x}) = \text{sign}\left(\sum_{j=1}^{M} \alpha_j \; H_j(\mathbf{x})\right)$.

Algorithm 8 Modest AdaBoost

Input: Dataset $Z = \{z_1, z_2, \ldots, z_N\}$, with $z_i = (\mathbf{x_i}, y_i)$, where $\mathbf{x}_i \in \mathscr{X}$ and $y_i \in \{-1, +1\}$.
M, the maximum number of classifiers.

Output: A classifier $H : \mathscr{X} \rightarrow \{-1, +1\}$.

1: Initialize the weights $w_i = 1/N, i \in \{1, \ldots, N\}$.
2: **for** $m = 1$ to M and while $H_m \neq 0$ **do**
3: Train $H_m(\mathbf{x})$ by weighted least-squares of y_i to $\mathbf{x_i}$, with weights w_i.
4: Compute "inverted" distribution $\overline{w}_i = (1 - w_i)$ and renormalize to $\sum_i \overline{w}_i = 1$.
5: Compute $P_m^{+1} = P_w(y = +1, H_m(\mathbf{x}))$, $\overline{P}_m^{+1} = P_{\overline{w}}(y = +1, H_m(\mathbf{x}))$.
6: Compute $P_m^{-1} = P_w(y = -1, H_m(\mathbf{x}))$, $\overline{P}_m^{-1} = P_{\overline{w}}(y = -1, H_m(\mathbf{x}))$,
7: Set $H_m(\mathbf{x}) = \left(P_m^{+1} \left(1 - P_m^{+1}\right) - P_m^{-1} \left(1 - P_m^{-1}\right)\right)$
8: Update $w_i \leftarrow w_i \exp(-y_i H_m(\mathbf{x_i}))$ and renormalize to $\sum_i w_i = 1$.
9: **end for**
10: Final classifier: $H(\mathbf{x}) = \text{sign}\left(\sum_{j=1}^{M} \alpha_j \; H_j(\mathbf{x})\right)$.

2.4.1.4 Modest AdaBoost

The Modest AdaBoost algorithm [119] is known to have lower GE and higher TE, as compared to Real and Gentle AdaBoost variants. Algorithm 8 shows the details of Modest AdaBoost, which, as compared to the previous variants, uses a different weighting scheme for the correctly and incorrectly classified patterns, using an "inverted" distribution.

The standard distribution w_i assigns high weights to training samples misclassified by earlier steps. On the contrary, $\overline{w}_i$ gives higher weights to samples that are already correctly classified by earlier steps.

Lines 5 and 6 deal with the direct and "inverted" distributions, using the expressions $P_m^{+1} = P_w(y = +1, H_m(\mathbf{x}))$ and $P_m^{-1} = P_w(y = -1, H_m(\mathbf{x}))$; these expressions compute how good is the current weak classifier at predicting class labels. On the other hand, the expressions $\overline{P}_m^{+1} = P_{\overline{w}}(y = +1, H_m(\mathbf{x}))$ and

$\overline{P}_m^{-1} = P_{\overline{w}}(y = -1, H_m(\mathbf{x}))$ estimate how well our current WL $H_m(\mathbf{x})$ is working on the data that has been correctly classified by previous steps.

The update $H_m(\mathbf{x}) = (P_m^{+1}(1-\overline{P}_m^{+1}) - P_m^{-1}(1-\overline{P}_m^{-1}))$ decreases weak classifiers contribution, if it works "too well" on data that has been already correctly classified with high margin. This way, the algorithm is named *Modest* because the classifiers tend to work only in their domain, as defined by w_i.

2.4.1.5 Float Boost

The Float Boost [72,73] variant is composed of the following stages: 1-initialization; 2-forward inclusion; 3-conditional exclusion; 4-output. All of these stages, with the exception of stage 3, are similar to those of AdaBoost and other variants as discussed so far. The novelty here is the conditional exclusion stage, in which the least significant weak classifier is removed from the set of classifiers, subject to the condition that the removal leads to an error below some threshold. The Float Boost algorithm details are described as Algorithm 9.

2.4.1.6 Emphasis Boost

The Emphasis Boost variant uses a *weighted emphasis* (WE) function [53]. Each input pattern is weighted according to a criterion (parameterized by λ), through the WE function, in such a way that the training process focuses on the "critical" patterns (near the classification boundary) or on the quadratic error of each pattern. Algorithm 10 presents the details of Emphasis Boost.

The WE function is defined by

$$w_i = \exp\left(\lambda\left(\sum_{j=1}^{m}\left(\alpha_j H_j(x_i) - y_i\right)^2\right) - (1-\lambda)\left(\sum_{j=1}^{m} H_j(x_i)\right)^2\right) \tag{2.8}$$

and controls where the emphasis is placed. This flexible formulation allows choosing how much to consider the *proximity* terms by means of a weighting parameter ($0 \leq \lambda \leq 1$). This way, we have a *boosting by weighting boundary and erroneous samples* technique. Regarding the value of λ, three particular cases are interesting enough to be considered:

- $\lambda = 0$, focus on the "critical" patterns because only the "proximity" to the boundary is taken into account

$$w_i = \exp\left[-\left(\sum_{j=1}^{m} H_j(\mathbf{x_i})\right)^2\right]. \tag{2.9}$$

Algorithm 9 Float Boost

Input: Dataset $Z = \{z_1, z_2, \ldots, z_N\}$, with $z_i = (\mathbf{x_i}, y_i)$, where $\mathbf{x}_i \in \mathscr{X}$ and $y_i \in \{-1, +1\}$.
M, the maximum number of classifiers.
N examples $N = a + b$; a examples have $y_i = +1$ and b examples have $y_i = -1$.
$J(H_M)$, the cost function and the maximum acceptable cost J^*.
Output: A classifier $H : \mathscr{X} \rightarrow \{-1, +1\}$.

1: {1 - Initialization stage.}
2: Initialize the weights $w_i^{(0)} = 1/2a$, for those examples with $y_i = +1$.
3: Initialize the weights $w_i^{(0)} = 1/2b$, for those examples with $y_i = -1$.
4: $J_m^{\min} = J^*$ $m = \{1, \ldots, M\max\}$.
5: $M = 0$, $\mathscr{H}_0 = \{\}$.
6: {2 - Forward inclusion stage.}
7: $M \leftarrow M + 1$.
8: Learn $H_m(x)$ and α_M.
9: Update $w_i^{(M)} \leftarrow w_i^{(M-1)} \exp(-y_i \alpha_M H_M(\mathbf{x_i}))$ and renormalize to $\sum_i w_i = 1$.
10: $\mathscr{H}_M = \mathscr{H}_{M-1} \bigcup \{H_M\}$.
11: **if** $J_M^{\min} > J(H_M)$ **then**
12: $J_M^{\min} = J(H_M)$.
13: **end if**
14: {3 - Conditional exclusion stage.}
15: $h' = \arg\min_{h \in \mathscr{H}_M} J(H_M - h)$.
16: **if** $J(H_M - h') < J_{M-1}^{\min}$ **then**
17: $\mathscr{H}_{M-1} = \mathscr{H}_M - h'$.
18: $J_{M-1}^{\min} = J(H_M - h')$
19: $M \leftarrow M - 1$
20: **if** $h' = h'_m$ **then**
21: Recalculate $w_i^{(j)}$ and h_j for $j = \{m', \ldots, M\}$.
22: Goto line 15.
23: **else**
24: **if** $M = M\max$ or $J(\mathscr{H}_M) < J^*$ **then**
25: Goto line 32.
26: **else**
27: Goto line 7.
28: **end if**
29: **end if**
30: **end if**
31: {4 - Output stage.}
32: Final classifier: $H(\mathbf{x}) = \text{sign}\left(\sum_{j=1}^{M} \alpha_j \; H_j(\mathbf{x})\right)$.

- $\lambda = 0.5$, we get the classical Real AdaBoost emphasis function

$$w_i = \exp\left[\left(\frac{\sum_{j=1}^{m}(\alpha_j H_j(\mathbf{x_i}) - y_i)^2}{2}\right) - \frac{(\sum_{j=1}^{m} H_j(\mathbf{x_i}))^2}{2}\right]. \tag{2.10}$$

Algorithm 10 Emphasis Boost

Input: Dataset $Z = \{z_1, z_2, \ldots, z_N\}$, with $z_i = (\mathbf{x_i}, y_i)$, where $\mathbf{x}_i \in \mathscr{X}$ and $y_i \in \{-1, +1\}$.
M, the maximum number of classifiers.
λ, weighting parameter $(0 \leq \lambda \leq 1)$.
Output: $H(\mathbf{x})$, a classifier suited for the training set.

1: Initialize the weights $w_i = 1/N, i \in \{1, \ldots, N\}$.
2: **for** $m = 1$ to M and while $H_m \neq 0$ **do**
3: Fit a classifier $H_m(\mathbf{x})$ to the training data using weights w_i.
4: Let $\text{err}_m = \sum_{i=1}^{N} w_i y_i H_m(\mathbf{x_i}) \Big/ \sum_{i=1}^{N} w_i$.
5: Compute $\alpha_m = 0.5 \log((1 + \text{err}_m)/(1 - \text{err}_m))$.
6: Set $w_i = \exp\left(\lambda \left(\sum_{j=1}^{m} (\alpha_j H_j(\mathbf{x_i}) - y_i)^2\right) - (1 - \lambda) \left(\sum_{j=1}^{m} H_j(\mathbf{x_i})\right)^2\right)$.
7: Renormalize to $\sum_i w_i = 1$.
8: **end for**
9: Final classifier: $H(\mathbf{x}) = \text{sign}\left(\sum_{j=1}^{M} \alpha_j \ H_j(\mathbf{x})\right)$.

- $\lambda = 1$, the emphasis function only pays attention to the quadratic error of each pattern

$$w_i = \exp\left[\sum_{j=1}^{m} (\alpha_j H_j(\mathbf{x_i}) - y_i)^2\right]. \tag{2.11}$$

The key issue with this algorithm is the choice of λ.

2.4.1.7 Reweight Boost

In the Reweight Boost variant [92], the weak classifiers are stumps (decision trees with a single node). The main idea is to consider as base classifier for boosting, not only the last weak classifier, but a classifier formed by the last r selected weak classifiers, using a classifier reuse technique. Algorithm 11 presents the details of the Reweight Boost variant.

2.4.1.8 Other Variants

For the sake of both completeness of this chapter and fairness to the many authors of AdaBoost variants, in this subsection we describe further variants for binary classification. We show the name of each variant as well as its main characteristics.

The *KLBoost* [74] variant uses *Kullback–Leibler* (KL) [22] divergence and operates as follows. First, classification is based on the sum of histogram divergences along corresponding global and discriminating linear features. Then,

Algorithm 11 Reweight Boost

Input: Dataset $Z = \{z_1, z_2, \ldots, z_N\}$, with $z_i = (\mathbf{x}_i, y_i)$, where $\mathbf{x}_i \in \mathcal{X}$ and $y_i \in \{-1, +1\}$.
M, the maximum number of classifiers.
r, the last r selected weak classifiers .

Output: $H(\mathbf{x})$, a classifier suited for the training set.

1: Initialize the weights $w_i = 1/N, i \in \{1, \ldots, N\}$.
2: **for** $m = 1$ to M and while $H_m \neq 0$ **do**
3: Fit a classifier $H_m(\mathbf{x})$ to the training data using weights w_i.
4: Get combined classifier H_t^r from $H_t, H_{t-1}, \ldots, H_{\max(t-r,1)}$.
5: Let $\mathrm{err}_m = \sum_{i=1}^{N} w_i y_i H_m(\mathbf{x_i}) \Big/ \sum_{i=1}^{N} w_i$.
6: Compute $\alpha_m = 0.5 \log((1 - \mathrm{err}_m)/\mathrm{err}_m)$.
7: Set $w_i \leftarrow w_i \exp\left(-\alpha_m y_i H_t^r(\mathbf{x_i})\right)$.
8: Renormalize to $\sum_i w_i = 1$.
9: **end for**
10: Final classifier: $H(\mathbf{x}) = \mathrm{sign}\left(\sum_{j=1}^{M} \alpha_j \; H_j(\mathbf{x})\right)$.

these linear KL features, are iteratively learned by maximizing the projected KL divergence in a boosting manner. Finally, the coefficients to combine the histogram divergences are learned by minimizing the recognition error, once a new feature is added to the classifier. This contrasts with conventional AdaBoost, in which the coefficients are empirically set. Because of these properties, KLBoosting classifier generalizes very well and has been applied to high-dimensional spaces of image data.

One of the experimental drawbacks of AdaBoost is that it can not improve the performance of *Naïve Bayes* (NB) [34,130] classifier as expected. *ActiveBoost* [124] overcomes this difficulty by using active learning to mitigate the negative effect of noisy data and introduce instability into the boosting procedure. Empirical studies on a set of natural domains show that ActiveBoost has clear advantages with respect to the increasing of the classification accuracy of NB when compared against AdaBoost.

The *Jensen–Shannon Boosting* [61] incorporates *Jensen–Shannon* (JS) divergence into AdaBoost. JS divergence is advantageous in that it provides a more appropriate measure of dissimilarity between two classes and it is numerically more stable than other measures such as KL divergence.

Infomax Boosting [76] is an efficient feature pursuit scheme for boosting. It is based on the infomax principle, which seeks optimal feature that achieves maximal mutual information with class labels. Direct feature pursuit with infomax is computationally prohibitive, so an efficient gradient ascent algorithm is proposed, based on the quadratic mutual information, nonparametric density estimation and fast Gauss transform. The feature pursuit process is integrated into a boosting framework as infomax boosting. It is similar to Real AdaBoost, but with the following differences:

- features are general linear projections;
- generates optimal features;
- uses KL divergence to select features;
- finer tuning on the coefficients.

Ent-Boost [68] uses entropy measures. The class entropy information is used to automatically subspace splitting and optimal weak classifier selection. The number of bins is estimated through a discretization process. KL divergence is applied to probability distribution of positive and negative samples, to select the best weak classifier in the weak classifier set.

The *MadaBoost* [30] algorithm consists on a modification of the weighting scheme of AdaBoost. This variant mitigates the problems that AdaBoost suffers from noisy data, improving its performance.

The *SoftBoost* algorithm [126] is a totally corrective algorithm which optimizes the soft margin and tries to produce a linear combination of hypotheses. The term *soft* means that the algorithm does not concentrate too much on outliers and hard to classify examples. It allows them to lie below the margin (with wrong predictions) but penalizes them linearly via slack variables. SoftBoost tries to avoid the problem of overfitting as in AdaBoost when using training data with high degree of noise.

The *linear programming boosting* (*LPBoost*) [27] algorithm maximizes the margin between training samples of different classes; this way, it belongs to the class of margin-maximizing supervised classification algorithms. The boosting task consists of constructing a learning function in the label space that minimizes misclassification error and maximizes the soft margin, formulated as a linear program which can be efficiently solved using column generation techniques, developed for large-scale optimization problems. Unlike gradient boosting algorithms, which may converge in the limit only, LPBoost converges in a finite number of iterations to a global solution, being computationally competitive with AdaBoost. The optimal solutions of LPBoost are very sparse in contrast with gradient-based methods. Empirical findings show that LPBoost converges quickly, often faster than other formulations.

LPBoost performs well on natural data, but there are cases where the number of iterations is linear in the number of training samples instead of logarithmic. By simply adding a relative entropy regularization to the linear objective of LPBoost, we get *entropy-regularized LPBoost* ERLPBoost [127], for which there is a logarithmic iteration bound. As compared to a previous algorithm, named SoftBoost, it has the same iteration bound and better GE. ERLPBoost does not suffer from this problem and has a simpler motivation. A detailed theoretical and experimental comparison between LPBoost and AdaBoost can be found in [69].

The *MarginBoost* algorithm [81] is a variant of the more general algorithm *AnyBoost* [81]. MarginBoost is also a general algorithm. It chooses a combination of classifiers to optimize the sample average of any cost function of the margin. MarginBoost performs gradient descent in function space, at each iteration choosing a base classifier to include in the combination so as to maximally reduce the cost function. As in AdaBoost, the choice of the base classifier corresponds to a minimization problem involving weighted classification error. That is, for a certain weighting of the training data, the base classifier learning algorithm attempts to return a classifier that minimizes the weight of misclassified training examples.

The general class of algorithms named AnyBoost consists of gradient descent algorithms for choosing linear combinations of elements of an inner product space so as to minimize some functional cost. Each component of the linear combination is chosen to maximize a certain inner product. In MarginBoost, this inner product corresponds to the weighted TE of the base classifier.

Brown Boost [45] uses a nonmonotonic weighting function such as examples far from the boundary decrease in weight, trying to achieve a given target error rate. It de-emphasizes outliers when it seems clear that they are too hard to classify correctly, being an adaptive version of Freund's boost-by-majority algorithm [44]. This variant reveals an intriguing connection between boosting and Brownian motion.

The *Weight Boost* algorithm [63] uses input-dependent weighting factors for WLs. It tries to cope with two possible problems of AdaBoost: suffer from overfitting, especially for noisy data; the assumption that the combination weights are fixed constants and therefore does not take particular input patterns into consideration. A learning procedure which is guaranteed to minimize TEs is devised. Empirical studies show that Weight Boost almost always achieves a considerably better classification accuracy than AdaBoost. Furthermore, experiments on data with artificially controlled noise indicate that the Weight Boost algorithm is more robust to noise than AdaBoost.

Asymmetric Boosting [79] is a cost-sensitive extension of boosting. It is derived from decision-theoretic principles, which exploit the statistical interpretation of boosting to determine a principled extension of the boosting loss. Similarly to AdaBoost, the cost-sensitive extension minimizes this loss by gradient descent on the functional space of convex combinations of WLs, and produces large margin detectors. Asymmetric boosting is fully compatible with AdaBoost, in the sense that it becomes the latter when errors are weighted equally.

In [59] we have an asymmetric boosting method, *Boosting with Different Costs*. The motivation is as follows; traditional boosting methods assume the same cost for misclassified instances from different classes, and in this way focus on good performance with respect to overall accuracy. This method is more generic than AdaBoost, and is designed to be more suitable for problems where the major concern is a low false positive (or negative) rate, such as SPAM filtering.

The *Quadratic Boost* [86] algorithm improves AdaBoost with a quadratic combination of base classifiers. It operates by constructing an intermediate learner on the combined linear and quadratic terms. A new method for iterative optimization is proposed; first a classifier is trained by randomizing the labels of the training examples. Subsequently, the input learner is called repeatedly with a systematic update of the labels of the training examples in each round. The quadratic-boosting algorithm converges under the condition that the given base learner minimizes the empirical error. The experimental results show that quadratic boosting compares favorably with AdaBoost on large datasets at the cost of the training time.

The *WaldBoost* [106] variant has near optimal time and error rate trade-off. It integrates the AdaBoost algorithm for measurement selection and ordering and the joint probability density estimation, with the optimal sequential probability ratio test decision strategy. It is suited for computer vision classification problems, in which both the error and time characterize the quality of a decision.

In [75] feature reweighting is integrated into the boosting scheme, which not only weights the samples but also weights the features iteratively; it is named *I.Boosting*. To avoid overfitting problems, a relevance feedback mechanism is applied into the boosting framework. I.Boosting is implemented using *adaptive discriminant analysis* (ADA) as base classifiers. The experimental results show the superior performance of I.Boosting over AdaBoost.

In [46] we have a new boosting algorithm, motivated by the large margins theory for boosting. The experimental results point out that the new algorithm is significantly more robust against label noise than existing boosting algorithms.

The algorithm proposed in [85] combines the base learners with symmetric functions. Among its properties of practical relevance, we have significant resistance against noise, and its efficiency even in an agnostic learning setting. Experimental results show the reliability of the classifiers built.

The *MilBoost* [123] variant uses cost functions from the *multiple instance learning* (MIL) literature combined with the AnyBoost framework. The feature selection criterion of MILBoost is modified to optimize the performance of the Viola–Jones cascade method for object detection (see Section 2.6.1). Experiments with this variant show improvement on the detection rate, as compared to previous approaches. This increased detection rate is a consequence of simultaneously learning the locations and scales of the objects in the training set along with the parameters of the classifier.

The *totally corrective boosting* [128], the weight update of each patterns is analyzed as the minimization of the relative entropy, subject to linear constraints. The algorithm is "totally corrective" in the sense that it takes into account the outputs of all the past WLs; the "corrective" versions only take into account the last WL results. A connection with margin maximization is also shown for totally corrective versions. The experimental results show that the totally corrective versions of AdaBoost attain smaller combinations of WLs than the corrective ones, being competitive with LPBoost (itself a totally corrective boosting algorithm with no regularization, for which there is no iteration bound known). An asymmetric totally corrective boosting approach for real-time object detection is proposed in [125].

In [105], the *BoostMetric* algorithm is proposed. The goal of this algorithm is to learn a semidefinite metric using boosting techniques. It is a generalization of AdaBoost in the sense that the WL is a matrix instead of a classifier, being simple and efficient. It attains better performance than many existing metric learning methods.

A boosting algorithm called *winner-take-all multiple category boosting* (*WTA-McBoost*) was proposed in [135]. On the learning process, the example subcategory labels are modified in order to make better object/nonobject decision. Multiple subcategory boosting classifiers are learned simultaneously with the assumption that the final classification of an example will only be determined by the highest score of all the subcategory classifiers (the winner will take all). The subcategory labels of the examples are dynamically assigned in this process, reducing the risk of having outliers in each subcategory. The WTA-McBoost algorithm uses confidence-rated prediction with asymmetric cost and is thus very efficient to train and test. The algorithm is successfully applied by building a multiview face detector.

The standard boosting procedure is extended to train a two-layer classifier dedicated to handwritten character recognition [43]. This learning scheme relies on a hidden layer and an output layer to obtain a final classification decision. The classical AdaBoost procedure is extended to train a multilayered structure by propagating the error through the output layer. This extension allows for the selection of optimal WLs by minimizing a weighted error, in both the output layer and the hidden layer.

2.4.2 Multiclass Variants

Since the first binary classification versions of AdaBoost, several generalizations of this algorithm to the multiclass case have been proposed. As a result of this direction of research, several multiclass AdaBoost variants have been proposed, as depicted in the timeline shown in Fig. 2.9. Similarly to what we did in Section 2.4.1.8, for each variant we will point out its main features as well as the connections among them. Many variants address multiclass classification problems as the concatenation of binary problems (a multiclass classification problem can be reformulated as a set of binary problems), while other (more recent) variants apply and combine multiclass classifiers directly.

AdaBoost.M1 and *AdaBoost.M2* [50] are multiclass extensions of (Discrete) AdaBoost (Algorithm 4). They differ between themselves in the way they treat each class. In the M1 variant, the weight of a base classifier is a function of the error rate. In M2, the sampling weights are increased for instances for which the pseudo-loss exceeds 0.5. The *AdaBoost.M1W* [39] algorithm changes AdaBoost.M1 as follows. In AdaBoost.M1, the weight of a base classifier is a function of the error rate. For AdaBoost.M1W this function is such that it gets positive, if the error rate is less than the error rate of random guessing. *BoostMA* [40] is also a simple modification of AdaBoost.M2 with the advantage that the base classifier minimizes the confidence-rated error, whereas for AdaBoost.M2 the base classifier should minimize the pseudo-loss. This makes BoostMA more easily applicable to already existing base classifiers; it also tends to converge faster than AdaBoost.M2.

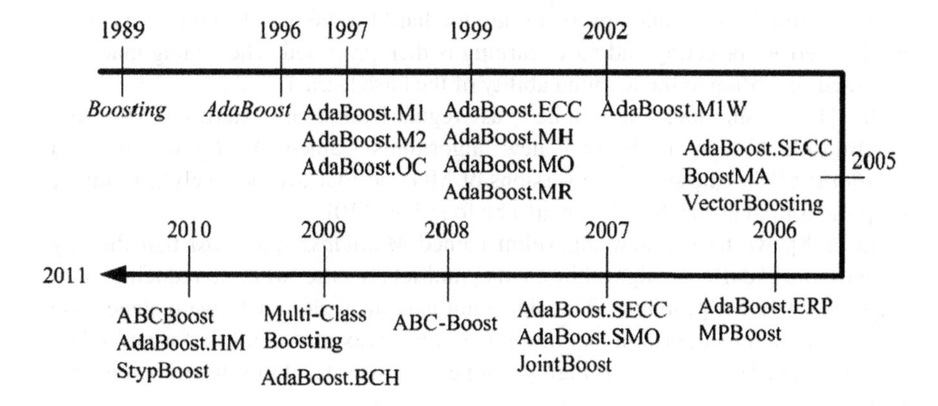

Fig. 2.9 A (possibly incomplete) timeline of AdaBoost variants for supervised learning, on multiclass problems, as of 2011

AdaBoost.MH [100] is a multiclass and *multilabel*[2] version of AdaBoost based on Hamming loss [100]. AdaBoost.MH generalizes AdaBoost, being tailored for multilabel text categorization tasks with decision stumps as WLs. *MPBoost* [41] further improves AdaBoost.MH augmenting its efficiency by performing a multiple pivot selection at each boosting iteration. Both these algorithms use binary features.

The *AdaBoost.MO* algorithm [100] performs a stage-wise functional gradient descent procedure on a given cost function. *AdaBoost.MR* [100] is a multiclass, multilabel version of AdaBoost based on ranking loss. *AdaBoost.OC* [96], where OC stands for output codes and *AdaBoost.ECC* [56], where ECC stands for error-correcting codes are similar algorithms; AdaBoost.OC is a shrinkage version of AdaBoost.ECC, which performs a stage-wise functional gradient descent procedure on an exponential loss cost function.

The *Vector Boosting* [60] algorithm is an extension of the Real AdaBoost in which both its WL and its final output are vectors rather than scalars. The idea of Vector Boosting comes from the *multiclass multilabel* (MCML) version of the Real AdaBoost, which assigns a set of labels for each sample and decomposes the original problem into k orthogonal binary ones. The major problem of this algorithm is that for each binary classification problem, a sample is regarded as either positive or negative. However, in many complicated cases, it is not tenable since some samples are neither positive nor negative for certain binary classification problems of which they are independent, which makes the MCML version of Real AdaBoost inapplicable.

AdaBoost.ERP [70] is AdaBoost.ECC with repartitioning. This algorithm improves two well-known issues of the quality of the ensemble learned by AdaBoost.ECC: the performance of the base learner; the error-correcting ability of the coding matrix. A coding matrix with strong error-correcting ability may not be

[2] A given instance can be classified into one or more classes.

overall optimal if the binary problems are too hard for the base learner. A trade-off between error-correcting and base learning is then proposed. The coding matrix is modified according to the learning ability of the base learner.

In [109], shrinkage is applied as regularization in AdaBoost.MO and AdaBoost.ECC and leads two new algorithms named *AdaBoost.SMO* and *AdaBoost.SECC* (the shrinkage versions of MO and ECC, respectively). A similar proposal for AdaBoost.SECC can also be found in [110].

In [138], we have a new algorithm named *Multiclass AdaBoost* that directly extends the AdaBoost algorithm to the multiclass case without reducing it to multiple two-class problems. The algorithm is equivalent to a forward stage-wise additive modeling algorithm that minimizes a novel exponential loss for multiclass classification. The algorithm is highly competitive in terms of misclassification error rate.

The *AdaBoost.BCH* algorithm [65] is a multiclass boosting algorithm which solves a C class problem by using $C - 1$ binary classifiers arranged by a hierarchy that is learned on the classes based on their closeness. AdaBoost is then applied to each binary classifier. AdaBoost.BCH requires less computation than AdaBoost.MH, with better or comparable generalization.

In [71], the concept of *adaptive base class boost* (*ABC-Boost*) for multiclass classification is addressed deriving *ABC-MART*, a concrete implementation of ABC-Boost. For binary classification, ABC-MART recovers MART and for multiclass classification, ABC-MART considerably improves MART, as evaluated on several public datasets.

AdaBoost.HM was proposed in 2010 [64]. It is based on *hypothesis margin* and directly combines multiclass weak classifiers, instead of learning binary WLs. The hypothesis margin maximizes the output about the positive class and minimizes the maximal outputs about the negative classes. Upper bounds on the TE of AdaBoost.HM are derived and compared against AdaBoost.M1 upper bounds. The WLs are feedforward NNs. AdaBoost.HM yields higher classification accuracies than both the AdaBoost.M1 and the AdaBoost.MH algorithms, being computationally efficient in training.

Recently, a totally corrective multiclass boosting was proposed [57]. After an analysis of some methods that extend two-class boosting to multiclass, a column-generation based totally corrective framework for multiclass boosting learning is derived, using the Lagrange dual problems. Experimental results show that the new algorithms have comparable generalization capability but converge much faster than their counterparts.

StypBoost [129] is a bilinear boosting algorithm, which extends the multiclass boosting framework of JointBoost to optimize a bilinear objective function. This allows style parameters to be introduced to aid classification, where style is any factor which the classes vary with systematically, modeled by a vector quantity. The algorithm allows learning with different styles. It is applied successfully to two object class segmentation tasks: road surface segmentation and general scene parsing.

JointBoost [107, 113] is a method where boosted one-versus-all classifiers are trained jointly and are forced to share features. It has been demonstrated to lead both to higher accuracy and smaller classification time, compared to using one-versus-all classifiers that were trained independently and without sharing features.

2.4.2.1 Analysis of Multiclass Boosting Algorithms

In [109] the AdaBoost.MO, AdaBoost.OC, and AdaBoost.ECC algorithms are studied. It is shown that MO and ECC perform stage-wise functional gradient descent on a cost function defined over margin values, and that OC is a shrinkage version of ECC. The AdaBoost.SMO and AdaBoost.SECC are the shrinkage versions of MO and ECC, respectively.

A unifying framework for studying the solution of multiclass categorization problems, by reducing them to multiple binary problems that are then solved using a margin-based binary learning algorithm is proposed in [2]. The proposed framework unifies some of the most popular approaches in which each class is compared against all others, or in which all pairs of classes are compared to each other, or in which output codes with error-correcting properties are used. A general method for combining the classifiers generated on the binary problems is proposed. A generic empirical multiclass loss bound given the empirical loss of the individual binary-learning algorithms is proven. The scheme and the corresponding bounds apply to many popular classification learning algorithms including SVM, AdaBoost, regression, logistic regression, and decision-tree algorithms. A multiclass GE analysis for general output codes with AdaBoost is provided.

The ability of boosting to achieve drastic improvements compared to the individual WLs has been noticed by several researchers. For two-class problems it has been observed that AdaBoost, is quite unaffected by overfitting. However, for the case of noisy data, it is also known that AdaBoost can be improved considerably by introducing some regularization technique. In speech-related problems one often considers multiclass problems and boosting formulations have been used successfully to solve them. Under this context, [90] reviews and extends the existing multiclass boosting algorithms to derive new boosting algorithms, which are more robust against outliers and noise in the data; these algorithms are also able to exploit prior knowledge about relationships between the classes.

In [110], a new interpretation of AdaBoost.ECC and AdaBoost.OC is presented. AdaBoost.ECC performs stage-wise functional gradient descent on a cost function, defined in the domain of margin values; AdaBoost.OC is a shrinkage version of AdaBoost.ECC. AdaBoostBCH has slower training and higher generalization ability as compared to AdaBoost.ECC [65].

2.5 Boosting for Semi-Supervised Learning

Semi-supervised learning (SSL) [15] has attracted considerable research efforts in the last few years. In many learning problems, we have a large amount of data available, but only a subset of it is labeled. In this section, we review several boosting algorithms for SSL, shown in the timeline of Fig. 2.10.

2.5.1 *MixtBoost*

The MixtBoost algorithm [55] was the first variant of AdaBoost to for SSL; the authors address the question: *can boosting be adapted for SSL learning?* The base classifiers are mixture models; thus, MixtBoost can be seen as boosting of mixture models.

The main ingredients of AdaBoost are the *loss* and the *margin*. The simplest way to generalize to SSL is to define these quantities for unlabeled data. This generalization to unlabeled data should not affect the labeled examples and should penalize inconsistencies between the classifier output and the available information. A loss definition is proposed for unlabeled data, from which margins are defined. The missing labels are interpreted as the absence of class information with the key idea that the pattern belongs to a class, but the class is unknown.

2.5.2 *SSMarginBoost*

The SSMarginBoost algorithm is an extension of MarginBoost to SSL [26] that explores the *clustering* assumption of SSL [15] and the large margin criterion.

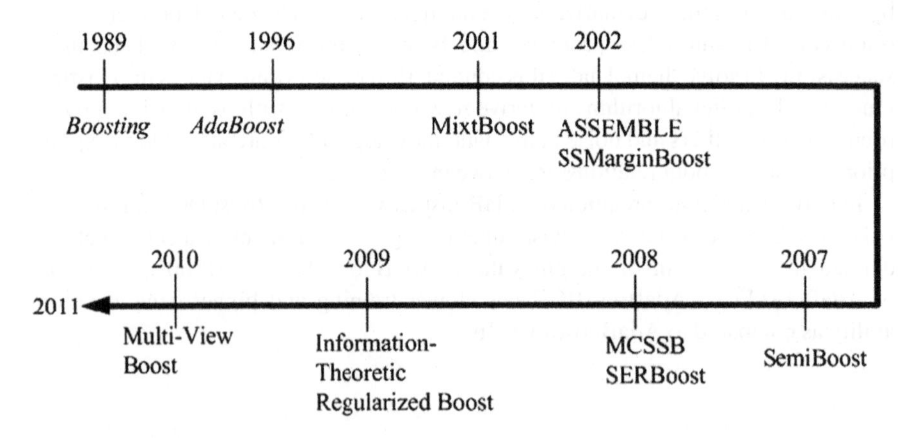

Fig. 2.10 A (possibly incomplete) timeline of AdaBoost variants for semi-supervised learning on binary and multiclass problems, as of 2011

The margin definition is extended to unlabeled data and the gradient descent algorithm for optimizing the resulting margin cost function is derived. SSMarginBoost can be applied with any base classifier able to handle unlabeled data, by means of mixture models trained with an *expectation-maximization* (EM) algorithm [28, 34].

2.5.3 ASSEMBLE

In [6], an adaptive semi-supervised ensemble method, named ASSEMBLE, was proposed. The method constructs ensembles based on both labeled and unlabeled data, by alternating between assigning "pseudo-classes" to the unlabeled data using the existing ensemble and constructing the next base classifier using both the labeled and pseudo-labeled data. This algorithm corresponds to maximizing the classification margin in hypothesis space as measured on both the labeled and unlabeled data. Unlike alternative approaches, ASSEMBLE does not require a SSL method for the base classifier. It can be used in conjunction with any cost-sensitive classification algorithm for both two-class and multiclass problems. As in SSMarginBoost, ASSEMBLE adopts the MarginBoost notation and strategy adapted to the margin measured on both the labeled and unlabeled data; the key difference is that ASSEMBLE assigns "pseudo-classes" to the unlabeled data.

Moreover, ASSEMBLE using decision trees won the Neural Information Processing Systems (NIPS) 2001 Unlabeled Data Competition. It achieves good results on several benchmark datasets using both decision trees and neural networks.

2.5.4 SemiBoost

The SemiBoost [78] algorithm was proposed as a boosting framework aiming at improving the classification accuracy of any given supervised learning algorithm by using the available unlabeled examples. The main advantages of SemiBoost over previous approaches are: (1) performance improvement of any supervised learning algorithm, using unlabeled data; (2) efficient computation by the iterative boosting algorithm and (3) exploiting both the SSL *manifold* and *cluster* assumptions [15].

An empirical study on 16 different datasets and text categorization demonstrates that SemiBoost improves the performance of several commonly used supervised learning algorithms, by using a large number of unlabeled examples.

2.5.5 MultiClass Semi-Supervised Boosting

The *multiclass semi-supervised boosting* (MCSSB) algorithm was proposed in [115]. Compared to the existing semi-supervised boosting methods, MCSSB has the advantage to exploit both classification confidence and similarities among examples, when deciding the pseudo-labels for unlabeled examples. This way, it overcomes the shortcoming of the multiclass approach *one-against-the-rest* applied on binary classifiers. Empirical evidence on several datasets shows that the MCSSB algorithm performs better than the previous algorithms for SSL.

2.5.6 SERBoost

The problem of bad scaling behavior of many SSL methods on large scale vision problems is addressed in [93]. Based on the *expectation regularization* (ER) principle, the SERBoost SSL boosting algorithm is proposed. It can be applied to large scale vision problems and its complexity is dominated by the base learners. The algorithm provides a margin regularizer for the boosting cost function and shows a principled way of utilizing prior knowledge. As compared to supervised and semi-supervised methods, SERBoost shows improvement both in terms of classification accuracy and computational speed.

2.5.7 Information Theoretic Regularization Boosting

An SSL boosting algorithm that incrementally builds linear combinations of weak classifiers through generic functional gradient descent, using both labeled and unlabeled training data, was proposed in [136]. The approach is based on extending the information regularization framework to boosting, bearing loss functions that combine log loss on labeled data with the information-theoretic measures to encode unlabeled data. Even though the information-theoretic regularization terms make the optimization nonconvex, a simple sequential gradient descent optimization algorithm is applied. This approach attains good results on synthetic, benchmark, and real world tasks as compared to supervised and semi-supervised boosting algorithms.

2.5.8 Multiview Boosting

A multiview boosting algorithm was proposed in [94], that, unlike other approaches, specifically encodes the uncertainties over the unlabeled samples in terms of given priors. Instead of ignoring the unlabeled samples during the training phase of each view, it uses the different views to provide an aggregated prior which is then used as a regularization term inside a semi-supervised boosting method, for multiclass problems. The algorithm uses priors as a regularization component over the unlabeled data. Since the priors may contain a significant amount of noise, a new loss function for the unlabeled regularization is introduced, being robust to noisy priors.

2.5.9 *Extensions on Semi-Supervised Boosting Algorithms*

Different strategies have been applied to extend boosting algorithms to SSL problems. Typically, these strategies do not take into account the local smoothness constraints among data into account during ensemble learning. A local smoothness regularizer to semi-supervised boosting algorithms based on the universal optimization framework of margin cost functionals was proposed in [17]. This regularizer is applicable to existing SSL boosting algorithms to improve their generalization and speed up their training.

In [18], the problem of using all the three SSL assumptions (*smoothness*, *cluster*, and *manifold*) during boosting is addressed. A novel cost functional consisting of the margin cost on labeled data and the regularization penalty based on unlabeled data is proposed. Thus, minimizing the proposed cost functional with a greedy stage-wise optimization procedure leads to a generic boosting framework for SSL.

A local smoothness regularizer for SSL boosting algorithm, based on the universal optimization framework of margin cost functionals, was proposed in [17]. This regularizer is applicable to existing SSL boosting algorithms to improve their generalization and speed up their training.

2.6 Experimental Evaluation

In this section, we discuss the application of AdaBoost and its variants on a wide variety of problems. We compare the boosting algorithms with other machine learning techniques, exploiting the theoretical properties of AdaBoost.

2.6.1 *Successful Applications*

Besides its nice theoretical properties, the AdaBoost algorithm and its variants have been found to work very well on problems from different domains. Empirical evidence from many researchers has shown the adequacy of boosting algorithms for real-world problems. This section outlines some successful applications of AdaBoost and its variants for binary and multiclass problems. There are many papers which evaluate AdaBoost and its variants for many types of problems; see for instance [5,29,33,62,77,89,104]. It has been shown empirically that AdaBoost with decision trees has excellent performance, being considered the best "off-the-shelf" classification algorithm [5, 58].

The first boosting algorithms were tested on a *optical character recognition* (OCR) problem of optical handwritten digits, with a set of 118,000 instances in boosting multilayer perceptrons [32].

Table 2.2 Summary of the use of boosting algorithms for face detection (adapted from [72])

Face Detector	*AdaBoost Variant*	*Weak Learner*
Viola–Jones [120]	Discrete AdaBoost	Stubs
Float Boost	Float Boost [72, 73]	1D Histograms
KLBoost	KLBoost [74]	1D Histograms
Schneiderman	Real AdaBoost [100]	One group of nD Histograms

In [122], a *pedestrian detection* system that integrates image intensity information with motion information is proposed. A detection-style algorithm scans a detector over two consecutive frames of a video sequence. The detector is trained using AdaBoost to take advantage of both motion and appearance information to detect a walking person. The detector combines two sources of information.

The breast cancer detection problem is addressed in [111]. A data preprocessing, feature selection and Modest AdaBoost algorithm, are applied to the breast cancer survival databases in Thailand. For this task, Modest AdaBoost outperforms Real and Gentle AdaBoost variants.

In [101], boosting is applied to *multiclass text categorization* tasks. The approach named BoosTexter has comparable results to other text-categorization algorithms, on a variety of tasks. The BoosTexter system is also applied to *speech categorization* to call-type identification from unconstrained spoken customer responses.

For *face detection*, boosting algorithms have been the most effective of all those developed so far, achieving the best results. They produce classifiers with about the same error rate than neural networks, with faster training [72]. Table 2.2 summarizes the use of boosting algorithms for face detection (a *stub* is a decision tree with a single decision node).

In [25] a fast and efficient face detection method has been devised, which relies on the AdaBoost algorithm and a set of Haar wavelet-like features. The face detection problem was been addressed also with *Asymmetric Boosting* [79], where it is shown to outperform a number of previous heuristic proposals for cost-sensitive boosting. For an updated literature on face detection and the use of boosting and other machine learning techniques, see [133].

A method for selecting edge-type features for *iris recognition* is proposed in [16]. The AdaBoost algorithm is used to select a filter bank from a pile of filter candidates. The decisions of the weak classifiers associated with the filter bank are linearly combined to form a strong classifier. The boosting algorithm can effectively improve the recognition accuracy at the cost of a slight increase on the computation time.

A new approach, proposing two *particle swarm optimization* (PSO) methods within AdaBoost for *object detection*, for constructing weak classifiers in AdaBoost is proposed in [83]. The experiments show that using PSO for selecting features and evolving associated weak classifiers in AdaBoost is more effective than for selecting features only for this problem.

In [131] a *face recognition* method using AdaBoosted low dimensional and discriminant Gabor features is proposed. AdaBoost is successfully applied to face recognition by introducing the intra-face and extra-face difference space in the Gabor feature space. By using the proposed method, only hundreds of Gabor features are selected. Experiments shown that these hundreds of Gabor features are enough to achieve good performance comparable to that of methods using the complete set of Gabor features.

A *feature selection approach* based on Gabor wavelets and AdaBoost is proposed in [137]. The features are first extracted by a Gabor wavelet transform. For each individual, a small set of significant features are selected by the AdaBoost algorithm from the pool of the Gabor wavelet features. In the feature selection process, each feature is the basis for a weak classifier. In each round of AdaBoost learning, the feature with the lowest error of weak classifiers is selected. The results from the experiments have shown that the approach successfully selects meaningful and explainable features for face verification. The experiments suggest that the feature selection algorithm for face verification selects the features corresponding to the unique characteristics rather than common characteristics, and a large example size statistically shows the benefits of AdaBoost feature selection.

The problem of *classifying music by genre* by partitioning songs into smaller pieces and classifying each one separately is addressed in [7]. The choice of features together with an AdaBoost.MH classifier proved to be the most effective method for genre classification at the MIREX 2005 international contest in music information extraction, and the second-best method for recognizing artists.

In [84], 2D cascaded AdaBoost, a novel classifier designing framework, is presented and applied to the *eye localization* problem. There are two cascade classifiers in two directions: the first one is a cascade designed by bootstrapping the positive samples; the second one, as the component classifiers of the first one, is cascaded by bootstrapping the negative samples. The proposed structure is applied to eye localization and evaluated on four public face databases, and extensive experimental results verified the effectiveness, efficiency, and robustness of the proposed method.

AdaBoost can also improve the performance of a strong learning algorithm as proposed in [103]: a NN based *online character recognition system*. AdaBoost can be used to learn automatically a great variety of writing styles even when the amount of training data for each style varies a lot. The system achieves about 1.4% error on a handwritten digit database of more than 200 writers.

In [87] the use of boosting and SVM is explored for the *segmentation of white-matter lesions in the MR scans of human brain*. Simple features are generated from proton density scans. Radial basis function-based AdaBoost technique and SVM are employed for this task. The classifiers are trained on severe, moderate, and mild cases. The results indicate that the proposed approach can handle MR field inhomogeneities quite well.

A *visual object detection* framework that is capable of processing images extremely rapidly while achieving high detection rates is proposed in [121]. The learning algorithm, based on AdaBoost, selects a small number of critical visual features and yields extremely efficient classifiers. The method combines classifiers in a cascade allowing background regions of the image to be quickly discarded while spending more computation on promising object-like regions. A set of experiments in the domain of *face detection* is presented.

A framework for *classifying face images* using AdaBoost and domain-partitioning based classifiers is addressed in [118]. The most interesting aspect of this framework is its ability to build classification systems with high accuracy in dynamical environments, which achieve, at the same time, high processing and training speed. This framework is applied to the specific problem of gender classification using different features, on standard face databases.

In [112] an approach for *image retrieval* using a very large number of highly selective features and efficient online learning is proposed. This approach is predicated on the assumption that each image is generated by a sparse set of visual "causes" and that images which are visually similar share causes between them. A mechanism for computing a very large number of highly selective features which capture some aspects of this causal structure (with over 45,000 highly selective features) is proposed. At query time a user selects a few example images, and boosting is used to learn a classification function in this feature space. The boosting procedure learns a simple classifier which only relies on 20 of the features. As a result, a very large database of images can be scanned rapidly.

The *boosting-based multimodal speaker detection* (BMSD) algorithm is proposed in [132]. It performs *speaker detection, identifying the active speaker* in a video, which can be very helpful for remote participants to understand the dynamics of the meeting. This algorithm fuses audio and visual information at feature level by using boosting to select features from a combined pool of both audio and visual features simultaneously. It achieves a very accurate speaker detector with extremely high efficiency.

2.6.1.1 Online Boosting

In the recent years, some attention has been given to *online boosting*, in which the training examples become available one at a time [4, 14, 23, 24].

A new family of topic-ranking algorithms for multilabeled documents is proposed in [23, 24]. The algorithms are simple to implement being both time and memory efficient. Experiments with the proposed family of topic-ranking algorithms on standard corpora, show that these algorithms attain adequate results, outperforming other topic-ranking adaptations of well-known classifiers.

The problem of *online adaptation of binary classifiers for tracking* is addressed in [54]. Online learning allows for simple classifiers since only the current view of the object from its surrounding background needs to be discriminated. However, online adaptation has one key problem: each update of the tracker may introduce an error which, finally, can lead to tracking failure (drifting). A novel online semi-supervised boosting method which significantly alleviates the drifting problem in tracking applications was proposed in [54]. This allows to limit the drifting problem while still staying adaptive to appearance changes. The main idea is to formulate the update process in a semi-supervised fashion as combined decision of a given prior and an online classifier without any parameter tuning.

A boosting framework that can be used to derive online boosting algorithms for various cost functions was proposed in [4]. Within this framework, online boosting algorithms for logistic regression, least squares regression, and multiple instance learning are derived.

In [14] a *real-time vision-based vehicle detection* system employing an online boosting algorithm is proposed. It is an online AdaBoost approach for a cascade of strong classifiers instead of a single strong classifier. The idea is to develop a cascade of strong classifiers for vehicle detection that is capable of being online trained in response to changing traffic environments. The proposed online boosting method can improve system adaptability and accuracy to deal with novel types of vehicles and unfamiliar environments.

TransientBoost [108] is an online learning algorithm, which is highly adaptive but still robust. It uses an internal multiclass representation and models reliable and unreliable data in separate classes. Unreliable data is considered transient, and thus highly adaptive learning parameters are applied to adapt to fast changes in the scene while errors fade out fast. In contrast, the reliable data is preserved completely and not harmed by wrong updates. The algorithm is applied successfully on the tasks of object detection and object tracking.

2.6.2 Comparison with Other Machine Learning Techniques

In this section, we show some detailed experimental results comparing AdaBoost variants, on well-known public domain datasets. We also describe some public domain software packages with code for AdaBoost and its variants.

Table 2.3 briefly describes the datasets used in the experiments, shown by increasing dimensionality. These datasets have several types of data and represent many different learning problems and are available from the UCI Repository [8].[3] We also have some datasets from bioinformatics (micro-array and gene expression data)[4] as well as datasets of the NIPS2003 FS Challenge[5] namely, Arcene, Madelon, Gisette, and Dexter.

The Leptograpsus Crabs dataset is from Ripley's book [91], and is publicly available at http://www.stats.ox.ac.uk/pub/PRNN/. The Crabs dataset is considered as a two-class problem for male/female detection. The Phoneme[6] dataset holds log-periodograms to represent speech phonemes as used in [58].

[3]http://archive.ics.uci.edu/ml/datasets.html

[4]http://www.gems-system.org/

[5]http://www.nipsfsc.ecs.soton.ac.uk

[6]http://orange.biolab.si/datasets/phoneme.htm

Table 2.3 Datasets with binary problems used in the experiments: P and N are the number of features and patterns, respectively. The datasets are shown by increasing dimensionality

Dataset	P	N	*Type of data / Classification problem*
Crabs	5	200	Classify crabs by gender
Phoneme	5	5404	Speech phoneme classification
Abalone	8	4177	Predict the age of abalone from physical measurements
Pima	8	768	The Pima Indians diabetes detection
Contraceptive	9	1473	Predict the current contraceptive method choice
Hepatitis	19	155	Detect if patients lived or died from hepatitis
WBCD	30	569	Wisconsin breast cancer diagnostic database
Ionosphere	34	351	Radar data–signals returned from the ionosphere
SpamBase	54	4601	Sparse BoW data/classify email as SPAM or not
Madelon	500	4400	Float data/artificial dataset, highly nonlinear and difficult
Colon	2000	62	Colon cancer detection
Gisette	5000	13500	Dense integer/distinguish handwritten digits "4" and "9"
DLBCL	5470	77	Dense integer/Lymphoma detection from medical analysis
Leukemia	7129	72	Cancer detection from medical analysis
Example 1	9947	2600	Sparse BoW (subset of Reuters)/text classification
Arcene	10000	900	Dense integer/detect cancer versus normal patterns
Prostate Tumor	10509	102	Cancer detection from medical analysis
Dexter	20000	2600	Same data as Example 1 with 10053 distractor features

The Abalone and Pima Indians are well-known datasets from the UCI Repository; their tasks is to predict the age of abalones from their shell measurements and to predict the presence of Diabetes in the Pima Indians population, respectively.

The Contraceptive dataset has the task to predict the current contraceptive method choice, for married women who were either not pregnant or do not know if they were at the time of interview. It is a subset of the 1987 National Indonesia Contraceptive Prevalence Survey.

The task of Hepatitis dataset is to classify if patients lived or died from hepatitis, given a set of medical analysis. The WBCD dataset is the well-known Wisconsin breast cancer database. The Ionosphere dataset is a binary classification problem on radar data; "good" radar returns are those showing evidence of some type of structure in the ionosphere and "bad" returns are those that do not; their signals pass through the ionosphere. The SpamBase dataset has sparse bag-of-words floating-point data; the task is to classify email messages as SPAM or nonSPAM. We have considered only the first 54 features which constitute a *bag-of-words* representation. The Madelon dataset is an artificial problem of the NIPS2003 FS challenge[7]; it is a difficult problem, because it is multivariate and highly nonlinear. The Colon, DLBCL, Leukemia, and Prostate Tumor datasets deal with the problem of cancer detection from microarray data.

[7]http://clopinet.com/isabelle/Projects/NIPS2003/#challenge

In the case of Example 1,[8] each pattern is a 9947-dimensional BoW vector. The Dexter dataset has the same data as Example 1 (with different train, test, and validation partitions) with 10053 additional distractor features, at random locations; it was created for the NIPS 2003 FS challenge. For both datasets, the task is learn to classify Reuters articles as being about "corporate acquisitions" or not.

The Arcene and Gisette datasets also belong to the NIPS2003 FS challenge, and their tasks are: to distinguish cancer versus normal patterns from mass-spectrometric data; to separate the highly confusable handwritten digits "4" and "9."

2.6.2.1 Software Packages for Boosting Algorithms

There are several software packages, freely available online that include implementations of boosting algorithms:

- The GML AdaBoost Matlab Toolbox[9] provides implementations of Real, Gentle, and Modest AdaBoost.
- The ENTOOL Matlab Toolbox http://www.j-wichard.de/entool/ which has many machine learning techniques and includes Real, Gentle, and Modest AdaBoost from the GML Toolbox.
- A Java implementation is available at http://jboost.sourceforge.net/, including AdaBoost, LogitBoost, RobustBoost, and BoosTexter.
- The well-known WEKA machine learning package includes AdaBoost.M1 and MultiBoost classifiers, and it is available at http://www.cs.waikato.ac.nz/~ml/weka/.
- A C++ implementation of the MPBoost algorithm, is available at this internet address http://www.esuli.it/mpboost.
- An efficient C++ implementation of various boosting algorithms can be found in http://www.stat.purdue.edu/~vishy/.
- An open-source implementation of BoosTexter[10] (see Section 2.6.1) can be found at http://code.google.com/p/icsiboost/.
- A BoostMetric implementation as well as other boosting algorithms are available at http://code.google.com/p/boosting/.
- In http://cseweb.ucsd.edu/~yfreund/adaboost/index.html, we have a Java Applet that shows how AdaBoost behaves during training.
- The *generalized boosted regression models* (GBM) implements extensions to AdaBoost and gradient boosting machine. It includes regression methods for least squares, absolute loss, quantile regression, logistic, Poisson, Cox proportional hazards partial likelihood, and AdaBoost exponential loss. It is available at http://cran.r-project.org/web/packages/gbm/index.html.

---

[8]http://svmlight.joachims.org/

[9]http://graphics.cs.msu.ru/ru/science/research/machinelearning/adaboosttoolbox

[10]http://www.cs.princeton.edu/~schapire/boostexter.html

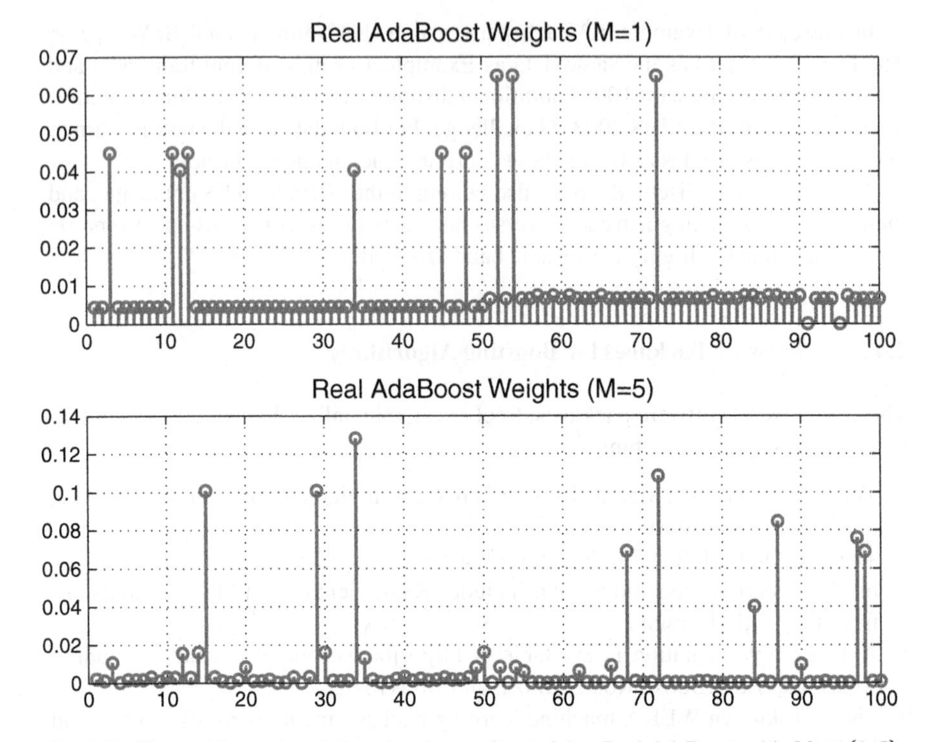

Fig. 2.11 The weights assigned to each pattern, after training Real AdaBoost with $M \in \{1, 5\}$ rounds (learners), on the Ionosphere dataset

The PRTools toolbox [36] available at http://www.prtools.org/prtools.html has many machine learning techniques, but it lacks implementations on boosting algorithms.[11]

2.6.2.2 Analysis of Training and Test Error

Figure 2.11 shows the weights of each pattern, after training Real AdaBoost with $M \in \{1, 5\}$ rounds, on the Ionosphere dataset using 100 training patterns. Notice that at the beginning of the first round, the weights have an uniform distribution with $1/N$.

On the first few iterations, the weight of many patterns is changed in such a way that we get a distribution which is quite different from the uniform.

[11]As of version PRTools 4.0, available at the time of this writing (July, 2011).

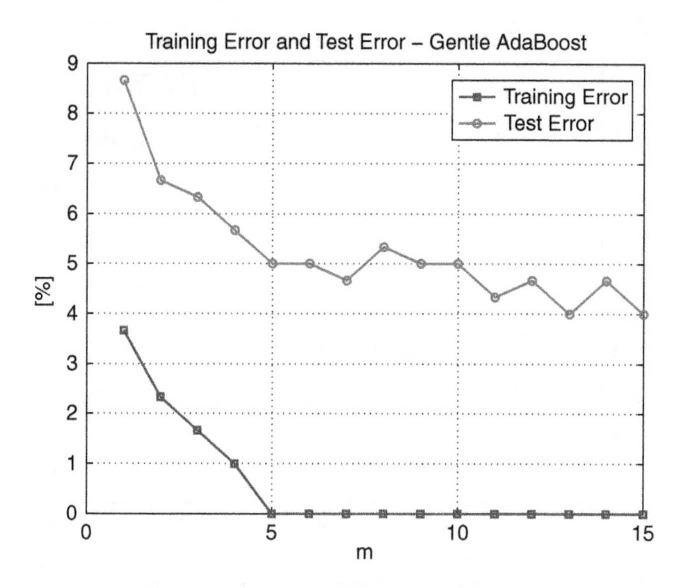

Fig. 2.12 The training error and the test error for Gentle AdaBoost on the WBCD dataset, with $M = 15$ learners

On Fig. 2.12, we have the TE and the test error of Gentle AdaBoost on the WBCD dataset, as a function of the number of WLs.

We see that the TE drops fast on the first few iterations. Even after the TE reaches zero, the test error continues to drop. Figure 2.13 shows the test error rates for Real, Gentle, and Modest AdaBoost classifiers, on the WBCD and Pima datasets, as a function of the number of WLs.

For these three classifiers, we have an adequate test set error rate on both datasets. On the WBCD dataset, the best performance is achieved by Gentle AdaBoost and for the Pima dataset Real AdaBoost attains the best results.

2.6.2.3 Comparison with Other Classifiers

The reported results in Table 2.4 are averages over ten different random replications of different training/testing partitions, for the standard datasets described in Table 2.3. We compare Real, Gentle, and Modest AdaBoost with linear SVM [10, 20, 116] and K-nearest neighbor (KNN) [1] classifiers from the PRTools toolbox. The linear kernel SVM classifiers are trained up to 20,000 iterations and the KNN classifier uses $K = 3$ neighbors. Regarding AdaBoost variants we use the ENTOOL toolbox with $M = 15$ WLs (tree nodes).

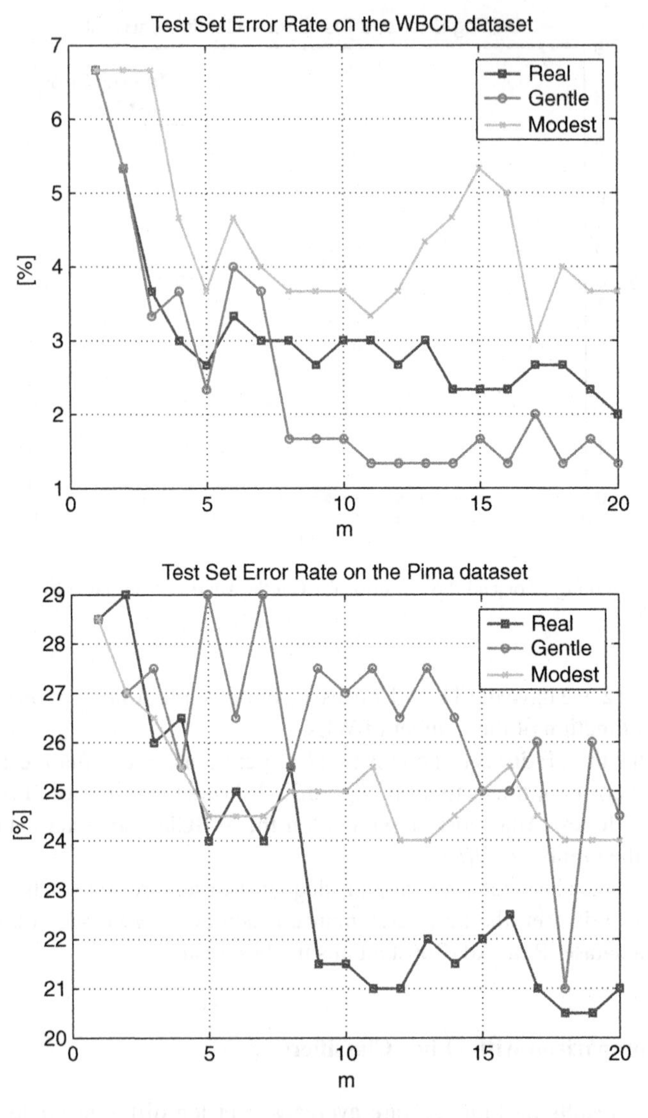

Fig. 2.13 Test error rates for Real, Gentle, and Modest AdaBoost classifiers, on the WBCD and Pima datasets

In many low and medium dimensional datasets, one of the AdaBoost variants attains adequate results being better than SVM and KNN. However, for the higher-dimensional datasets SVM and KNN tend to perform slightly better than these AdaBoost variants.

Table 2.4 Experimental comparison of three AdaBoost variants with linear SVM and 3-NN classifiers. We show the average ± standard deviation of the test set error rate, for ten runs with different random training/test partitions on standard datasets of Table 2.3. The best results are in bold face

Dataset	*Real*	*Gentle*	*Modest*	*SVM*	*3-NN*
Crabs	23.50 ± 11.56	23.50 ± 12.26	19.50 ± 12.12	**5.50 ± 4.97**	33.50 ± 15.28
Phoneme	24.45 ± 4.50	23.70 ± 3.25	23.90 ± 2.63	26.70 ± 2.49	**22.90 ± 1.73**
Abalone	26.25 ± 3.26	26.15 ± 3.12	25.65 ± 3.98	**23.05 ± 4.56**	28.55 ± 3.24
Pima	23.05 ± 3.35	**22.20 ± 3.28**	23.60 ± 2.17	25.85 ± 2.52	28.25 ± 3.50
Contraceptive	**34.35 ± 4.45**	34.50 ± 4.29	35.50 ± 3.06	37.95 ± 3.61	40.05 ± 3.20
Hepatitis	**19.00** ± 5.80	22.25 ± 7.77	20.50 ± 9.92	25.25 ± 7.77	41.25 ± 7.38
WBCD	3.57 ± 1.81	**2.40 ± 1.64**	3.83 ± 2.24	4.80 ± 1.10	7.10 ± 1.28
Ionosphere	9.60 ± 5.02	7.60 ± 2.99	**7.30 ± 2.36**	12.70 ± 4.42	18.20 ± 2.86
SpamBase	14.60 ± 2.68	14.60 ± 3.11	14.10 ± 3.04	**13.57 ± 2.39**	18.33 ± 1.52
Madelon	50.12 ± 2.29	49.78 ± 1.98	**49.45 ± 1.85**	50.97 ± 1.85	50.58 ± 1.97
Colon	**1.67 ± 5.27**	**1.67 ± 5.27**	**1.67 ± 5.27**	13.33 ± 4.10	15.56 ± 6.70
Gisette	13.67 ± 1.59	11.06 ± 0.90	10.62 ± 1.51	7.79 ± 0.67	11.39 ± 1.81
DLBCL	14.00 ± 10.16	19.00 ± 9.94	17.00 ± 10.59	**4.67 ± 4.22**	18.67 ± 7.73
Leukemia	37.00 ± 17.51	37.00 ± 17.51	37.00 ± 17.51	**11.00 ± 7.75**	13.50 ± 9.44
Example 1	12.07 ± 2.62	9.62 ± 1.57	11.63 ± 1.42	**4.30 ± 0.61**	10.98 ± 1.13
Arcene	31.10 ± 6.26	32.00 ± 5.60	29.70 ± 6.36	31.00 ± 0.94	**20.90 ± 6.06**
Prost. Tumor	5.75 ± 4.87	7.25 ± 3.11	4.75 ± 4.03	**3.50 ± 2.49**	15.63 ± 4.87
Dexter	18.13 ± 3.07	15.30 ± 1.06	14.57 ± 1.66	**10.10 ± 0.80**	25.97 ± 4.63

2.7 Summary and Discussion

The AdaBoost algorithm (and its variants) has many practical advantages, which, combined with theoretical guarantees, makes it a very attractive general purpose learning method. On the practical side, boosting algorithms are simple to implement and debug. The base learner and the number of iterations (learners) are the only two important choices to be made.

AdaBoost had been shown to be resistant to overfitting, despite the fact that it can produce combinations involving very large numbers of base classifiers. However, recent studies have shown that this is not the case, even for base classifiers as simple as decision stumps. The success of the boosting algorithms depends on the amount of data available for training as well as on the type of WL.

There are dozens of variants for binary and multiclass problems that have been proven successful on many problems. In recent years, the research on boosting algorithms has been focused mainly on multiclass and semi-supervised problems. The study of online boosting algorithms, in which the training examples arrive one at a time, as contrary to the batch mode, is also a fruitful field of research with many successful algorithms. Many of these algorithms are applied to real-time computer vision problems, such as detection or tracking. This is a focus of intensive current research.

AdaBoost and its variants have also been combined with different machine learning techniques, such as random subspaces [66], genetic algorithms [19], and rotation forest [134].

2.8 Bibliographical and Historical Remarks

There are many papers and tutorials addressing the many boosting algorithms, proposing variants for the multiclass case and/or SSL problems. In this section, we point out some of these elements that can be found in the literature.

For the origins of boosting algorithms, one can be interested in reading about the bootstrapping [37, 38] and bagging (bootstrap aggregation) [11]) techniques. Bootstrap was initially proposed in 1982, but it regained interest in the decade of 1990–1999, in which bagging was proposed. The use of bagging for classification problems is addressed in many papers, see [12, 13, 95, 139, 140] for many applications. The seminal paper of Schapire [95], proposing the first provable polynomial time boosting procedure is a must read. For a comparison of the effectiveness of randomization, bagging, and boosting see [29].

The seminal papers in the middle of the decade of 1990–1999 introducing *adaptive boosting* (AdaBoost) algorithm [47, 48, 50], with the idea that we can *weight* the data instead of resampling it, are also a must read. In the second half of the decade 1990–1999, the AdaBoost algorithm was also extended for regression tasks, as addressed in [3, 31], for instance.

From 1999 until this date, there are dozens of extensions and variants of AdaBoost for supervised and semi-supervised binary and multiclass problems, covering a wide range of successful applications. In this chapter, rather than the theoretical aspects of boosting algorithms, we tried to cover as many variants and successful applications as possible. Section 2.4 covers many variants for supervised learning, whereas Section 2.5 addresses the semi-supervised variants on binary and multiclass problems. In Section 2.6, we have described a wide range of applications. The vast majority of these variants and successful applications were published in the decade of 2000–2009. Online boosting, in contrast with batch boosting, has received a great deal of attention in recent research; we covered many approaches for online boosting on Section 2.6.1.1.

For further reading on adaptive boosting algorithms, please see [82], which complements well this chapter. Whereas we have aimed at covering a wide range of variants, [82] focuses more on theoretical and practical aspects of boosting and ensemble learning. The webpage http://cbio.mskcc.org/~aarvey/boosting_papers.html has many papers, tutorials, and links to software on boosting algorithms. There are also some tutorials about boosting available on-line.[12,13]

---

[12] http://www.site.uottawa.ca/~stan/csi5387/boost-tut-ppr.pdf

[13] http://www.stat.purdue.edu/~vishy/

The web pages of boosting and adaptive boosting pioneers R. Schapire[14] and Y. Freund[15] have many useful information about boosting algorithms, being useful for both experienced researchers as well as to the new researchers entering this exciting field.

Appendix A: Proofs

A.1: Proof of (2.4)

Consider the final weights, $w_i^{(M+1)}$, and explicitly write the recursion that starts at $w_i^{(1)} = 1/N$ (recall that S_j is the normalizing constant used in line 7 of Algorithm 4, AdaBoost at iteration j)

$$w_i^{(M+1)} = w_i^{(M)} \frac{\exp(-\alpha_M y_i H_M(\mathbf{x}_i))}{S_M} = \frac{\prod_{j=1}^{M} \exp\left(-\alpha_j y_i H_j(\mathbf{x}_i)\right)}{N \prod_{j=1}^{M} S_j}. \quad (2.12)$$

This can be re-written as

$$w_i^{(M+1)} = \frac{\exp\left(-y_i \sum_{j=1}^{M} \alpha_j H_j(\mathbf{x}_i)\right)}{N \prod_{j=1}^{M} S_j} = \frac{\exp(-y_i \ f(\mathbf{x}_i))}{N \prod_{j=1}^{M} S_j}, \quad (2.13)$$

where $f(\mathbf{x}) = \sum_{j=1}^{M} \alpha_j H_j(\mathbf{x})$, from which we can conclude that

$$\exp(-y_i \ f(\mathbf{x}_i)) = w_i^{(M+1)} N \prod_{j=1}^{M} S_j. \quad (2.14)$$

Now, noticing that $H(\mathbf{x}) = \text{sign}(f(\mathbf{x}))$, and recalling that h denotes the Heaviside function (defined above), we have

$$h(-y_i \ H(\mathbf{x}_i)) = h(-y_i \ f(\mathbf{x}_i)) \leq \exp(-y_i \ f(\mathbf{x}_i)) = w_i^{(M+1)} N \prod_{j=1}^{M} S_j. \quad (2.15)$$

We can now write and bound the TE rate of H,

$$\text{TE} = \frac{1}{N} \sum_{i=1}^{N} h(-y_i \ H(\mathbf{x}_i)) \leq \frac{1}{N} \sum_{i=1}^{N} w_i^{(M+1)} N \prod_{j=1}^{M} S_j = \prod_{j=1}^{M} S_j, \quad (2.16)$$

because $\sum_{i=1}^{N} w_i^{(M+1)} = 1$, thus concluding the proof of (2.4). □

---

[14] http://www.cs.princeton.edu/~schapire/boost.html

[15] http://cseweb.ucsd.edu/~yfreund/papers/index.html

A.2: Proof of (2.5)

Let us plug the expression for α_m (line 5 of AdaBoost) into the expression of the normalizing factor S_m, and use the fact that $y_i\, H_m(\mathbf{x}_i) = 1$, if and only if $H_m(\mathbf{x}_i) = y_i$, while $y_i\, H_m(\mathbf{x}_i) = -1$, if and only if $H_m(\mathbf{x}_i) \neq y_i$,

$$S_m = \sum_{i=1}^{N} w_i^{(m)} \exp\left(y_i\, H_m(\mathbf{x}_i) \log \sqrt{\frac{\mathrm{err}_m}{1-\mathrm{err}_m}} \right) \tag{2.17}$$

$$= \sqrt{\frac{\mathrm{err}_m}{1-\mathrm{err}_m}} \sum_{i:y_i=H(\mathbf{x}_i)} w_i^{(m)} + \sqrt{\frac{1-\mathrm{err}_m}{\mathrm{err}_m}} \sum_{i:y_i\neq H(\mathbf{x}_i)} w_i^{(m)} \tag{2.18}$$

$$= 2\sqrt{\mathrm{err}_m(1-\mathrm{err}_m)}. \tag{2.19}$$

Recalling that $\mathrm{err}_m = 1/2 - \gamma_m$, we have

$$S_m = 2\sqrt{\left(\frac{1}{2} - \gamma_m\right)\left(\frac{1}{2} + \gamma_m\right)} \tag{2.20}$$

$$= \sqrt{1 - 4\gamma_m^2} \tag{2.21}$$

$$= \exp\left(\frac{1}{2}\log\left(1 - 4\gamma_m^2\right)\right) \tag{2.22}$$

$$\leq \exp\left(-2\,\gamma_m^2\right), \tag{2.23}$$

where in (2.23) we have used the inequality $\log u \leq u - 1$ (often referred to as the Gibbs inequality). Plugging this inequality into (2.4), and invoking the assumption $\gamma_m > \gamma$, yields

$$\mathrm{TE} \leq \exp\left(-2\sum_{m=1}^{M} \gamma_m^2\right) \leq \exp\left(-2\,M\,\gamma^2\right) \tag{2.24}$$

thus proving inequality (2.5).

References

1. D. Aha, D. Kibler, and M. Albert. Instance-based learning algorithms. In *Machine Learning*, pages 37–66, 1991.
2. E. Allwein, R. Schapire, and Y. Singer. Reducing multiclass to binary: A unifying approach for margin classifiers. *Journal of Machine Learning Research*, 1:113–141, 2000.
3. R. Avnimelech and N. Intrator. Boosting regression estimators. *Neural Computation*, 11:491–513, 1999.

4. B. Babenko, M. Yang, and S. Belongie. A family of online boosting algorithms. In *Learning09*, pages 1346–1353, 2009.
5. E. Bauer and R. Kohavi. An empirical comparison of voting classification algorithms: Bagging, boosting, and variants. *Machine Learning*, 36:105–139, 1999.
6. K. Bennett, A. Demiriz, and R. Maclin. Exploiting unlabeled data in ensemble methods. In *Proceedings of the Eighth ACM SIGKDD International Conference on Knowledge Discovery and Data Mining*, KDD '02, pages 289–296, New York, NY, USA, 2002. ACM.
7. J. Bergstra and B. Kégl. Meta-features and adaboost for music classification. In *Machine Learning Journal : Special Issue on Machine Learning in Music*, 2006.
8. C. Blake and C. Merz. UCI repository of machine learning databases. Technical report, University of California, Irvine, Department of Informatics and Computer Science, 1999.
9. A. Blumer, A. Ehrenfeucht, D. Haussler, and M. Warmuth. Learnability and the vapnik-chervonenkis dimension. *Journal of the ACM*, 36:929–965, 1989.
10. B. Boser, I. Guyon, and V. Vapnik. A training algorithm for optimal margin classifiers. In *Proc. of the 5th Annual ACM Workshop on Computational Learning Theory*, pages 144–152, New York, NY, USA, 1992. ACM Press.
11. L. Breiman. Bagging predictors. *Machine Learning*, 24(2):123–140, 1996.
12. L. Breiman. Bias, variance, and arcing classifiers. Technical report, UC Berkeley, CA, 1996.
13. P. Bühlmann and B.Yu. Analyzing bagging. *Annals of Statistics*, 30:927–961, 2002.
14. W.-C. Chang and C.-W. Cho. Online Boosting for Vehicle Detection. *IEEE Transactions on Systems, Man, and Cybernetics, Part B (Cybernetics)*, 40(3):892–902, June 2010.
15. O. Chapelle, B. Schalkopf, and A. Zien. *Semi-Supervised Learning*. The MIT Press, Cambridge, MA, 2006.
16. K. Chen, C. Chou, S. Shih, W. Chen, and D. Chen. Feature selection for iris recognition with adaboost. *Third International Conference on Intelligent Information Hiding and Multimedia Signal Processing, 2007. IIHMSP 2007*, 2:411–414, 2007.
17. K. Chen and S. Wang. Regularized boost for semi-supervised learning. In *Neural Information Processing Systems*, 2007.
18. K. Chen and S. Wang. Semi-supervised Learning via Regularized Boosting Working on Multiple Semi-supervised Assumptions. *IEEE Transactions on Pattern Analysis and Machine Intelligence*, 99(1), 2010.
19. H. Chouaib, O. Terrades, S. Tabbone, F. Cloppet, and N. Vincent. Feature selection combining genetic algorithm and adaboost classifiers. In *ICPR08*, pages 1–4, 2008.
20. N. Christiani and J. Shawe-Taylor. *An Introduction to Support Vector Machines and other kernel based learning methods*. Cambridge University Press, Cambridge, MA, 2000.
21. M. Collins, R. Schapire, and Y. Singer. Logistic regression, adaboost and bregman distances. In *Machine Learning*, volume 48, pages 158–169, 2000.
22. T. Cover and J. Thomas. *Elements of Information Theory*. John Wiley & Sons, 1991.
23. K. Crammer and Y. Singer. A new family of online algorithms for category ranking. In *Proceedings of the 25th Annual International ACM SIGIR Conference on Research and Development in Information Retrieval*, SIGIR '02, pages 151–158, New York, NY, USA, 2002. ACM.
24. K. Crammer, Y. Singer, J. K., T. Hofmann, T. Poggio, and J. Shawe-Taylor. A family of additive online algorithms for category ranking. *Journal of Machine Learning Research*, 3:2003, 2003.
25. D. Cristinacce and T. Cootes. Facial feature detection using adaboost with shape constraints. In *British Machine Vision Conference (BMVC)*, pages 231–240, 2003.
26. F. d'Alché-Buc, Y. Grandvalet, and C. Ambroise. Semi-supervised MarginBoost. In *Advances in Neural Information Processing Systems* 14, pages 553–560, MIT Press, Cambridge, MA, 2001.
27. A. Demiriz, K. P. Bennett, and J. S. Taylor. Linear Programming Boosting via Column Generation. *Machine Learning*, 46(1-3):225–254, 2002.
28. A. Dempster, N. Laird, and D. Rubin. Maximum likelihood from incomplete data via the EM algorithm (with discussion). *Journal of the Royal Statistical Society, B*, 39:1–38, 1977.

29. T. Dietterich. An experimental comparison of three methods for constructing ensembles of decision trees: Bagging, boosting, and randomization. *Machine Learning*, 40(2):139–157, 2000.
30. C. Domingo and O. Watanabe. MadaBoost: A modification of AdaBoost. In *Proceedings of the 13th Annual Conference on Computational Learning Theory (COLT)*, pages 180–189, Palo Alto, CA, 2000.
31. H. Drucker. Improving regressors using boosting techniques. In *Proceedings of the Fourteenth International Conference on Machine Learning* ICML 97, pages 107–115, San Francisco, CA, USA, 1997. Morgan Kaufmann Publishers Inc.
32. H. Drucker, R. Schapire, and P. Simard. Improving performance in neural networks using a boosting algorithm. In *Advances in Neural Information Processing Systems 5, [NIPS Conference]*, pages 42–49, San Francisco, CA, USA, 1993. Morgan Kaufmann Publishers Inc.
33. H. Drucker and C. Tortes. Boosting decision trees. In *Advances in Neural Information Processing Systems*, volume 8, pages 479–485. MIT Press, 1996.
34. R. Duda, P. Hart, and D. Stork. *Pattern Classification*. John Wiley & Sons, 2nd edition, 2001.
35. N. Duffy and D. Helmbold. Potential boosters? In *Advances in Neural Information Processing Systems 12*, pages 258–264. MIT Press, New York, NY, 2000.
36. R. Duin, P. Juszczak, P. Paclik, E. Pekalska, D. Ridder, D. Tax, and S. Verzakov. PRTools4.1, a Matlab Toolbox for Pattern Recognition. Technical report, Delft University of Technology, 2007.
37. B. Efron. The jackknife, the bootstrap and other resampling plans. *Society for Industrial and Applied Mathematics (SIAM)*, 1982.
38. B. Efron and R. Tibshirani. *An Introduction to the Bootstrap*. Chapman & Hall, New York, 1993.
39. G. Eibl and K. Pfeiffer. How to make adaboost.M1 work for weak classifiers by changing only one line of the code. In *Machine Learning: Thirteenth European Conference*, volume 1, pages 109–120, 2002.
40. G. Eibl and K. Pfeiffer. Multiclass boosting for weak classifiers. *Journal of Machine Learning Research*, 6:189–210, 2005.
41. A. Esuli, T. Fagni, and F. Sebastiani. MP-Boost: A multiple-pivot boosting algorithm and its application to text categorization. In *Proceedings of the 13th International Symposium on String Processing and Information Retrieval (SPIRE'06)*, 2006.
42. A. Ferreira and M. Figueiredo. Boosting of (very) weak classifiers. In *6th Portuguese Conference on Telecommunications, Conftele'07*, Peniche, Portugal, 2007.
43. F. Fleuret. Multi-layer boosting for pattern recognition. *Pattern Recognition Letters*, 30:237–241, February 2009.
44. Y. Freund. Boosting a Weak Learning Algorithm by Majority. *Information and Computation*, 121(2):256–285, 1995.
45. Y. Freund. An adaptive version of the boost by majority algorithm. In *Proceedings of the Twelfth Annual Conference on Computational Learning Theory*, pages 102–113, 2000.
46. Y. Freund. A more robust boosting algorithm. *http://arxiv.org/abs/0905.2138*, 2009.
47. Y. Freund and R. Schapire. A decision-theoretic generalization of on-line learning and application to boosting. In *European Conference on Computational Learning Theory – EuroCOLT*. Springer, 1994.
48. Y. Freund and R. Schapire. Experiments with a new boosting algorithm. In *Thirteenth International Conference on Machine Learning*, pages 148–156, Bari, Italy, 1996.
49. Y. Freund and R. Schapire. Game theory, on-line prediction and boosting. In *Proceedings of the Ninth Annual Conference on Computational Learning Theory*, pages 325–332. ACM Press, 1996.
50. Y. Freund and R. Schapire. A decision-theoretic generalization of on-line learning and an application to boosting. *Journal of Computer and System Sciences*, 55(1):119–139, 1997.
51. J. Friedman. Greedy function approximation: A gradient boosting machine. *Annals of Statistics*, 29:1189–1232, 2000.
52. J. Friedman, T. Hastie, and R. Tibshirani. Additive logistic regression: a statistical view of boosting. *The Annals of Statistics*, 28(2):337–374, 2000.

53. V. Gómez-Verdejo, M. Ortega-Moral, J. Arenas-Gárcia, and A. Figueiras-Vidal. Boosting of weighting critical and erroneous samples. *Neurocomputing*, 69(7–9):679–685, 2006.
54. H. Grabner, C. Leistner, and H. Bischof. Semi-supervised On-Line Boosting for Robust Tracking. In D. Forsyth, P. Torr, and A. Zisserman, editors, *Computer Vision ECCV 2008*, volume 5302, *Lecture Notes in Computer Science*, chapter 19, pages 234–247. Springer, Berlin, Heidelberg, 2008.
55. Y. Grandvalet, F. Buc, and C. Ambroise. Boosting mixture models for semi-supervised learning. In *ICANN International Conference on Artificial Neural Networks*, volume 1, pages 41–48, Vienna, Austria, 2001.
56. V. Guruswami and A. Sahai. Multiclass learning, boosting, and error-correcting codes. In *12th Annual Conference on Computational Learning Theory (COLT-99)*, Santa Cruz, USA, 1999.
57. Z. Hao, C. Shen, N. Barnes, and B. Wang. Totally-corrective multi-class boosting. In *ACCV10*, volume 6495/2011, *Lecture Notes in Computer Science*, pages 269–280, 2011.
58. T. Hastie, R. Tibshirani, and J. Friedman. *The Elements of Statistical Learning*. Springer, 2nd edition, New York, NY, 2001.
59. J. He and B. Thiesson. Asymmetric gradient boosting with application to SPAM filtering. In *Fourth Conference on Email and Anti-Spam (CEAS) 2007*, August 2–3, Mountain View, California, USA.
60. C. Huang, H. Ai, Y. Li, and S. Lao. Vector boosting for rotation invariant multi-view face detection. In *International Conference on Computer Vision (ICCV)*, volume 1, pages 446–453, 2005.
61. X. Huang, S. Li, and Y. Wang. Jensen–Shannon boosting learning for object recognition. In *International Conference on Computer Vision and Pattern Recognition (CVPR)*, volume 2, pages 144–149, 2005.
62. J. Jackson and M. Craven. Learning sparse perceptrons. In *Advances in Neural Information Processing Systems*, volume 8, pages 654–660. MIT Press, 1996.
63. R. Jin, Y. Liu, L. Si, J. Carbonell, and A. Hauptmann. A new boosting algorithm using input-dependent regularizer. In *Proceedings of Twentieth International Conference on Machine Learning (ICML 03)*. AAAI Press, 2003.
64. X. Jin, X. Hou, and C.-L. Liu. Multi-class adaboost with hypothesis margin. In *Proceedings of the 2010 20th International Conference on Pattern Recognition*, ICPR '10, pages 65–68, Washington, DC, USA, 2010. IEEE Computer Society.
65. G. Jun and J. Ghosh. Multi-class boosting with class hierarchies. *Multiple Classifier Systems*, 5519:32–41, 2009.
66. H. Kong and E. Teoh. Coupling adaboost and random subspace for diversified fisher linear discriminant. In *International Conference on Control, Automation, Robotics and Vision (ICARCV) 06*, pages 1–5, 2006.
67. L. Kuncheva. *Combining Pattern Classifiers: Methods and Algorithms*. Wiley, Hoboken, NJ, 2004.
68. D. Le and S. Satoh. Ent-boost: Boosting using entropy measures for robust object detection. *Pattern Recognition Letters*, 2007.
69. H. Li and C. Shen. Boosting the minimum margin: LPBoost vs. AdaBoost. *Digital Image Computing: Techniques and Applications*, 0:533–539, 2008.
70. L. Li. Multiclass boosting with repartioning. In *23rd International Conference on Machine Learning (ICML 07)*, Pennsylvania, USA, 2006.
71. P. Li. ABC-Boost: adaptive base class boost for multi-class classification. In *International Conference on Machine Learning (ICML)*, pages 79–632, 2009.
72. S. Li and A. Jain. *Handbook of Face Recognition*. Springer, New York, NY, 2005.
73. S. Li and Z. Zhang. Floatboost learning and statistical face detection. *Transactions on Pattern Analysis and Machine Intelligence*, 26(9):23–38, 2004.
74. C. Liu and H. Shum. Kullback–Leibler boosting. In *International Conference on Computer Vision and Pattern Recognition (CVPR)*, volume 1, pages 587–594, Madison, Wisconsin, USA, 2003.

75. Y. Lu, Q. Tian, and T. Huang. Interactive boosting for image classification. In *Proceedings of the 7th International Conference on Multiple Classifier Systems*, MCS'07, pages 180–189, Berlin, Heidelberg, 2007. Springer-Verlag.
76. S. Lyu. Infomax boosting. In *International Conference on Computer Vision and Pattern Recognition (CVPR)*, volume 1, pages 533–538, 2005.
77. R. Maclin. An empirical evaluation of bagging and boosting. In *Proceedings of the Fourteenth National Conference on Artificial Intelligence*, pages 546–551. AAAI Press, 1997.
78. P. Mallapragada, R. Jin, A. Jain, and Y. Liu. SemiBoost: Boosting for Semi-Supervised Learning. *IEEE Transactions on Pattern Analysis and Machine Intelligence*, 31(11):2000–2014, 2009.
79. H. Masnadi-Shirazi and N. Vasconcelos. Asymmetric boosting. In *Proceedings of the 24th International Conference on Machine Learning, (ICML)*, pages 609–619, New York, NY, USA, 2007. ACM.
80. L. Mason, J. Baxter, P. Bartlett, and M. Frean. Boosting algorithms as gradient descent. In *Advances in Neural Information Processing Systems 12*, pages 512–518, MIT Press, Cambridge, MA, 1999.
81. L. Mason, J. Baxter, P. Bartlett, and M. Frean. Functional gradient techniques for combining hypotheses. *Advances in Large Margin Classifiers*, 1:109–120, 2000.
82. R. Meir and G. Rätsch. An introduction to boosting and leveraging. In S. Mendelson and A. Smola, editors, *Advanced Lectures on Machine Learning*. Springer Verlag, 2006.
83. A. Mohemmed, M. Zhang, and M. Johnston. A PSO Based Adaboost Approach to Object Detection. *Simulated Evolution and Learning*, pages 81–90, 2008.
84. Z. Niu, S. Shan, S. Yan, X. Chen, and W. Gao. 2d cascaded adaboost for eye localization. In *18th International Conference on Pattern Recognition*, 2006.
85. R. Nock and P. Lefaucheur. A Robust Boosting Algorithm. In T. Elomaa, H. Mannila, and H. Toivonen, editors, *Machine Learning: ECML 2002*, volume 2430 of *Lecture Notes in Computer Science*, pages 319–331, 2002. Springer Berlin/Heidelberg.
86. T. Pham and A. Smeulders. Quadratic boosting. *Pattern Recoginition*, 41:331–341, January 2008.
87. A. Quddus, P. Fieguth, and O. Basir. Adaboost and Support Vector Machines for White Matter Lesion Segmentation in MR Images. In *2005 IEEE Engineering in Medicine and Biology 27th Annual Conference*, pages 463–466. IEEE, 2005.
88. J. Quinlan. *C4.5: Programs for Machine Learning*. Morgan Kaufmann, San Mateo, CA, 1993.
89. J. Quinlan. Bagging, Boosting, and C4.5. In *Proceedings of the Thirteenth National Conference on Artificial Intelligence*, pages 725–730, 1996.
90. G. Rätsch. Robust multi-class boosting. In *Eurospeech 2003, 8th European Conference on Speech Communication and Technology*, page 9971000, Geneva, Switzerland, 2003.
91. B. Ripley. *Pattern Recognition and Neural Networks*. Cambridge University Press, Cambridge, MA, 1996.
92. J. Rodriguez and J. Maudes. Boosting recombined weak classifiers. *Pattern Recognition Letters*, 29(8):1049–1059, 2007.
93. A. Saffari, H. Grabner, and H. Bischof. Serboost: Semi-supervised boosting with expectation regularization. In D. Forsyth, P. Torr, and A. Zisserman, editors, *Computer Vision European Conference on Computer Vision (ECCV) 2008*, volume 5304 *Lecture Notes in Computer Science*, pages 588–601. Springer Berlin/Heidelberg, 2008.
94. A. Saffari, C. Leistner, M. Godec, and H. Bischof. Robust multi-view boosting with priors. In *11th European Conference on Computer Vision (ECCV)*, pages 776–789, Berlin, Heidelberg, 2010. Springer-Verlag.
95. R. Schapire. The strength of weak learnability. In *Machine Learning*, volume 5, pages 197–227, 1990.
96. R. Schapire. Using output codes to boost multiclass learning problems. In *14th International Conference on Machine Learning (ICML)*, pages 313–321, Tennessee, USA, 1997.

97. R. Schapire. Theoretical views of boosting. In *Proceedings of the 4th European Conference on Computational Learning Theory*, EuroCOLT '99, pages 1–10, London, UK, 1999. Springer-Verlag.
98. R. Schapire. The boosting approach to machine learning: An overview. In *Nonlinear Estimation and Classification*, Berkeley, 2002. Springer.
99. R. Schapire, Y. Freund, P. Bartlett, and W. Lee. Boosting the margin: A new explanation for the effectiveness of voting methods. In *Proceedings of the 14th International Conference on Machine Learning (ICML)*, pages 322–330, Nashville, TN, 1997.
100. R. Schapire and Y. Singer. Improved boosting algorithms using confidence-rated predictions. *Machine Learning*, 37(3):297–336, 1999.
101. R. Schapire and Y. Singer. BoosTexter: A Boosting-based System for Text Categorization. *Machine Learning*, 39(2/3):135–168, 2000.
102. B. Schölkopf and A. Smola. *Learning with Kernels*. MIT Press, 2002.
103. H. Schwenk and Y. Bengio. Adaboosting neural networks: Application to on-line character recognition. In *International Conference on Artificial Neural Networks'97, LNCS, 1327, 967–972*, pages 967–972. Springer, 1997.
104. H. Schwenk and Y. Bengio. Boosting Neural Networks. *Neural Comp.*, 12(8):1869–1887, 2000.
105. C. Shen, J. Kim, L. Wang, and A. van den Hengel. Positive semidefinite metric learning with boosting. In Y. Bengio, D. Schuurmans, J. Lafferty, C. Williams, and A. Culotta, editors, *Advances in Neural Information Processing Systems (NIPS'09)*, pages 1651–1659, Vancouver, BC, Canada, December 2009. MIT Press.
106. J. Sochman and J. Matas. "Waldboost" learning for time constrained sequential detection. In *Proceedings of the 2005 IEEE Computer Society Conference on Computer Vision and Pattern Recognition (CVPR'05) – Volume 2*, pages 150–156, Washington, DC, USA, 2005. IEEE Computer Society.
107. A. Stefan, V. Athitsos, Q. Yuan, and S. Sclaroff. Reducing jointboost-based multiclass classification to proximity search. In *Computer Vision and Pattern Recognition (CVPR)*, pages 589–596. IEEE, 2009.
108. S. Sternig, M. Godec, P. Roth, and H. Bischof. Transientboost: On-line boosting with transient data. *IEEE Computer Society Conference on Computer Vision and Pattern Recognition Workshops (CVPRW), 2010*, pages 22–27, San Francisco, CA.
109. Y. Sun, S. Todorovic, and J. Li. Unifying multi-class adaboost algorithms with binary base learners under the margin framework. *Pattern Recognition Letters*, 28:631–643, 2007.
110. Y. Sun, S. Todorovic, J. Li, and D. Wu. Unifying the error-correcting and output-code adaboost within the margin framework. In *Proceedings of the 22nd International Conference on Machine Learning (ICML)*, pages 872–879, New York, NY, USA, 2005. ACM.
111. J. Thongkam, O. Xu, Y. Zhang, F. Huang, and G. Adaboosts. Breast cancer survivability via adaboost algorithms. In *Proceedings of the second Australasian workshop on Health data and knowledge management – Volume 80*, HDKM '08, pages 55–64, Darlinghurst, Australia, 2008. Australian Computer Society, Inc.
112. K. Tieu and P. Viola. Boosting image retrieval. In *Proceedings of the IEEE Conference on Computer Vision and Pattern Recognition – CVPR*, volume 1, pages 228–235, 2000.
113. A. Torralba, K. Murphy, and W. Freeman. Sharing visual features for multiclass and multiview object detection. *IEEE Transactions on Pattern Analysis and Machine Intelligence*, 29(5):854 – 869, March 2007.
114. L. Valiant. A theory of the learnable. *Communications of the ACM*, 27(11):1134–1142, 1984.
115. H. Valizadegan, R. Jin, and A. K. Jain. Semi-Supervised Boosting for Multi-Class Classification. In *ECML PKDD '08: Proceedings of the European Conference on Machine Learning and Knowledge Discovery in Databases – Part II*, pages 522–537, Berlin, Heidelberg, 2008. Springer-Verlag.
116. V. Vapnik. *The Nature of Statistical Learning Theory*. Springer-Verlag, New York, NY, 1999.

117. V. Vapnik and A. Chervonenkis. On the uniform convergence of relative frequencies of events to their probabilities. *Theory of Probability and its Applications*, 16(2):264–280, 1971.
118. R. Verschae, J. Ruiz-del-solar, and M. Correa. Gender classification of faces using adaboost. In *Lecture Notes in Computer Science (CIARP 2006) 4225*, page 78. Springer, 2006.
119. A. Vezhnevets and V. Vezhnevets. Modest AdaBoost - teaching AdaBoost to generalize better. *Graphicon*, 12(5):987–997, 2005.
120. P. Viola and M. Jones. Rapid object detection using a boosted cascade of simple features. In *International Conference on Computer Vision and Pattern Recognition (CVPR)*, volume 1, pages 511–518, Hawaii, 2001.
121. P. Viola and M. Jones. Robust real-time face detection. *International Journal of Computer Vision*, 57:137–154, 2004.
122. P. Viola, M. Jones, and D. Snow. Detecting pedestrians using patterns of motion and appearance. In *International Conference on Computer Vision – ICCV*, pages 734–741, 2003.
123. P. Viola, J. Platt, and C. Zhang. Multiple instance boosting for object detection. In Y. Weiss, B. Schölkopf, and J. Platt, editors, *Advances in Neural Information Processing Systems 18*, pages 1417–1424, Cambridge, MA, 2006. MIT Press.
124. L. Wang, S. Yuan, L. Li, and H. Li. Boosting naïve Bayes by active learning. In *Third International Conference on Machine Learning and Cybernetics*, volume 1, pages 41–48, Shanghai, China, 2004.
125. P. Wang, C. Shen, N. Barnes, H. Zheng, and Z. Ren. Asymmetric totally-corrective boosting for real-time object detection. In *Asian Conference on Computer Vision (ACCV)*, pages I: 176–188, 2010.
126. M. Warmuth, K. Glocer, and G. Rätsch. Boosting algorithms for maximizing the soft margin. In *Advances in Neural Information Processing Systems NIPS*, pages 1–8, MIT Press, 2007.
127. M. Warmuth, K. Glocer, and S. Vishwanathan. Entropy regularized LPBoost. In *Proceedings of the 19th International Conference on Algorithmic Learning Theory*, ALT '08, pages 256–271, Springer-Verlag, Berlin, Heidelberg, 2008.
128. M. Warmuth, J. Liao, and G. Rätsch. Totally corrective boosting algorithms that maximize the margin. In *Proceedings of the 23rd International Conference on Machine Learning (ICML)*, pages 1001–1008, New York, NY, USA, 2006. ACM.
129. J. Warrell, P. Torr, and S. Prince. Styp-boost: A bilinear boosting algorithm for learning style-parameterized classifiers. In *British Machine Vision Conference (BMVC)*, 2010.
130. J. Webb, J. Boughton, and Z. Wang. Not so naïve Bayes: Aggregating one-dependence estimators. *Machine Learning*, 58(1):5–24, 2005.
131. P. Yang, S. Shan, W. Gao, S. Z. Li, and D. Zhang. Face recognition using ada-boosted gabor features. In *Proceedings of the 16th International Conference on Face and Gesture Recognition*, pages 356–361, 2004.
132. C. Zhang, P. Yin, Y. Rui, R. Cutler, P. Viola, X. Sun, N. Pinto, and Z. Zhang. Boosting-based multimodal speaker detection for distributed meeting videos. *IEEE Transactions on Multimedia*, 10(8):1541–1552, December 2008.
133. C. Zhang and Z. Zhang. *Boosting-Based Face Detection and Adaptation*. Morgan and Claypool Publishers, 2010.
134. C. Zhang and J. Zhang. Rotboost: A technique for combining rotation forest and adaboost. *Pattern Recognition Letters*, 29(10):1524–1536, July 2008.
135. C. Zhang and Z. Zhang. Winner-take-all multiple category boosting for multi-view face detection. Technical report, One Microsoft Way, Redmond, WA 98052, USA, 2010.
136. L. Zheng, S. Wang, Y. Liu, and C.-H. Lee. Information theoretic regularization for semi-supervised boosting. In *Proceedings of the 15th ACM SIGKDD International Conference on Knowledge Discovery and Data Mining*, KDD '09, pages 1017–1026, New York, NY, USA, 2009. ACM.
137. M. Zhou, H. Wei, and S. Maybank. Gabor wavelets and AdaBoost in feature selection for face verification. In *Proceedings of the Workshop on Applications of Computer Visions*, pages 101–109, Graz, Austria, 2006.

138. J. Zhu, H. Zou, S. Rosset, and T. Hastie. Multi-class adaboost. *Statistics and Its Interface*, 2:349–360, 2009.
139. X. Zhu, C. Bao, and W. Qiu. Bagging very weak learners with lazy local learning. In *International Conference on Pattern Recognition (ICPR)*, pages 1–4, 2008.
140. X. Zhu and Y. Yang. A lazy bagging approach to classification. *Pattern Recognition*, 41:2980–2992, 2008.

Chapter 3
Boosting Kernel Estimators

Marco Di Marzio and Charles C. Taylor

3.1 Introduction

A boosting algorithm [1, 2] could be seen as a way to improve the fit of statistical models. Typically, M predictions are operated by applying a base procedure—called a *weak learner*—to M reweighted samples. Specifically, in each reweighted sample an individual weight is assigned to each observation. Finally, the output is obtained by aggregating through majority voting. Boosting is a *sequential* ensemble scheme, in the sense the weight of an observation at step m depends (only) on the step $m-1$. It appears clear that we obtain a specific boosting scheme when we choose a loss function, which orientates the data re-weighting mechanism, and a weak learner.

In statistical inference kernel estimators can be regarded as the most used and studied locally weighted learning procedures. They constitute a sound means to address the three main inferential problems, i.e., density estimation—which can be categorized as unsupervised learning, discrimination (where we predict labels) and regression (where we predict real values)—which can be categorized as supervised learning.

In the sequel, we will discuss boosting algorithms for density estimation, regression, and classification; all of them use kernel estimators as weak learners. To obtain the properties of our algorithms, we will see them as multistep estimators and will derive some statistical properties of them. The main conclusion will be that boosting has the potential to reduce the bias of kernel methods at the cost of a slight variance inflation. Notice that we will treat point estimation problems, but not

M. Di Marzio
Dipartimento di Metodi Quantitativi e Teoria Economica (DMQTE),
University of Chieti-Pescara, Viale Pindaro 42, 65127 Pescara, Italy
e-mail: mdimarzio@unich.it

C.C. Taylor (✉)
Department of Statistics, University of Leeds, Leeds LS2 9JT, UK
e-mail: charles@maths.leeds.ac.uk

C. Zhang and Y. Ma (eds.), *Ensemble Machine Learning: Methods and Applications*,
DOI 10.1007/978-1-4419-9326-7_3, © Springer Science+Business Media, LLC 2012

Algorithm 1 Discrete AdaBoost

1. Initialize $w_i(1) = 1/N, \quad 1 = 1, \ldots, N$
2. For $m = 1, \ldots, M$

   a. Estimate the classifier $\delta_m(x)$ using weights $w(m)$
   b. Compute the (weighted) error rate ξ_m and $b_m = \log \frac{1-\xi_m}{\xi_m}$
   c. Update the weights

$$w_i(m+1) = \begin{cases} w_i(m) & \text{if } Y_i = \delta_m(x_i) \\ w_i(m)(1-\xi_m)/\xi_m & \text{otherwise} \end{cases}$$

   d. Normalize $w_i(m)$ so that $\sum_i w_i(m) = 1$

3. Output sign $\sum_m b_m \delta_m(x)$

confidence regions or hypothesis tests. In fact, Machine Learning does not focus on these latter two. Nevertheless, because of boosting reduces bias, an usage of it also for improving the coverage of pointwise confidence intervals appears promising when, as it usually happens, the nominal coverage is reduce by a biased pointwise estimator. For more on this, see [3]. For a more general treatment of boosting from a statistical perspective, including existing software, see [4].

3.2 The Boosting Mechanism for Discrimination

Consider data $(x_i, Y_i), i = 1, \ldots, N$ in which $Y_i \in \{-1, 1\}$ denotes the class, and $\delta(x)$ denotes the classification of a (weak) learner. A proper boosting algorithm—called Discrete AdaBoost [5]—follows.
Note that misclassified data are given higher weights, and that the final vote is weighted in favor of the more successful classifiers.

We illustrate the above algorithm in two examples. Initially, we focus on the concept of a *weak* learner. This, in the classification setting, is defined as a method which has an expected error rate which is smaller than a random guess (in the case of equal priors) or a default classifier (in the case of unequal priors). In the following toy example, the *distribution* of the population (rather than weights assigned to observations) is boosted.

3.2.1 *Example 1: Data Without Noise*

Suppose we have a categorical variable X which can take the values A, B, C and that there are two classes with priors $\pi_i, i = 1, 2$. We have 2^3 possible partitions for the classification rule (writing as class 1|class 2):

$$A|BC,\ BC|A,\ B|CA,\ CA|B,\ C|AB,\ AB|C,\ ABC|\phi,\ \phi|ABC \tag{3.1}$$

An example of a "weak learner" is one which must select a rule from the reduced set of partitions:

$$A|BC,\ BC|A,\ C|AB,\ AB|C,\ ABC|\phi,\ \phi|ABC, \tag{3.2}$$

i.e., omitting the partitions which allocate only outcome B to one of the two classes. Then, for any distribution of the two classes, on A, B, C we can always obtain an error rate of at most $1/3$. To see this, denote $p_{1C} = P(X = C|x \in \text{class1})$, etc., so that $p_{kA} + p_{kB} + p_{kC} = 1$ for $k = 1, 2$. The worst case is when class 1 has support AC and class 2 has support B (or vice versa). Then

$$p_{1A} + p_{1C} = 1, \quad p_{2B} = 1, \quad p_{1B} = p_{2A} = p_{2C} = 0 \tag{3.3}$$

and at least one of the following partitions will give error rate less than $1/3$:

Partition	Error rate
$A\|BC$	$\pi_1 p_{1C}$
$BC\|A$	$\pi_1 p_{1A} + \pi_2$
$AB\|C$	$\pi_1 p_{1C} + \pi_2$
$C\|AB$	$\pi_1 p_{1A}$
$ABC\|\phi$	π_2
$\phi\|ABC$	π_1

For example (first row), partition $A|BC$ allocates $x = C$ incorrectly. This occurs with probability

$$P(X = C \mid X \in \text{class1})\pi_1 = \pi_1 p_{1C}. \tag{3.4}$$

Similar calculations and check for each of the other rows verify the assertion that we have constructed a weak learner, since the smallest error will be one of the first, fourth, or fifth partitions.

To complete the illustration we use some numerical values for the distribution to show the effect of boosting on the distribution weights. Suppose in our example that $p_{1A} = 0.8$ (so that $p_{1C} = 0.2$) and $\pi_1 = 0.6$ (so that $\pi_2 = 0.4$). Then

$$p(x = A) = 0.48,\ p(x = B) = 0.4,\ p(x = C) = 0.12. \tag{3.5}$$

Then the best first split is $A|BC$ with error $\xi_1 = 0.12$ which gives $b_1 = 1.99$. These values are displayed in the first row of the table below. The *distributions* (instead of w) can be reweighted, with the next two partitions given in the table, with consequent boosting constants, and classifiers. Thus, in the first row we allocate A to class 1 (+) and B and C to class 2 (−) and the boosting constant 1.99 is multiplied by $(1, -1, -1)$. In the second row, 1.22 multiplied by $(1, 1, 1)$ is added to the current classifier to yield $(3.21, -0.77, -0.77)$, and so on.

m	Partition	Error	b_m	$\sum b_m \delta_m(x)$	ξ_m
1	$A\|BC$	0.12	1.99	$(1.99, -1.99, -1.99)$	.12
2	$CAB\|\phi$	0.23	1.22	$(3.21, -0.77, -0.77)$	.12
3	$C\|AB$	0.18	1.54	$(1.68, -2.31, 0.77)$	0

After three iterations, the classifier $\sum b_m \delta_m(x)$ has signs $+, -, +$, so that A and C are classified to class 1 and B to class 2, as desired.

3.2.2 Example 2: Data with Noise

We continue to work with population (distribution)—corresponding to infinite data—in a second illustration in which the goal is to learn a nonlinear boundary using only linear splits. Consider two populations occupying squares on a mini-chess board, where some of the squares are partially occupied by an amount E $(0 \leq E < 1)$; so that some squares can be shared by the two populations, representing an overlap, or noise.

Class 1					Class 2				
1	1	1	1	1	0	0	E	0	0
1	1	E	1	1	E	0	1	0	E
E	1	0	1	E	1	E	1	E	1
0	E	0	E	0	1	1	1	1	1

In this example, we consider a weak learner which can only split in the horizontal or vertical direction, i.e., one of the following 30 splits shown as lines—each line rep resents a possible split (though these are not all distinct in their effect):

*	*	*	*	*
*	*	*	*	*
*	*	*	*	*
*	*	*	*	*

We can see the development of the boosting algorithm (using a straightforward implementation of Discrete AdaBoost, in which the distribution weights are updated) for overlapping amount $E = 0$. The individual (step-wise) classifiers are shown in Fig. 3.1, and the cumulative effect (after taking a weighted sum according to the boosting coefficients b_m, and taking the sign of the output) is also shown. Hence, when $E = 0$ we can see that the exact partition is recovered after 12 boosting iterations.

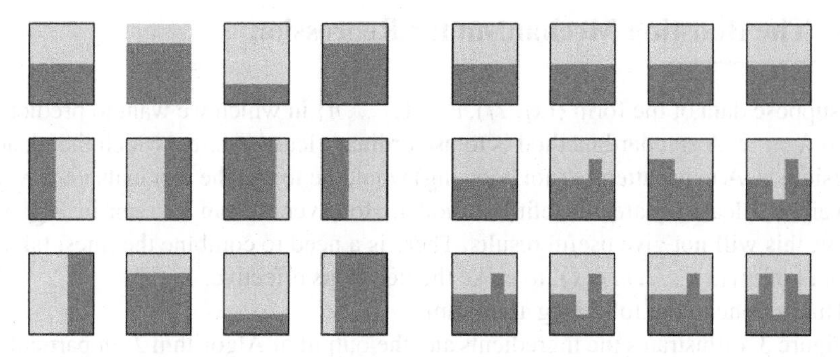

Fig. 3.1 *Left*: Individual (step-wise) classifiers of Discrete AdaBoost. *Right*: The cumulative effect of Discrete AdaBoost

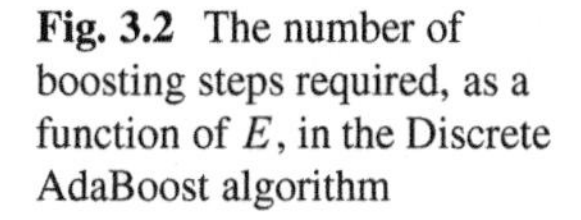

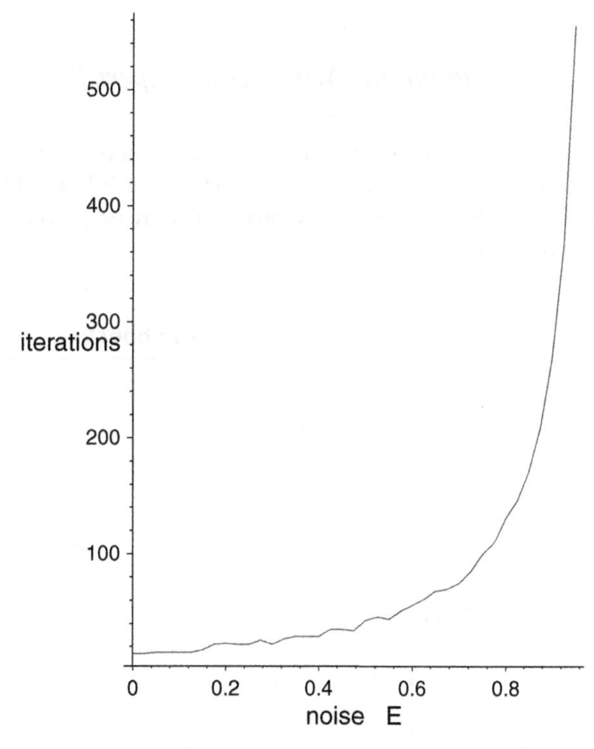

Fig. 3.2 The number of boosting steps required, as a function of E, in the Discrete AdaBoost algorithm

This example can be repeated for $E > 0$ and it is of interest to note the effect that this has on the number of iterations required before the true partition is found. Figure 3.2 shows the number of partitions as a function of E, before the Bayes classifier is obtained. It can be seen there is very rapid growth in the number of splits as $E \to 1$ since this represents squares with almost equal proportions of the two classes.

3.3 The Boosting Mechanism for Regression

We suppose data of the form $\{(x_i, Y_i), i = 1, \ldots, n\}$ in which we want to predict Y given $X = x$. A standard method is to use ordinary least squares, which then leads to residuals. A naive attempt (not boosting) would be to use the residuals iteratively in weighted least squares to refit the model. However, as can be seen in Fig. 3.3 below, this will not give useful results. There is a need to combine the lines, taking account of $m_1(x), \ldots, m_B(x)$, to make the iterations effective.

This is done in the following algorithm

Figure 3.4 illustrates the ingredients and the output of Algorithm 2. In particular, the left panel shows the components of each iteration, whereas the right panel shows their combination into the final prediction, which is very nonlinear.

3.3.1 *Relation to "Mixtures of Experts"*

As a generalization of standard mixture models in the regression context, the *mixtures of experts* model was introduced in [6]. The basic idea for regression is to use the EM algorithm to estimate a mixture proportion for each data point x_i, i.e., for B mixtures:

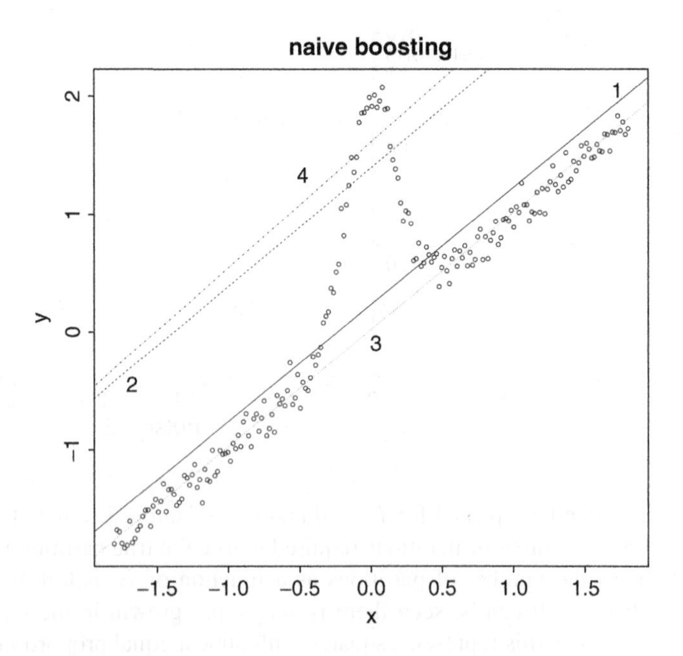

Fig. 3.3 Iteratively re-weighted least squares (3 iterations) of linear model

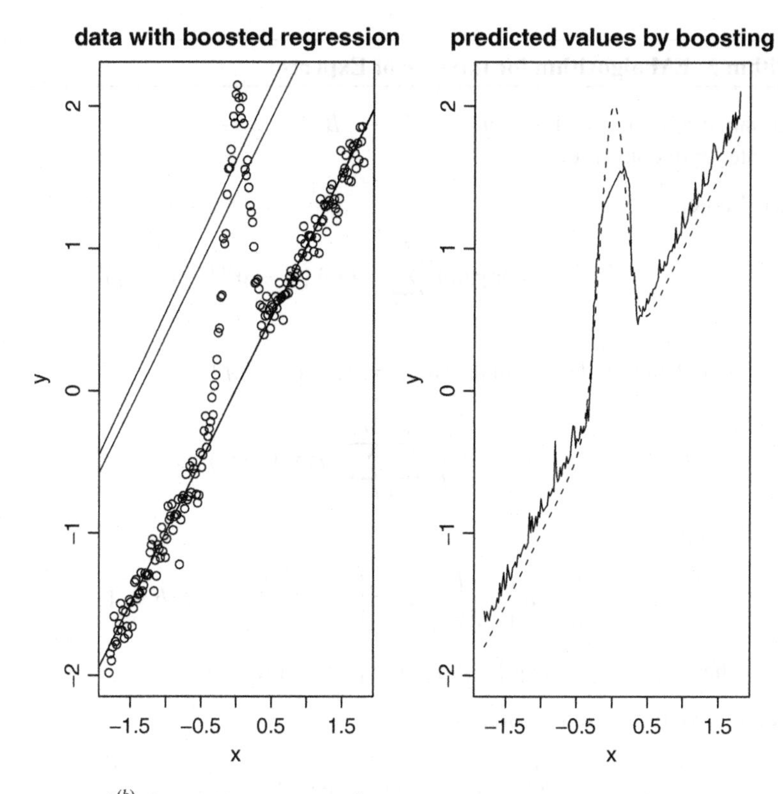

Fig. 3.4 *Left*: $\hat{y}_i^{(b)}, b = 1, \ldots, 5$. *Right*: $\hat{m}(x_i)$ (continuous line) and $m(x)$ (dashed line) (see Algorithm 2)

Algorithm 2 Boosting linear regression

---

1. Set $b = 1$ and $w_b(x_i) = 1/n, i = 1, \ldots, n$
2. Fit $\hat{y}_i^{(b)} = \alpha^{(b)} + \beta^{(b)} x_i$ by weighted least squares (with weights w_b).
3. $e_i = y_i - \hat{y}_i^{(b)} \quad w_{b+1}(x_i) = e_i^2$; renormalize weights.
4. increment $b := b + 1$ and return to step 2 (until $b = B$).
5. Summarize the overall prediction

$$\hat{m}(x_i) = \frac{\sum_{b=1}^{B} w_b(x_i) \hat{y}_i^{(b)}}{\sum w_b(x_i)}$$

- estimate $\alpha^{(b)}, \beta^{(b)}, b = 1, \ldots, B$ by weighted least squares
- estimate the mixing proportions $\pi_b(x_i), b = 1, \ldots, B,\ i = 1, \ldots, n$, such that

$$\sum_b \pi_b(x_i) = 1, \quad i = 1, \ldots, n. \tag{3.6}$$

Algorithm 3 EM algorithm for mixture of Experts

1. Initialize $\pi_b(x_i)$, $i = 1, \ldots, n$, $b = 1, \ldots, B$
2. Iterate, until convergence:

   2.1 M-step:

$$\hat{\alpha}^{(b)}, \hat{\beta}^{(b)} = \arg\min \sum_{i=1}^{n} \pi_b(x_i) \left(y_i - \alpha^{(b)} - \beta^{(b)} x_i\right)^2$$

   then, denoting the residuals $e_b(x_i) = y_i - \hat{\alpha}^{(b)} - \hat{\beta}^{(b)} x_i$

$$\sigma^2 = \frac{1}{nB} \sum_{b=1}^{B} \sum_{i=1}^{n} \pi_b(x_i) e_b(x_i)^2$$

   2.2 E-step:

$$\pi_b(x_i) = \frac{1}{S_i} \exp\left\{ -\frac{1}{2\oe^2} e_b(x_i)^2 \right\}, \quad i = 1, \ldots, n, \; b = 1, \ldots, B$$

   where $S_i = \sum_{b=1}^{B} \exp\left\{-\frac{1}{2\sigma^2} e_b(x_i)^2\right\}$, $i = 1, \ldots, n$

3. Summarize the overall prediction

$$\hat{m}(x_i) = \sum_{b=1}^{B} \pi_b(x_i) \left(\hat{\alpha}^{(b)} + \hat{\beta}^{(b)} x_i\right), \; i = 1, \ldots, n$$

This can be done via an iterative procedure, which maximizes the likelihood, giving the parameter estimates $\alpha^{(b)}, \beta^{(b)}, b = 1, \ldots, B$, and estimates the mixing proportions $\pi_b(x_i), b = 1, \ldots, B, i = 1, \ldots, n$. using the EM algorithm as follows.

Convergence using the EM algorithm is usually fast but the starting point is critical. Using the previous example we obtain the result in Fig. 3.5. An obvious problem, in general, is how to choose the number of mixtures, B. Moreover, the figure illustrates the effect of different choices of starting values—a well known issue for EM algorithms.

3.4 Main Definitions in Kernel Methods

Kernel methods can be regarded as an established way to nonparametrically face the three main inferential problems, i.e., density estimation, (which can be categorized as unsupervised learning), discrimination, and regression (which can be categorized as supervised learning). Each of these themes has had its own

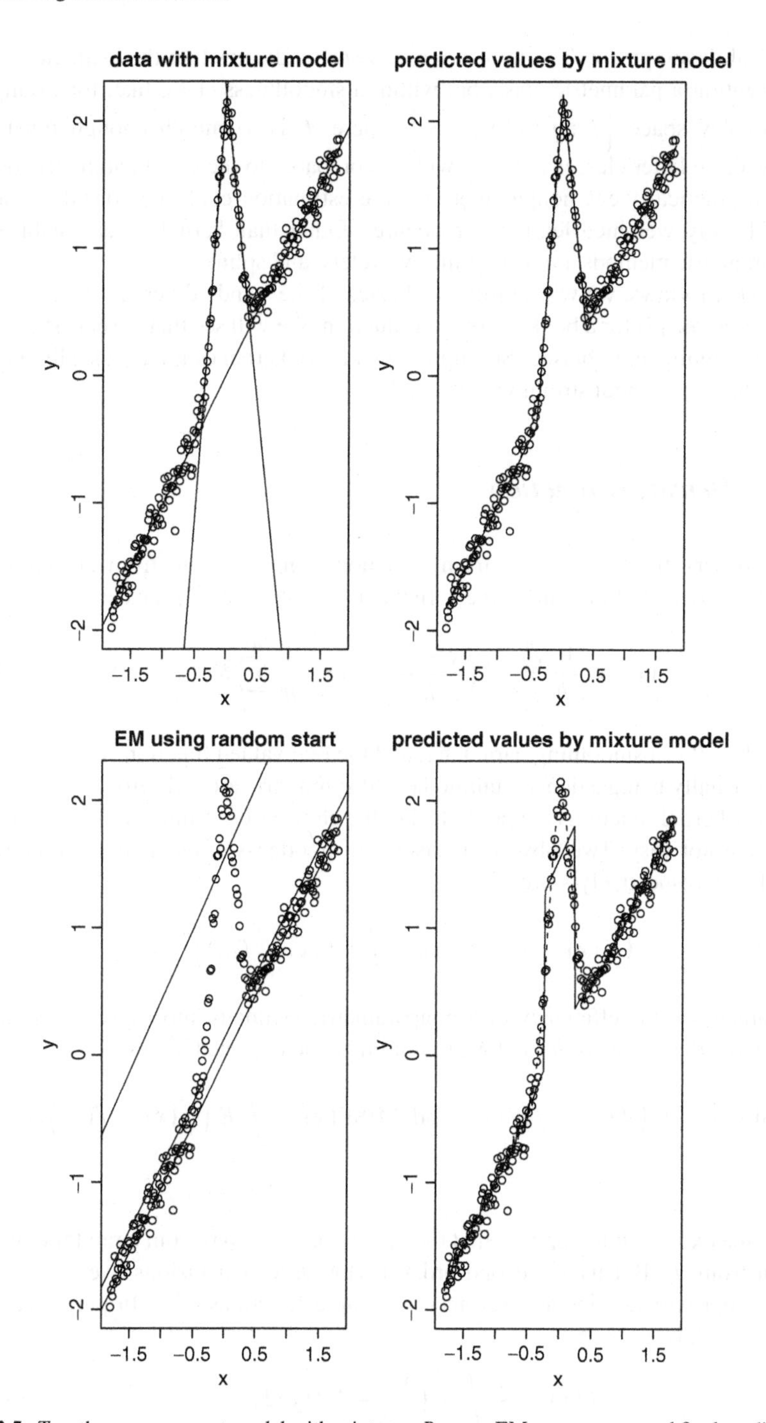

Fig. 3.5 *Top*: three-component model with mixtures. *Bottom*: EM components and final prediction

historical development. The term *nonparametric* indicates that the estimate is not found within a parametric class, but within a smoothness class, like, for example, the Sobolev space $\left\{f : \int (f'')^2 < \infty\right\}$ where f is a smooth enough function. Surely, these latter classes are very wide if compared to the usual parametric ones. From a practical view, nonparametric curve estimation could be considered as a set of locally weighted learning procedures. Other than kernels, well-established nonparametric methods are histograms, wavelets, and splines.

In what follows, we will outline the basics of the standard kernel theory in one dimension, see [7] for a book-length introduction. We will see that kernel estimators could be significantly biased. So improving a naive kernel estimator usually means bias reduction without strong variance inflation.

3.4.1 Density Estimation

Given observations $x_1, \ldots, x_n$ from a random sample drawn from an unknown probability density function f, the kernel density estimate of f at x is

$$\hat{f}(x) = \frac{1}{nh}\sum_{i=1}^{n} K\left(\frac{x - x_i}{h}\right) \quad \text{or} \quad \frac{1}{n}\sum_{i=1}^{n} K_h(x - x_i) \tag{3.7}$$

where $h > 0$ is a smoothing parameter, and the real valued function K is called the kernel, usually being a density unimodal and symmetric around zero.

Many kernel functions are possible, but the choice of h is much more important, and often not easy. Two obvious measures of goodness of estimation—pointwise and global, respectively—are:

$$\left(\hat{f}(x) - f(x)\right)^2 \quad \text{and} \quad \int \left(\hat{f}(x) - f(x)\right)^2 \mathrm{d}x \tag{3.8}$$

to characterize the efficiency of a nonparametric estimator most researchers have considered the choice of h (and K) to minimize the expected losses

$$\text{MSE}(x) = E\left(\hat{f}(x) - f(x)\right)^2 \quad \text{and} \quad \text{MISE}(x) = \int E\left(\hat{f}(x) - f(x)\right)^2 \mathrm{d}x \tag{3.9}$$

where the expectations suppose that the x_is are observations from a random sample drawn from f. But using second-order Taylor series expansions, we are able to obtain approximate bias and variance as simple functions of h. In particular, for small h we have:

$$E\hat{f}(x) = \frac{1}{h}\int K\left(\frac{x - y}{h}\right) f(y)\mathrm{d}y \tag{3.10}$$

$$= f(x) + \frac{h^2}{2} f''(x)\mu_2(K) + O(h^3) \tag{3.11}$$

(given $\int zK(z)dz = 0$) where $\mu_2(K) = \int z^2 K(z)dz$. Similarly, for large n and such that $h \to 0, n \to \infty, nh \to \infty$ we have:

$$\operatorname{var} \hat{f}(x) = \frac{1}{n}\left\{\frac{1}{h^2}\int K^2\left(\frac{x-y}{h}\right) f(y)dy - (f(x) + o(h))^2\right\} \tag{3.12}$$

$$= \frac{f(x)}{nh}\int K^2(z)dz + o\left(\frac{1}{nh}\right). \tag{3.13}$$

So, as $h \to 0$ the bias vanishes, but the variance becomes infinite, and as $h \to \infty$ the variance vanishes, but the bias becomes infinite. An asymptotic mean squared error is therefore (omitting smaller terms in the Taylor series):

$$\text{AMSE}(x) = \operatorname{var} \hat{f}(x) + \left(\operatorname{bias} \hat{f}(x)\right)^2 \tag{3.14}$$

$$= \frac{f(x)}{nh} R(K) + \frac{h^4}{4} f''(x)^2 \mu_2(K)^2 \tag{3.15}$$

where $R(K) = \int K^2(z)dz$. We can minimize this (over h) to find the optimal value

$$h = \left(\frac{f(x)R(K)}{f''(x)^2 \mu_2(K)^2 n}\right)^{1/5} \quad (f''(x) \neq 0) \tag{3.16}$$

which depends on the known quantities K and, n and on the unknown quantities $f(x)$ and $f''(x)$. In a similar way, we can get the optimal value with respect to the AMISE, which is defined as the integral of AMSE over the density support

$$h = \left(\frac{R(K)}{\int f''(x)^2 dx \; \mu_2(K)^2 n}\right)^{1/5}, \tag{3.17}$$

which depends on the known quantities K and n, and on the unknown quantity $f''(x)$. A common objection is that kernel density estimation is an ill-posed problem because the estimation of $f''(x)$ is an intermediate step to estimate f. A common approach is to use a normal reference rule whereby f is assumed to be gaussian. If we also use a gaussian kernel this leads to the choice $h = 1.06sn^{1/5}$ where s is the sample standard deviation.

Various methods have been suggested by which the bias can be reduced. In the above calculations, the bias is of order $O(h^2)$. This term arises because $\int z^2 K(z)dz > 0$. Selecting K so that $\int z^2 K(z)dz = 0$ would lead to bias of order $O(h^4)$, and such kernels are known as higher-order kernels. The inconvenience of these kernels are that they are not densities, and so they lead to $\hat{f}$ which are not pdfs (since they can be negative). Moreover, although they have small *asymptotic* bias, various research indicates that the gain is rather modest unless n is very large.

A so-called *multiplicative* approach to bias reduction can also be used. Jones et al. [7] proposed the following variable-kernel estimator and this gives nonnegative estimates, which, however, do not integrate to 1. The name multiplicative originates from the fact that the overall estimator is the product of the standard kernel (here $\tilde{f}$) and a second estimate. The second estimate is a *weighted* sum, with weights given by the inverse of the first estimate. Regarding the smoothing setting, Jones et al. [7] say that h should be the same in $\tilde{f}(x)$ as in $\hat{f}(x)$.

3.4.2 Density Classification

The discrimination problem could be regarded as the task of predicting a categorical variable, i.e., the label j of a *class* Π_j, to which an observations is supposed to belong. We indicate the jth class density as f_j. Given $G \geq 2$ classes with respective prior membership probabilities $\pi_1, \ldots, \pi_G$, the posterior probability of the observation x_i being from the jth class, is (using Bayes' rule):

$$P\left(x_i \in \Pi_j | x_i = x\right) = \pi_j f_j(x) \Big/ \sum_{j=1}^{G} \pi_j f_j(x) . \tag{3.18}$$

Bayes' classifier (which is optimal) allocates an observation to the class with highest posterior probability.

Typically, the priors are estimated by the sample proportions or other external information. But the main difficulty in using this optimal rule is that the $f_j(x)$ are unknown. An obvious method is to operate a kernel density estimate for each $f_j(x)$.

In the sequel we will focus on the case of two classes ($G = 2$), which amounts to the estimation of the intersection points of the population densities. It should be remarked that intersection estimation does not involve the usual zero-one loss, but a real-valued one. In the case of equal priors, it is sufficient to find all x_0 for which $f_1(x_0) = f_2(x_0)$. Therefore, we are aimed to estimate the difference $g(x) = f_1(x) - f_2(x)$ by using $\hat{f}_1(x) - \hat{f}_2(x)$. Here, once more, the crucial question is to set a smoothing degree specific to the discrimination problem, for more details see [8].

3.4.3 Regression

Given a double random variable (X, Y), our aim is to predict the continuous variable Y by observing X. Therefore, given data $\{(x_1, Y_1), \ldots, (x_n, Y_n)\}$, the kernel estimator—also called Nadaraya–Watson (NW) estimator—of the regression function m at x is so defined:

$$\hat{m}(x) = \frac{\frac{1}{n}\sum K_h(x - X_i)\,Y_i}{\frac{1}{n}\sum K_h(x - X_i)} = \frac{\hat{r}(x)}{\hat{f}(x)}, \quad \text{say} \tag{3.19}$$

the idea is to operate a kernel estimate of the conditional mean

$$m(x) = E(Y \mid X = x) = \frac{\int y f(x, y)\, dy}{f(x)} \tag{3.20}$$

where $f(x, y)$ denotes the joint density of the distribution of (X, Y) at (x, y). According to a different interpretation, kernel regression consists of fitting constants using a locally weighted training criterion. In both cases, local learning here means that the estimated regression will tend to match the responses as $h \to 0$.

Using Taylor series expansions again we have

$$E\,\hat{r}(x) = \iint K_h(x-u) y f(u, y)\, du dy = \int K_h(x-u) r(u)\, du \tag{3.21}$$

$$\approx r(x) + \frac{h^2}{2} r''(x) \mu_2(K) \tag{3.22}$$

which leads to

$$E\,\hat{m}(x) = \frac{\hat{r}(x)}{\hat{f}(x)} \approx \left(r(x) + \frac{h^2}{2} r''(x)\mu_2(K) \right) \left(f(x) + \frac{h^2}{2} f''(x)\mu_2(K) \right)^{-1} + o(h^2) \tag{3.23}$$

and so the bias in $\hat{m}(x)$ is

$$\frac{h^2 \mu_2(K)}{2} \left(m''(x) + \frac{2m'(x) f'(x)}{f(x)} \right) + o(h^2). \tag{3.24}$$

Similar calculations give the variance as

$$\frac{1}{nh} \frac{\sigma^2(x)}{f(x)} R(K) + o\left(\frac{1}{nh} \right), \tag{3.25}$$

where $\sigma^2(x)$ is the conditional variance. We easily see that the regression curve is more stable (lower variance) when there are more observations, and that the bias-squared is dominated by the second derivative $m''(x)$ (close to a turning point) or by $m'(x)$ when there are few observations.

3.4.4 Cumulative Distribution Function Estimation

Define the kernel cumulative distribution function (CDF) estimator as the integral of the kernel density estimate:

$$\hat{F}_h(x) = \int_{-\infty}^{x} \hat{f}_h(u)\mathrm{d}u \tag{3.26}$$

and, if we use a normal kernel, this is simply

$$\hat{F}_h(x) = \frac{1}{n}\sum_{i=1}^{n} \Phi\left(\frac{x - x_i}{h}\right), \tag{3.27}$$

where $\Phi(\cdot)$ is the CDF of a standard normal. If $h \to 0$, the kernel density estimate becomes a sum of Dirac delta functions placed on the data, which then yields the empirical CDF

$$\hat{F}_0(x) := \lim_{h \to 0} \hat{F}_h(x) = \frac{1}{n}\sum_{i=1}^{n} I[x_i \leq x]. \tag{3.28}$$

If f is smooth enough and $K(\cdot)$ is symmetric, the expectation admits the second order expansion

$$E\hat{F}_h(x) = F(x) + \frac{h^2}{2}\mu_2 f'(x) + O\left(h^4\right), \tag{3.29}$$

where $\mu_k = \int K(x)x^k \mathrm{d}x$, and this shows that $\hat{F}_0(x)$ is unbiased.

Incidentally, note that

$$\mathrm{var}\left(\hat{F}_0(x)\right) = \frac{F(x)(1 - F(x))}{n} \tag{3.30}$$

which means that, we do not get infinite variance (unlike the pdf estimation case) if $h \to 0$. In fact, it is known that $\hat{F}_0(x)$ is the unbiased estimator of $F(x)$ with the smallest variance. What about $\mathrm{var}\left(\hat{F}_h(x)\right)$ for $h > 0$? If we use a Normal kernel, we can calculate

$$\mathrm{var}\left(\hat{F}_h(x)\right) = \frac{1}{n}\left[E\Phi\left(\frac{x - X}{h}\right)^2 - E^2\Phi\left(\frac{x - X}{h}\right)\right]. \tag{3.31}$$

Since

$$E\int K\left(\frac{x-X}{h}\right) \approx F(x) + \frac{h^2}{2}\mu_2(K)f'(x) \tag{3.32}$$

and, using a Normal kernel we have

$$E\Phi\left(\frac{x-X}{h}\right)^2 \approx F(x) - \frac{hf(x)}{\sqrt{\pi}} \tag{3.33}$$

we then obtain

$$\mathrm{var}\left(\hat{F}_h(x)\right) \approx \frac{F(x)(1-F(x))}{n} - \frac{hf(x)}{n\sqrt{\pi}}. \tag{3.34}$$

In conclusion, $\hat{F}_h$ can have smaller mean squared error than $\hat{F}_0$. Notice that an alternative way to obtain (non*bona fide*) estimates of the CDF by regressing $\hat{F}_0(x_i)$s on x_is.

3.5 Boosting Kernel Estimators

In this section we will present various boosting algorithms, one for each kernel method discussed in the previous section. We will focus on the unidimensional case, for the multidimensional setting see [9].

In the cases of density estimation and discrimination, our algorithms repeatedly call, M times, a kernel estimator to iteratively estimate using reweighted kernels. The first weighting distribution is uniform, whilst the mth distribution $\{w_m(i), i = 1, \ldots, n\}$ with $m \in [2, \ldots, M]$ is determined on the basis of the estimation resulting from the $(m-1)$th call. The final sequence of estimates is summarized into a single prediction rule which should have superior standards of accuracy. The weighting distribution is designed to associate more importance to misclassified data (for discrimination) or with a poor density estimate through a proper *loss function*. Consequently, as the number of iterations increases the "hard to classify" observations receive an increasing weight.

Concerning regression, the boosting scheme is different because we iteratively add smooth of residuals.

A number of case studies will practically confirm that the typical boosting iteration will reduce the bias but slightly inflate the variance. In our simulations, we will explore the optimal smoothing parameter for each choice of M, In practical applications, it is common to use cross-validation to select the pair (h, M).

Algorithm 4. Boosting a kernel density estimate

1. *Given* $\{x_i, i = 1, \ldots, n\}$, *initialise* $w_1(i) = 1/n, \quad i = 1, \ldots, n,$ *Select* h.
2. *For* $m = 1, \ldots, M$, *obtain a weighted kernel estimate,*

$$\hat{f}_m(x) = \sum_{i=1}^{n} \frac{w_m(i)}{h} K\left(\frac{x - x_i}{h}\right),$$

and update weights according to $w_{m+1}(i) = w_m(i) + \log\left(\hat{f}_m(x_i) \Big/ \hat{f}_m^{(i)}(x_i)\right)$.

3. *Provide as output*

$$\prod_{m=1}^{M} \hat{f}_m(x),$$

renormalised to integrate to unity.

3.5.1 Boosting Density Estimation

Standard kernel density estimator could be seen as a single-step boosting algorithm where all of the kernels have the same weight $w(i) = 1/n$.

$$\hat{f}(x) = \frac{1}{h} \sum_{i=1}^{n} w(i) K\left(\frac{x - x_i}{h}\right). \tag{3.35}$$

In order to boost, we propose a loss function which compares $\hat{f}(x_i)$ with the leave-one-out estimate

$$\hat{f}^{(i)}(x_i) = \frac{n}{n-1}\left(\hat{f}(x_i) - \frac{K(0)}{nh}\right). \tag{3.36}$$

Boosting then reweights the kernels using a log-odds ratio, i.e., $\log\left(\hat{f}(x_i) / \hat{f}^{(i)}(x_i)\right)$.

Our pseudocode for boosting a kernel density estimate is given in Algorithm 4.

To formally appreciate how the algorithm works, it is sufficient to analyze the first iteration. Recalling the definition of $\hat{f}^{(i)}(x_i)$, we then have

$$w_2(i) = w_1(i) + \log\left(\frac{\hat{f}_1(x_i)}{\hat{f}_1^{(i)}(x_i)}\right) \tag{3.37}$$

$$\approx \frac{1}{n} + \frac{K(0)}{nh\hat{f}_1(x_i)} + \log\left(\frac{n-1}{n}\right) \approx \frac{K(0)}{nh\hat{f}_1(x_i)} \tag{3.38}$$

and so $w_2(i) \sim \hat{f}_1(x_i)^{-1}$. Hence, after one step we have $\hat{f}(x) = c\hat{f}_1(x)\hat{f}_2(x)$.

This is essentially equivalent to the variable-kernel estimator of [7] in which the bias is reduced to $O(h^4)$. So boosting (one-step) a kernel density estimate reduces the bias. For more details, see [10].

3.5.1.1 Simulation Examples

In this simple example, we investigate how the choice of M affects the optimal choice of h, and how the average integrated squared error changes with these two choices. We use the average over all simulations of the integrated squared error, MISE, as a criterion.

Figure 3.6 shows MISE(h) for various values of M and for standard normal and beta populations. We can see that larger smoothing parameters are required for boosting to be beneficial; this corresponds to the "weak learner" concept in which boosting was originally proposed. The optimal value is $M = 13$ with corresponding optimal smoothing parameter $h = 3.47$. This represents a reduction in MISE of more than 71% compared with the optimal value (using $h = 0.51$) for the usual estimator, with $M = 1$. In the beta example, the optimal value is $M = 23$ with corresponding optimal smoothing parameter $h = 1.22$. This represents a reduction in MISE of more than 15% of the optimal value (using $h = 0.11$) for the usual estimator, with $M = 1$. In both cases improvement due to boosting is diminishing with subsequent iterations, with most of the benefit being obtained at the second iteration. The inferior performance in the beta case is due to the fact that the beta density has a limited support, and, as is well known, local estimates are less efficient at the boundaries. This, in turn, causes a bad estimate at the first boosting step. Then in the subsequent steps the algorithm has usually failed to recover those inefficiencies, being often diverted toward different shapes. We can conclude that when a density is difficult to estimate a bigger sample size is needed to make boosting worthwhile. In the extreme case of the *claw* density (the results are not reported here), we have observed that boosting starts to be beneficial only for sample sizes bigger than 500. For examples on thick-tailed, bimodal, and skewed populations see [10].

3.5.2 *Boosting Classification*

Assume a two-class discrimination problem, the first problem in which the boosting algorithm has been conceived. Friedman et al. [11] propose Real AdaBoosting in which the weak classifier yields a membership probability, not a binary response. Its loss system gives x_i a weight proportional to

$$V_i = \sqrt{\frac{\min\left(p(x_i \in \Pi_1), p(x_i \in \Pi_2)\right)}{\max\left(p(x_i \in \Pi_1), p(x_i \in \Pi_2)\right)}} \tag{3.39}$$

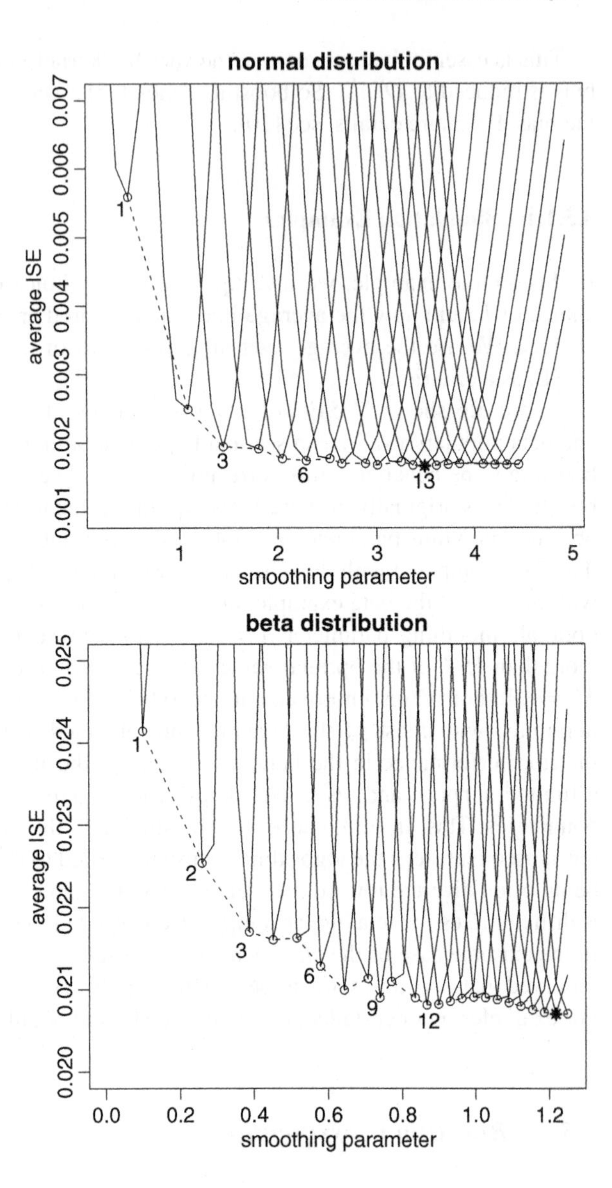

Fig. 3.6 For 500 samples of size $n = 50$, the average integrated squared error is shown as a function of the smoothing parameter h for various values of the boosting iteration M. The dashed line joins the points corresponding to the optimal smoothing parameters for each boosting iteration. Underlying distributions: top N(0, 1); bottom Beta(2, 2)

if x_i is correctly classified, and proportional to V_i^{-1} if x_i is misclassified. Because kernel classification (KC) estimates densities in order to classify, Real AdaBoost seems the natural framework for boosting KC, whereas discrete mappings do not employ the whole information generated by a KC, but only the resulting sign.

Our pseudocode for Real AdaBoost KC is given in Algorithm 5.

Algorithm 5. Boosting a kernel density classifier

1. *Given* $\{(x_i, Y_i), i = 1, \ldots, N\}$, *initialize* $w_1(i) = 1/N, \quad i = 1, \ldots, N.$
2. *Select* $h_J, J = 1, 2.$
3. *For* $m = 1, \ldots, M$

   (i) *Obtain a weighted kernel estimate using*

$$\widehat{f}_{J,m}(x) = \sum_{i:Y_i=J} \frac{w_m(i)}{h_J} K\left(\frac{x - x_i}{h_J}\right) \quad \textit{for } J = 1, 2.$$

   (ii) *Calculate*

$$\delta_m(x) = \frac{1}{2}\log\left(p_m/(1 - p_m)\right).$$

   where $p_m(x) = \widehat{f}_{2,m}(x) \Big/ \left(\widehat{f}_{1,m}(x) + \widehat{f}_{2,m}(x)\right)$

   (iii) *Update:*

$$w_{m+1}(i) = w_m(i) \times \begin{cases} \exp\left(\delta_m(x_i)\right) & \text{if } Y_i = 1 \\ \exp\left(-\delta_m(x_i)\right) & \text{if } Y_i = 2 \end{cases}$$

4. *Output*

$$H(x) = \operatorname{sign}\left[\sum_{m=1}^{M} \delta_m(x)\right]$$

Note that $\widehat{f}_{J,m}(x)$ does not integrate to 1 (even for $m = 1$); so in effect, we are considering $\pi_j f_j(x)$ (with $\pi_j = n_j/N$) in our estimation. Note also that we do not need to re-normalise the weights because we consider the ratio $\hat{f}_{2,m}(x)/\hat{f}_{1,m}(x)$ so any normalization constant will cancel.

It is possible to prove that this algorithm gives an estimator of a frontier point (i.e., the point where $f_1 = f_2$) which is smaller biased. The initial classifier is (assuming equal priors):

$$\delta_1 = \frac{1}{2}\left[\log \hat{f}_{2,1} - \log \hat{f}_{1,1}\right] \tag{3.40}$$

$$= \frac{1}{2}\left\{\log\left[\frac{f_2}{f_1}\right] + \frac{h_2^2 f_2''}{2f_2} - \frac{h_1^2 f_1''}{2f_1} + O(h_1^4) + O(h_2^4)\right\} \tag{3.41}$$

so when $f_1(x) = f_2(x) = f(x)$, say, we have

$$\Delta_1(x) = \frac{h_2^2 f_2''(x) - h_1^2 f_1''(x)}{4f(x)}. \tag{3.42}$$

Then at the next iteration,

$$\hat{f}_{1,2}(x) = \int \frac{1}{h_1} K\left(\frac{x-y}{h_1}\right) \sqrt{\frac{\hat{f}_{2,1}(y)}{\hat{f}_{1,1}(y)}} f_1(y)\mathrm{d}y, \tag{3.43}$$

$$\hat{f}_{2,2}(x) = \int \frac{1}{h_2} K\left(\frac{x-y}{h_2}\right) \sqrt{\frac{\hat{f}_{1,1}(y)}{\hat{f}_{2,1}(y)}} f_2(y)\mathrm{d}y. \tag{3.44}$$

This leads to (up to terms of order h^2):

$$\delta_2(x) = \frac{1}{2}\left\{\frac{(h_1^2 - h_2^2)}{8}\left(\frac{f_1'(x)}{f_1(x)} - \frac{f_2'(x)}{f_2(x)}\right)^2 + \frac{(h_2^2 + h_1^2)}{4}\left(\frac{f_1''(x)}{f_1(x)} - \frac{f_2''(x)}{f_2(x)}\right)\right\} \tag{3.45}$$

which gives an updated classifier $\delta_1(x) + \delta_2(x)$.

Noting that, near the point of interest, $f_1(x) \approx f_2(x) = f(x)$, say, we have

$$\Delta_2(x) = \frac{\Delta_1(x)}{2} + \frac{h_2^2 f_1'' - h_1^2 f_2''}{8f} + (h_1^2 - h_2^2)\left(\frac{f_1' - f_2'}{4f}\right)^2. \tag{3.46}$$

Now letting $h_1 = h_2$, we see that $\Delta_2(x) = O(h^4)$. That boosting reduces the bias comes as no surprise, but it is somewhat counterintuitive that the bias reduction is enhanced by taking equal smoothing parameters. For more details, see [8].

3.5.2.1 A Simulation Example

We use $f_1 = N(0, 4^2)$, $f_2 = N(4, 1^2)$ with $n_1 = n_2 = 50$ and 500 simulations. For these densities, we focus on the primary intersection point: $x_0 = 2.243$ where $f_1(x_0) = f_2(x_0) = .0852$. We compare various methods in which the strategy is to estimate $\hat{x}_0$, and then use this as a decision boundary (ignoring the other solution in the other tail). We consider LDA, and various density estimates for the two groups, in which the smoothing parameter is obtained by a Normal reference rule (Fixed plug-in), an adaptive smoother [12], a fourth-order kernel, and the multiplicative estimator of [7]. The results are given in Table 3.1. It can be seen that LDA behaves poorly since this method assumes equal variances (which would give an expected bias of −0.243), and that the higher-order kernels all perform similarly. The Bayes

Table 3.1 Left: Estimation of main intersection boundary between $N(0, 4^2)$ and $N(5, 1^2)$ by various kernel methods and linear discriminant, based on 500 samples of size $n_1 = n_2 = 50$. Right: Resulting error rates, given as an exceedence over Bayes rule × 100

Method	Bias	SD	MSE	Method	Mean	SD
LDA	−0.2882	0.3021	0.1743	LDA	1.1315	1.2100
Fixed plug-in	−0.1630	0.2354	0.0820	Fixed plug-in	0.5774	0.7116
Abramson	0.0190	0.2425	0.0592	Abramson	0.4856	0.7182
Fourth-order	−0.0793	0.2332	0.0607	Fourth-order	0.4542	0.6080
Jones	0.0154	0.2389	0.0573	Jones	0.4675	0.7165

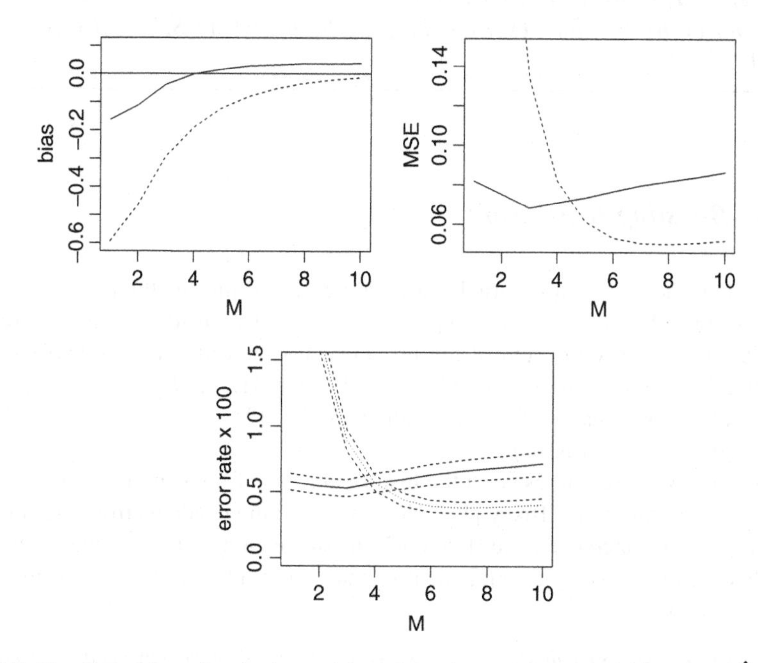

Fig. 3.7 *Top*: Estimation of x_0 by boosted kernel density estimates, where $\hat{x}_0 = \arg \hat{f}_1(x) = \hat{f}_2(x)$, versus number of boosting iterations M. Continuous line uses Normal plug-in rule; dashed line is plug-in rule ×2. *Bottom*: Consequent error rate for the classifier as a function of boosting iteration

rule gives error rate $\times 100 = 16.3$, and Table 3.1 shows the inflation in error rate for the various methods. Again, the higher-order kernels all perform in a similar way, and substantially better than LDA.

For the same example, we again consider estimation of x_0, and consequent error rate, for a standard kernel density estimator in which the smoothing parameter is based on the normal-reference plug-in rule. However, we now consider boosting the kernels as described above. Figure 3.7 shows the dependence on the number of boosting iterations, and illustrate the fact that the performance of boosting is affected by the choice of smoothing parameter. In this case, using larger smoothing parameters (and so making a weaker learner) can lead to an improved performance.

Algorithm 6. Boosting kernel regression

1. *(Initialization)* Given data $S = \{(x_i, y_i), i = 1, \ldots, n\}$ and $h > 0$, let

$$\widehat{m}_0(x; h) := \widehat{m}_{\mathrm{NW}}(x; S, h).$$

2. *(Iteration)* Repeat for $b = 1, \ldots, B$
   (i) $e_i := Y_i - \widehat{m}_{b-1}(X_i; h) \quad i = 1, \ldots, n;$
   (ii) $\widehat{m}_b(x; h) := \widehat{m}_{b-1}(x; h) + \widehat{m}_{\mathrm{NW}}(x; S_e, h)$, where $S_e = \{(X_i, e_i), i = 1, \ldots, n\}$.

3.5.3 Boosting Regression

Our strategy here is to use kernel regression as the fitting method within an L_2 boosting algorithm. L_2 boosting is a procedure of iterative residual fitting where the final output is simply the sum of the fits. Formally, consider a *weak learner* $\mathcal{M}$. An initial least squares fit is denoted by $\mathcal{M}_0$. For $b \in [1, \ldots, B]$, $\mathcal{M}_b$ is the sum of $\mathcal{M}_{b-1}$ and a least squares fit of the residuals $S_e := \{X_i, e_i := Y_i - \mathcal{M}_{b-1}(X_i)\}$. The L_2 boost estimator is $\mathcal{M}_B$.

For those who are interested in the historical roots of boosting in the regression context it is worth noting Tukey [13]. He considered the residuals from the smooth of the data, and coined the name "twicing." His suggestion for an iterative procedure (which we can see today as equivalent to many boosting iterations) is somewhat obscure at first read:

> *"then we get a final smooth whose rough is precisely the rough of the rough of the original sequence—hence the name* Reroughing*"* … "Twicing" *or use of some specified smoother twice, is just a special case of reroughing. … One natural thing is to smooth again*

The scheme is visualized in Fig. 3.8. Our pseudocode for L_2 boosting of kernel regression is given in Algorithm 6.

To formally study the algorithm properties, some preliminary notation is needed, as follows. The NW estimates at the observation points are compactly denoted as

$$\widehat{\boldsymbol{m}}_0 = \boldsymbol{NKy} \tag{3.47}$$

where

$$\widehat{\boldsymbol{m}}_0^T := (\widehat{m}_0(x_1; h), \ldots, \widehat{m}_0(x_n; h)),$$

$$\boldsymbol{N}^{-1} := \mathrm{diag}\left(\left\{\sum_{i=1}^n K_h(x_1 - x_i)\right\}, \ldots, \left\{\sum_{i=1}^n K_h(x_n - x_i)\right\}\right),$$

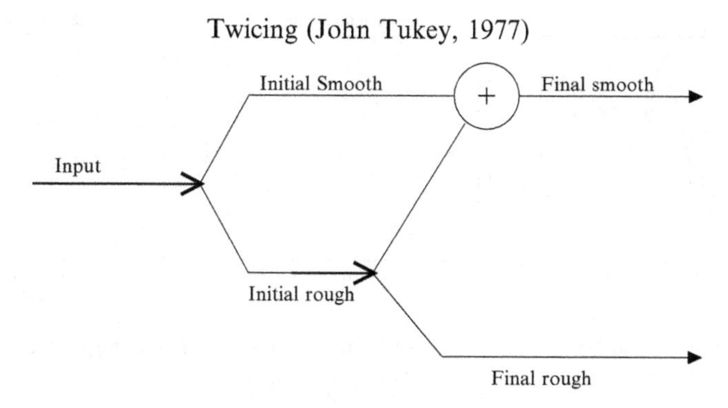

Fig. 3.8 Twicing. From Tukey (1977) *Exploratory Data Analysis*

$$\begin{aligned} \boldsymbol{y}^T &:= (y_1, \ldots, y_n) \qquad \text{and} \\ (\boldsymbol{K})_{ij} &:= K_h(x_i - x_j). \end{aligned}$$

Concerning the algorithm accuracy, Di Marzio and Taylor [14] show what follows.

$$\text{ave-bias}^2 = \frac{1}{n}\boldsymbol{m}^T \left(\boldsymbol{U}^{-1}\right)^T \text{diag}\left((1-\lambda_k)^{b+1}\right) \boldsymbol{U}^T \boldsymbol{U} \text{diag}\left((1-\lambda_k)^{b+1}\right) \boldsymbol{U}^{-1}\boldsymbol{m}, \tag{3.48}$$

$$\text{ave-var} = \frac{\sigma^2}{n}\text{trace} \left\{\boldsymbol{U}\text{diag}\left(1-(1-\lambda_k)^{b+1}\right)\boldsymbol{U}^{-1}\left(\boldsymbol{U}^{-1}\right)^T \text{diag}\left(1-(1-\lambda_k)^{b+1}\right)\boldsymbol{U}^T\right\}. \tag{3.49}$$

where $\lambda_1, \ldots, \lambda_n$ are the characteristic roots of $\boldsymbol{NK}$, $b \geq 0$, with at least one smaller than 1, and $\boldsymbol{U}$ is a $n \times n$ invertible matrix of real numbers. Moreover, if $\text{spec}(\boldsymbol{NK}) \subset (0, 1]$, then

$$\begin{aligned} \lim_{b\to\infty} \text{ave-bias}^2 &= 0, \\ \lim_{b\to\infty} \text{ave-var} &= \sigma^2, \\ \lim_{b\to\infty} \text{ave-MSE} &= \sigma^2. \end{aligned}$$

Quite remarkably, not all of the kernels match above condition about the characteristic roots. Gaussian and triangular kernels meet the condition, whereas Epanechnikov, biweight, and triweight do not.

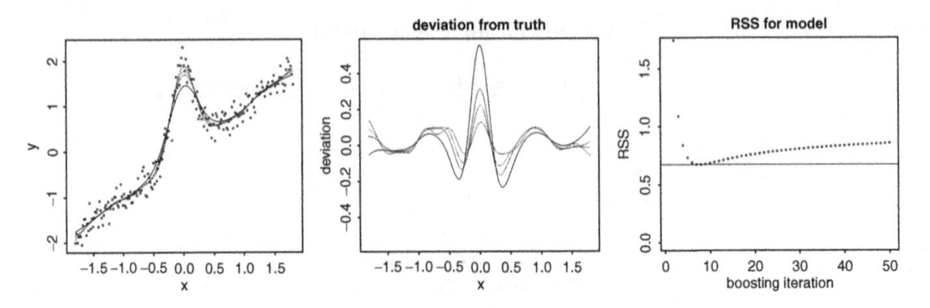

Fig. 3.9 *Left*: Data and NW estimate, boosted for three iterations. *Middle*: respective errors from true model. *Right*: Residual sum of squares to expected value of Y

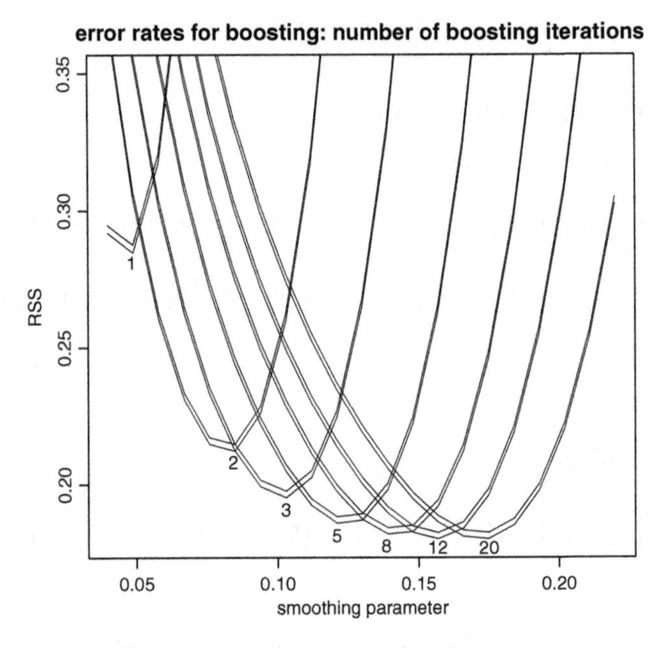

Fig. 3.10 Average residual sum of squares from boosted NW estimate with approximate 95% confidence intervals

3.5.3.1 Simulated Example

We used $n = 200$ fixed design points from $[-1.8, 1.8]$, and simulate $y_i = x_i + 2\exp(-16x_i^2) + \varepsilon_i$, with $\varepsilon_i \sim N(0, 0.2^2)$. Using a bandwidth of 0.16, in Fig. 3.9 we illustrate the effect of boosting NW estimate over three iterations.

Over 10,000 simulated datasets from the above model, we can find the residual sum of squares to the expected value of Y as a function of smoothing parameter and number of boosting iterations. The results are shown in Fig. 3.10. It can be seen that the optimal number of boosting iterations depends on the smoothing parameter. We again note that the optimal setting is to weaken the NW estimate by using a larger smoothing value, together with many boosting iterations.

3.5.4 Generalizations

Let $H = (h_0, \ldots, h_B), \Phi = (\phi_0, \ldots, \phi_B)$ be vectors of smoothing parameters and weights, respectively, then

$$\mathscr{K}_H^{\Phi} = \sum_{j=0}^{B} \phi_j K_{h_j} \tag{3.50}$$

is a weighted sum of kernel functions. The kernel used in Algorithm 2 when $B = 1$ is a special case of this formulation with $\Phi = (2, -1)$ and $H = (h, \sqrt{2}h)$. We can thus generalize $\widehat{m}_B$ to

$$\widehat{m}_g(x; S, \Theta) := \sum_{j=0}^{B} \phi_j \frac{\sum K_{h_j}(x - X_i) Y_i}{\sum K_{h_j}(x - X_i)} = \sum_{j=0}^{B} \phi_j \widehat{m}_{\mathrm{NW}}(x; S, h_j) \tag{3.51}$$

in which $\Theta = (\Phi, H) = (\phi_0, \ldots, \phi_B, h_0, \ldots, h_B)$ includes all the required parameters. This is simply a linear combination of N–W estimators, each with its own bandwidth. Similar proposals were considered by [15] and [16] who used weighted combinations of kernels to improve estimators at the boundary. In order for $\widehat{m}_g$ to be asymptotically unbiased, we require $\sum \phi_j = 1$. Given a set of bandwidths $H = (h_0, \ldots, h_B)$ we can choose the ϕ_j to eliminate the bias terms which arise as a consequence of $\mu_k = \int v^k K(v)\, dv$.

Since we have used a Normal kernel, for a given bandwidth h we have $\mu_{2k} \propto h^{2k}$, and $\mu_{2k-1} = 0$ so a simple approach to obtain the weights ϕ_j would be to set $H = \left(h, \sqrt{c}h, \ldots, c^{B/2}h\right)$ for some c, and then to solve $C\Phi = (1, 0, \ldots, 0)^T$ for Φ, where (for $B \geq 1$)

$$C = \begin{pmatrix} 1 & 1 & \cdots & 1 \\ 1 & c & \cdots & c^B \\ \vdots & \vdots & \vdots & \vdots \\ 1 & c^B & \cdots & c^{2B-1} \end{pmatrix} \tag{3.52}$$

and this simplification requires the selection of only two parameters (c and h), for a given B. Note that the above convolution kernel K_h^* uses $c = 2$, and that the solution for Φ gives the desired value $(2, -1)$.

As an alternative approach, we could consider obtaining the ϕ_j by estimation through an ordinary least squares regression, i.e., obtain Φ from $\widehat{\Phi} = (X^T X)^{-1} X^T Y$ where $Y = (Y_1, \ldots, Y_n)$ is the vector of responses, and the jth

Algorithm 7. L_2 boosting a kernel CDF estimate

1. *(Initialization) Given* $x_1, \ldots, x_n$,
   (i) *define the responses* $Y_i = \widehat{F}_0(x_i)\,, \quad i = 1, \ldots, n$;
   (ii) *select* $h > 0$;
   (iii) *calculate* ${}_0\widehat{F}_h(x) = \hat{m}(x, \{Y_i\})$.
2. *(Iteration) For* $m = 1, \ldots, M$
   (i) *compute the residuals* $e_i = Y_i - {}_{m-1}\widehat{F}_h(x_i)\,, \quad i = 1, \ldots, n$;
   (ii) *calculate* $\hat{m}(x, \{e_i\})$;
   (iii) *update* ${}_m\widehat{F}_h(x) = {}_{m-1}\widehat{F}_h(x) + \hat{m}(x)\,.$

column of the matrix X is given by $\left(\widehat{m}_{\mathrm{NW}}(X_1; S, h_j), \ldots, \widehat{m}_{\mathrm{NW}}(X_n; S, h_j)\right)^T$. This approach could also allow for the selection of B through standard techniques in stepwise regression. Also, note the connection between (3.51) and a radial basis function (RBF) representation. In this framework the ϕ_j are the weights, and $m_{\mathrm{NW}}(x; S, h_j)$ act as "basis functions" which are themselves a weighted sum of basis functions. So this formulation is equivalent to a generalized RBF network, in which an extra layer is used to combine estimates, but with many of the weights being fixed.

Many questions arise. How similar will $\widehat{m}_g$ be to the boosting estimate $\widehat{m}_B$, and for which value of c is the correspondence closest? How would we choose H (or h and c) in practice, i.e., from the data? Which estimator is "best"? Which is the best way to obtain Φ? etc. Similar work by [17] on the use of higher-order kernels could provide some of these answers for the fixed, equispaced design. However, in several simulation experiments, we have not been able to obtain any method of selection of $\phi_j, h_j, j = 0, \ldots, B$ for which $\widehat{m}_g$ performs better than the boosting estimates $\widehat{m}_k$ for equivalent $B \geq 2$.

3.5.5 Boosting Cumulative Distribution Function Estimation

As said before, we could also estimate the CDF by regression on empirical CDF values. So, it theoretically becomes feasible to employ L_2 boosting also to get improved CDF estimates. We do this in the algorithm below.

Because of the weak learner is a basic regression estimator, we notice that Algorithm 7 may give estimates of $F(x)$ which are not *bona fide*, i.e., they can locally decrease or violate $0 \leq \hat{F}(x) \leq 1$. Surely for sample sizes big enough this problem tends to disappear.

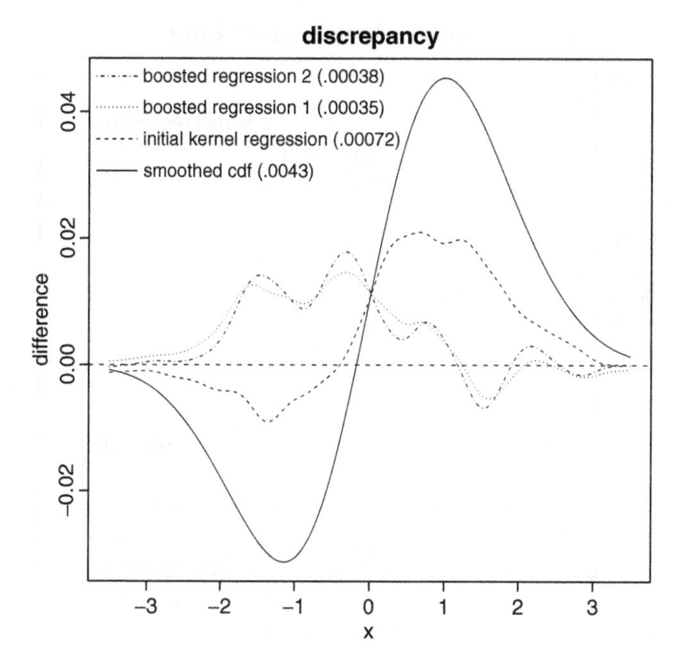

Fig. 3.11 Discrepancies between various estimates of the cumulative distribution function, based on 1,000 observations from a standard normal distribution

3.5.5.1 Simulation Example

In the following simulation experiment, we will compare $\hat{F}_0(x)$, $\hat{F}_h(x)$, and ${}_m\hat{F}_h(x)$ for $m = 0, 1, 2, \ldots$ (see Fig. 3.11).

In Fig. 3.12, we simulate 100 datasets of size $n = 1{,}000$ from a standard Normal distribution and compare MSE for: (1) empirical CDF; (2) smoothed CDF; (3)–(5) kernel regression; and two boosted versions. The results are shown in Fig. 3.12. Once again it can be seen that boosting can be effective when the smoothing parameter is larger (and hence the learner is weaker). Also of interest, we note that the regression estimates perform slightly better than the more standard (unbiased) empirical CDF and smoothed versions of it.

3.6 Bibliographical and Historical Remarks

Two surprising links to previous work have arisen. The first one concerns boosting and density estimation, in which we have noted that the multiplicative estimator introduced by [7] can be described as a boosting algorithm, for more details see Sect. 3.5.1. The second one concerns L_2-boosting applied to regression, whereas

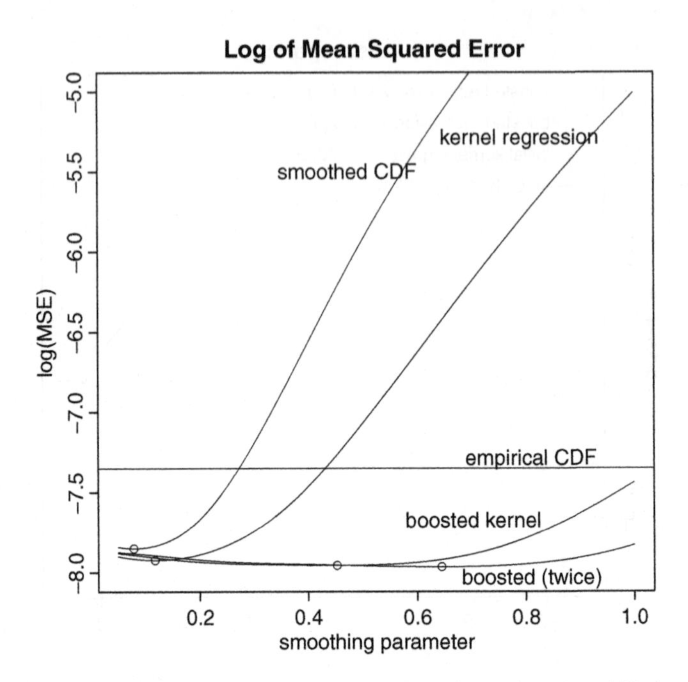

Fig. 3.12 Log mean squared error from a standard Normal CDF based on 100 simulations of sample size $n = 1{,}000$ using an empirical CDF, a smoothed CDF, and boosted iterations of a kernel regression estimate, for different smoothing parameters

the basic strategy of iterative adding of smooth of residuals has been proposed as late as 1977 by [13], in his classical book *Exploratory Data Analysis*; for more details, see Sect. 3.5.3.

References

1. R. E. Schapire, "The strength of weak learnability," *Machine Learning*, vol. 5, pp. 197–227, 1990.
2. Y. Freund, "Boosting a weak learning algorithm by majority," *Information and Computation/information and Control*, vol. 121, pp. 256–285, 1995.
3. M. D. Marzio and C. C. Taylor, "Using small bias nonparametric density estimators for confidence interval estimation using small bias nonparametric density estimators for confidence interval estimation," *Journal of Nonparametric Statistics*, vol. 21, pp. 229–240, 2009.
4. P. Buhlmann and T. Hothorn, "Boosting Algorithms: Regularization, Prediction and Model Fitting," *Statistical Science*, vol. 22, no. 4, pp. 477–505, 2007.
5. Y. Freund and R. E. Schapire, "A decision-theoretic generalization of on-line learning and an application to boosting," in *European Conference on Computational Learning Theory*, pp. 23–37, 1995.
6. R. A. Jacobs, M. I. Jordan, S. J. Nowlan, and G. E. Hinton, "Adaptative mixture of local experts," *Neural Computation*, vol. 3, pp. 1–12, 1991.

7. M. C. Jones, O. Linton, and J. P. Nielsen, "A simple bias reduction method for density estimation," *Biometrika*, vol. 82, pp. 327–338, 1995.
8. M. D. Marzio and C. C. Taylor, "Kernel density classification and boosting: an l2 analysis," *Statistics and Computing*, vol. 15, pp. 113–123, 2005.
9. M. D. Marzio and C. C. Taylor, "On boosting kernel density methods for multivariate data: density estimation and classification," *Statistical Methods and Applications*, vol. 14, pp. 163–178, 2005.
10. M. D. Marzio and C. C. Taylor, "Boosting kernel density estimates: A bias reduction technique?," *Biometrika*, vol. 91, pp. 226–233, 2004.
11. J. Friedman, T. Hastie, and R. Tibshirani, "Additive logistic regression: a statistical view of boosting (with discussion and a rejoinder by the authors)," *Annals of Statistics*, vol. 28, pp. 337–407, 2000.
12. I. S. Abramson, "On Bandwidth Variation in Kernel Estimates-A Square Root Law," *The Annals of Statistics*, vol. 4, pp. 1217–1223, 1982.
13. J. W. Tukey, *Exploratory Data Analysis*. Addison-Wesley, Philippines, 1977.
14. M. D. Marzio and C. C. Taylor, "On boosting kernel regression," *Journal of Statistical Planning and Inference*, vol. 138, pp. 2483–2498, 2008.
15. J. A. Rice, "Boundary modifications for kernel regreession," *Comm. Statist. Theory Meth.*, vol. 13, pp. 893–900, 1984.
16. M. C. Jones, "Simple boundary correction for kernel density estimation," *Statistics and Computing*, vol. 3, pp. 135–146, 1993.
17. T. Gasser, H.-G. Müller, and V. Mammitzsch, "Kernels for nonparametric curve estimation," *Journal Royal Statist. Soc. B*, vol. 47, pp. 238–252, 1985.

Chapter 4
Targeted Learning

Mark J. van der Laan and Maya L. Petersen

4.1 Introduction

Suppose we observe n i.i.d. copies $O_1, \ldots, O_n$ of a random variable O with probability distribution P_0, and assume that it is known that $P_0 \in \mathscr{M}$ for some set of probability distributions $\mathscr{M}$. One refers to $\mathscr{M}$ as the statistical model for P_0. We consider so called semiparametric models that cannot be parameterized by a finite dimensional Euclidean vector. In addition, suppose that our target parameter of interest is a parameter $\Psi : \mathscr{M} \rightarrow \mathscr{F} = \{\Psi(P) : P \in \mathscr{M}\}$, so that $\psi_0 = \Psi(P_0)$ denotes the parameter value of interest. We wish to estimate ψ_0 based on the dataset $O_1, \ldots, O_n$. In the first part of this article, we consider the case that Ψ is not pathwise differentiable; in the typical applications $\Psi(P_0)$ is itself a whole function, so that ψ_0 is an infinite dimensional parameter. For example, $\Psi(P_0) = P_0$ itself, or some conditional density or conditional mean identified by P_0. Let $\mathscr{M}_{\mathrm{NP}}$ be the nonparametric model that also includes the realizations of the empirical probability distribution P_n of $O_1, \ldots, O_n$. An estimator $\hat{\Psi} : \mathscr{M}_{\mathrm{NP}} \rightarrow \mathscr{F}$ will be viewed as a mapping from the empirical distribution P_n of $O_1, \ldots, O_n$ into the parameter space $\mathscr{F}$. In this chapter, we discuss estimator selection for the target parameter $\Psi(P_0)$. We use this terminology over "model selection", since the formal meaning of a (statistical) model in the field of statistics is the set of possible probability distributions, and most algorithms are not indexed by a statistical model choice.

An example is provided by $O = (W, A, Y)$, where W denotes baseline covariates, A a subsequent dose of a drug, Y a final outcome, $\mathscr{M}$ a nonparametric model, and where $\Psi(P)(a) = E_P E_P(Y \mid A = a, W)$ maps a probability distribution P

M.J. van der Laan
Division of Biostatistics, School of Public Health, University of California, Berkeley, CA, USA
e-mail: laan@berkeley.edu

M.L. Petersen (✉)
Division of Biostatistics, School of Public Health, University of California, Berkeley, CA, USA
e-mail: mayaliv@berkeley.edu

C. Zhang and Y. Ma (eds.), *Ensemble Machine Learning: Methods and Applications*,
DOI 10.1007/978-1-4419-9326-7_4, © Springer Science+Business Media, LLC 2012

into a function of dose level a. If one wishes to make the additional nontestable causal assumptions that there exists an underlying collection of counterfactual random variables $Y(a)$ so that $Y = Y(A)$, and that A is independent of $(W, (Y(a) : a))$, given W (i.e., that W is sufficient to control for confounding of the effect of A on Y), we have that $\Psi(P)(a) = EY(a)$. Thus under these additional causal assumptions, $\Psi(P)$ can be interpreted as a causal dose–response curve summarizing the causal effect of setting the drug dose at level a on outcome Y.

Due to the dimension of the model $\mathcal{M}$, the maximum likelihood estimator is often ill behaved or not defined. As a consequence, a careful trade-off of bias and variance is required to make a choice between a large class of candidate estimators, and thereby construct sensible well-behaved estimators of ψ_0. We define an appropriate risk function $R(\psi, P_0)$ of a candidate value ψ that is minimized at ψ_0, and whose risk-based dissimilarity $R(\psi, P_0) - R(\psi_0, P_0)$ measures a desired dissimilarity between a candidate ψ and truth ψ_0. If the risk function is linear, then $R(\psi, P_0) = E_{P_0} L(\psi)(O)$ for some loss function $(\psi, O) \rightarrow L(\psi)(O)$. Given an estimator $\hat{\Psi}$, an estimate of the conditional risk $R\left(\hat{\Psi}(P_n), P_0\right)$ can then be used as a criterion to select among candidate estimators of ψ_0.

In the first part of this chapter, we propose a template for construction of an estimator of ψ_0 that is defined by (1) a library of initial estimators, which themselves may be data adaptive algorithms; (2) a parametric family of weighted combinations of these estimators that provides the set of candidate estimators; and, (3) a risk function. A cross-validated estimated risk is computed for each of the candidate estimators and the cross-validation selector selects the optimal candidate estimator by minimizing the cross-validated estimated risk over all candidate estimators in the set. We refer to this template as super learning, and to the resulting estimator as a super learner. We then consider the case that the risk function is defined by a loss function indexed by a relatively easy to estimate nuisance parameter (relative to ψ_0). In this case, the risk can be estimated with an empirical mean of the loss using a simple plug-in estimator for the nuisance parameter. We present a finite sample oracle inequality that demonstrates the asymptotic optimality of the super learner under this scenario. Specifically, if none of the candidate estimators behave as an estimator based on a correctly specified parametric model (i.e., converge to ψ_0 at rate $1/\sqrt{n}$), then our finite sample inequality for the cross-validation selector implies that the super learner asymptotically outperforms all the candidate algorithms in the library as long as the number of candidate algorithms converges to infinity at a rate that is polynomial in sample size.

In the second part of this chapter, we consider the case that Ψ is a pathwise differentiable parameter at $P \in \mathcal{M}$ with canonical gradient $D^*(P)$, for each $P \in \mathcal{M}$. That is, for a rich collection of one-dimensional parametric submodels $\{P(\varepsilon) : \varepsilon\}$ we have that $\frac{d}{d\varepsilon}\Psi(P(\varepsilon))\big|_{\varepsilon=0} = E_P D^*(P)(O)S(O)$, where S is the score of the one-dimensional parametric model, and the components of $D^*(P)$ are an element of the closure of the linear span of all the scores generated by this family of one-dimensional models. Due to this pathwise differentiability, the target parameter $\Psi(P)$ is a smooth enough function of P so that one can construct an asymptotically linear estimator of ψ_0 under some regularity conditions. An estimator $\hat{\Psi}(P_n)$ is asymptotically linear at P_0 if

$$\hat{\Psi}(P_n) - \psi_0 = (P_n - P_0)\mathrm{IC}(P_0) + o_P\left(1/\sqrt{n}\right).$$

Here, we used the notation $Pf = \int f(o)\mathrm{d}P(o)$. The function $\mathrm{IC}(P_0)$ is called the influence curve of the estimator at P_0. If $\hat{\Psi}$ is asymptotically linear, then (by the CLT) $\sqrt{n}\left(\hat{\Psi}(P_n) - \psi_0\right)$ converges to a multivariate normal distribution with mean zero and covariance matrix $\Sigma_0 = P_0\mathrm{IC}(P_0)\mathrm{IC}(P_0)^{\top}$. Statistical inference in terms of confidence intervals and tests can now be based on an estimator Σ_n of Σ_0.

When the target parameter Ψ is pathwise differentiable, it is often not possible to construct an appropriate risk function that directly targets ψ_0 such that the super learner achieves its desirable optimality properties. In particular, the finite sample inequality established for the cross-validation selector is now not of as much interest since the rate of convergence of different candidate estimators is now $1/\sqrt{n}$. In this case one can still use super learning to obtain an estimator Q_n of a nonpathwise differentiable parameter Q_0 for which $\psi_0 = \Psi(Q_0)$. However, one now needs to carry out a subsequent targeting step that removes residual bias (and possibly variance) w.r.t. the lower-dimensional ψ_0 to correct for the fact that the super learner was targeting Q_0, and not directly targeting ψ_0. The update of the estimator of Q_0 to directly target ψ_0 results in a substitution estimator $\Psi(Q_n^*)$, where Q_n^* is the targeted estimate of Q_0. We review this template for targeted learning of pathwise differentiable parameters, which we refer to as Targeted Minimum Loss Based Estimation (TMLE) [84].

Finally, with TMLE in our toolbox, we return to the case that our target parameter is nonpathwise differentiable, but now consider general risk functions, not necessarily expressed as an expectation of a loss function. We note that estimation of the risk as the empirical mean of a plug-in estimator of the loss function does not necessarily satisfy the constraints implied by the model and the definition of risk, such as that the risk is positive valued and bounded from above. We present an alternative approach to estimation of the risk function at a candidate estimator that instead makes use of a cross-validated targeted maximum likelihood estimator. This provides a generalized approach to super learning that utilizes TMLE for the purpose of estimation of the risk function.

4.1.1 Organization

In Sect. 4.2 we present the template for super learning of a nonpathwise differentiable (infinite dimensional) parameter in terms of a risk function, library of estimators, parametric family for generating weighted combinations of these estimators, and cross-validation as a tool for estimator selection. The risk function is assumed to be either linear in the probability distribution of the data, or linear in the probability distribution of the data up to a dependence on a relatively easy to estimate nuisance parameter. Section 4.3 presents an oracle inequality of the cross-validation estimator selector for risk functions that are linear in the probability distribution of the data, possibly up to a relatively easy to estimate nuisance parameter. These oracle inequalities imply the asymptotic optimality of the cross-validation estimator selector and thereby the super learner whenever none of the candidate estimators converges at the rate $1/\sqrt{n}$; if, by luck (e.g., by guessing a correctly specified parametric model), one of the estimators achieves this rate of convergence, then the estimator selected by the cross-validation selector will also converge at this rate. In Sect. 4.4 we demonstrate the super learner in a number of examples. In Sect. 4.5, we consider estimation of a pathwise differentiable parameter. For that purpose, we augment the super learning template with an additional targeted loss-based estimation update, resulting in a template for targeted minimum loss-based estimation (TMLE). In Sect. 4.6, we demonstrate TMLE using variable importance analysis as an example. Finally, in Sect. 4.7 we return to the problem of estimation of a nonpathwise differentiable parameter for general risk functions, now addressing risk functions for which simple plug in estimators of the risk might be inappropriate and result in a super learner with poor performance. We assume that the risk function at a candidate parameter value is pathwise differentiable. We propose the application of a cross-validated-TMLE (CV-TMLE) to estimate the risk function at a candidate parameter value. This provides a general super learner of a nonpathwise differentiable parameter. We demonstrate this general super learning template using the estimation of the causal dose–response curve for a continuous valued exposure/dose as an example. Section 4.8 provides some summary remarks. This chapter represents an overview of previously established results on unified loss-based cross-validation for nonpathwise differentiable parameters, and targeted minimum loss-based estimation of pathwise differentiable parameters (e.g., [81,84]), as well as presents new methods for general risk functions. The work presented builds on an extensive literature on data-adaptive estimation, ensemble learning, semiparametric estimation, and efficiency theory. Rather than providing exhaustive citations in the main text, we conclude in Sect. 4.9 with bibliographic remarks that provide an overview of some of the relevant literature.

4.2 Targeted Super Learning

In this section we define the super learning algorithm in terms of a risk function, library of estimators, and family of weighted combinations of these estimators. A cross-validated estimator of risk is computed for each candidate estimator, and the optimal weighted combination, defined as the minimizer of the cross-validated estimated risk, is selected as the final estimator.

4.2.1 Loss and Risk Functions

Let $O \sim P_0 \in \mathcal{M}$. Define a parameter $\Psi : \mathcal{M} \to \mathcal{F} \equiv \{\Psi(P) : P \in \mathcal{M}\}$ in terms of a risk function R as follows:

$$\Psi(P) = \arg\min_{\psi \in \mathcal{F}} R(\psi, P).$$

Thus, $\psi_0 = \Psi(P_0) = \arg\min_\psi R(\psi, P_0)$ represents the true parameter value. We wish to estimate ψ_0 based on n i.i.d. copies $O_1, \ldots, O_n$ of $O \sim P_0$. This estimation problem is defined by the statistical model $\mathcal{M}$ and the parameter mapping $\Psi : \mathcal{M} \to \mathcal{F}$.

In typical applications, ψ_0 will be a high dimensional function such as a conditional density, conditional mean, or classifier. If $R(\psi, P) = PL(\psi) \equiv \int L(\psi)(o)dP(o)$ for some function L, then we refer to R as a linear risk function, and $(O, \psi) \to L(\psi)(O)$ is a loss function that assigns a loss to a candidate ψ and observation O.

If R depends on P in a nonsmooth manner, then we aim to determine a representation $R(\psi, P) = R_{\Gamma(P)}(\psi, P)$, where $P \to R_\gamma(\psi, P)$ is smooth (e.g, linear) in P, and $\Gamma : \mathcal{M} \to \mathcal{F}_\gamma$ is a nuisance parameter. It is common that $R_{\Gamma(P)}(\psi, P) = PL_{\Gamma(P)}(\psi)$ for a loss function L that is now also indexed by a nuisance parameter $\Gamma : \mathcal{M} \to \mathcal{F}_\gamma$. That is, the loss function L for ψ_0 is now unknown, and is itself a parameter of P_0 through γ_0. In this case, for a given value γ of the nuisance parameter, the risk function is linear in P. Such risk functions defined by generalized loss functions $(O, \psi, \gamma) \to L_\gamma(\psi)(O)$ indexed by a nuisance parameter Γ represent an important class of nonlinear risk functions.

If γ_0 is a relatively easy to estimate nuisance parameter and $\gamma \to L_\gamma$ is reasonably smooth, then one may be willing to assume that the estimated loss L_{γ_n} would perform as well as L_{γ_0} for the purpose of estimation of ψ_0. The underlying assumption in such a case is that estimation of $L_{\gamma_0}(\psi)$ as a function of γ_0 is an easier estimation problem than estimation of ψ_0. Such an assumption might be warranted, for example, if one has particular knowledge about γ_0 such that one can confidently obtain good well-behaved estimators of γ_0. For risk functions of the two types mentioned above (increasing in generality, linear, or linear for fixed nuisance parameter), one expects the empirical risk $R_{\gamma_n}(\psi, P_n)$ (for good estimators of the nuisance parameter) to behave as an empirical mean, so that empirical process inequalities can be utilized to prove important results of interest about the cross-validation selector defined below.

4.2.2 Library of Algorithms

Let $\hat{\Psi}_j : \mathcal{M}_{NP} \to \mathcal{F}$ be a j-specific estimator of ψ_0, where $\mathcal{M}_{NP}$ denotes a nonparametric model so that it includes the empirical probability distribution P_n

of a sample $O_1, \ldots, O_n$ with probability 1, $j = 1, \ldots, J$. One particular type of estimator is defined as the minimizer of the empirical risk $R(\psi, P_n)$ over a j-specific parametric or semiparametric subspace $\mathcal{F}_j \subset \mathcal{F}$, such as one implied by a submodel $\mathcal{M}_j$ of $\mathcal{M}$. Other estimators are obtained using sieve (sequence of subspaces approximating $\mathcal{F}$)-based estimation based on this risk function and an algorithm for minimizing the empirical risk function over these subspaces, using internal fine-tuning of a variety of fine-tuning parameters. Some estimators involve resampling or ensemble learning, thereby combining algorithms into a new algorithm: for example, in prediction, bagging involves bootstrapping, computing an estimator on each bootstrap sample, and averaging the bootstrap-specific estimators.

A library of estimators can be further expanded by mapping an initial algorithm in the library into a number of new algorithms. For example, an initial algorithm can be indexed by a set of different prior dimension reductions, or can be stratified on a discrete variable. In this manner, a rich library of diverse algorithms can be constructed, including state of the art algorithms in machine learning as well as classical parametric model-based estimators. Each algorithm $\hat{\Psi}_j$ results in an estimate $\hat{\Psi}_j(P_n)$ when applied to a dataset $O_1, \ldots, O_n$.

4.2.3 Family for Combining Algorithms

Given the library of algorithms $\hat{\Psi}_j$, $j = 1, \ldots, J$, we consider a family of algorithms $\hat{\Psi}_\alpha = f\left(\left(\hat{\Psi}_j : j\right), \alpha\right)$ indexed by a Euclidean vector α for some function f. For example,

$$\hat{\Psi}_\alpha = \sum_{j=1}^{J} \alpha(j) \hat{\Psi}_j.$$

If ψ_0 is known to be a function with values in [0, 1], then one might use the family defined by

$$\log \frac{\hat{\Psi}_\alpha}{1 - \hat{\Psi}_\alpha} = \sum_{j=1}^{J} \alpha(j) \log \frac{\hat{\Psi}_j}{1 - \hat{\Psi}_j}.$$

One could also consider parametric families that combine cross-product terms $\hat{\Psi}_{j_1} \hat{\Psi}_{j_2}$ for pairs j_1, j_2 beyond the main algorithms. The vector α may be constrained in a user-supplied way, such as $\alpha(j) \geq 0$ and $\sum_j \alpha(j) = 1$.

4.2.4 Cross-Validated Estimator of Risk of Candidate Estimator

Let $B_n \in \{0, 1\}^n$ be a random variable defining a split of $\{O_1, \ldots, O_n\}$ into a training sample $\{O_i : B_n(i) = 0\}$ and validation sample $\{O_i : B_n(i) = 1\}$

with corresponding empirical probability distributions P^0_{n,B_n} and P^1_{n,B_n}, respectively. Let $\hat{\Gamma} : \mathcal{M}_{\mathrm{NP}} \to \mathcal{F}_\gamma$ be an estimator of the nuisance parameter in the risk function (if there is such a nuisance parameter). The cross-validated estimator of risk of a candidate estimator $\hat{\Psi} : \mathcal{M}_{\mathrm{NP}} \to \mathcal{F}$ is defined as

$$E_{B_n} R_{\hat{\Gamma}(P^0_{n,B_n})} \left(\hat{\Psi} \left(P^0_{n,B_n} \right), P^1_{n,B_n} \right).$$

For linear risk functions defined by loss function L this reduces to

$$E_{B_n} P^1_{n,B_n} L \left(\hat{\Psi} \left(P^0_{n,B_n} \right) \right).$$

For loss functions indexed by a nuisance parameter, this reduces to

$$E_{B_n} P^1_{n,B_n} L_{\hat{\Gamma}(P^0_{n,B_n})} \left(\hat{\Psi} \left(P^0_{n,B_n} \right) \right).$$

The cross-validated estimator of risk of the estimator $\hat{\Psi}$ represents a measure of performance. Specifically, the cross-validated estimator or risk targets the conditional risk $E_{B_n} P_0 L_{\gamma_0}(\hat{\Psi}(P^0_{n,B_n}))$ of the estimator when applied to samples of size $n(1-p)$, where $p = \sum_i B_n(i)/n$ is the proportion of observations that fall into the validation sample.

4.2.5 Super Learning Algorithm

The class of candidate estimators is given by $\hat{\Psi}_\alpha : \mathcal{M}_{\mathrm{NP}} \to \mathcal{F}$ indexed by a Euclidean parameter α. The cross-validation selector of α is defined as

$$\alpha_n = \arg\min_\alpha E_{B_n} R_{\hat{\Gamma}(P^0_{n,B_n})} \left(\hat{\Psi}_\alpha \left(P^0_{n,B_n} \right), P^1_{n,B_n} \right).$$

The super learner of ψ_0 is defined by

$$\hat{\Psi}_{\mathrm{SL}}(P_n) = \hat{\Psi}_{\alpha_n}(P_n).$$

4.2.6 Evaluation of Performance of Super Learner

The performance of the Super Learner $\hat{\Psi}_{\mathrm{SL}}$ itself can be assessed with its cross-validated estimator of risk:

$$E_{B_n} R_{\hat{\Gamma}(P^0_{n,B_n})} \left(\hat{\Psi}_{\mathrm{SL}} \left(P^0_{n,B_n} \right), P^1_{n,B_n} \right).$$

4.3 Oracle Inequality for Cross-Validation Selector

We present a finite sample oracle inequality for the cross-validation selector for the linear risk function. We focus on loss functions that imply a quadratic dissimilarity and refer to [81] and subsequent articles for the analogue results for loss functions that are nonquadratic. We also present a generalized oracle inequality for loss functions that are indexed by a nuisance parameter. We conclude with a corollary representing the implication of this oracle inequality for the super learner.

4.3.1 *Oracle Inequality for Linear Risk Function*

We now present the oracle inequality for this cross-validation selector α_n, as presented originally in [81]. Let $d(\psi, \psi_0) = P_0\{L(\psi) - L(\psi_0)\}$ denote the loss-function-based dissimilarity. Assume that the loss function is bounded: $M_1 \equiv \sup_{\psi,O} \mid L(\psi)(O) - L(\psi_0)(O) \mid < \infty$. In addition, we assume that $P_0\{L(\psi) - L(\psi_0)\}^2 \leq M_2 P_0\{L(\psi) - L(\psi_0)\}$. As explained in [81], the latter assumption corresponds with the loss-based dissimilarity being quadratic in the difference between ψ and ψ_0. These two properties of the loss function allow us to apply the oracle inequality for the cross-validation selector as presented in [81]: if the cross-validation selector α_n is defined as a minimizer over a grid with $K(n)$ α-values, then for any $\delta > 0$,

$$Ed\left(\hat{\Psi}_{\alpha_n}\left(P^0_{n,B_n}\right), \psi_0\right) \leq (1+2\delta)E \min_{\alpha} E_{B_n} d\left(\hat{\Psi}_\alpha\left(P^0_{n,B_n}\right), \psi_0\right)$$
$$+C(M_1, M_2, \delta)\frac{\log K(n)}{n},$$

where $C(M_1, M_2, \delta)$ is a specified constant. The $\tilde{\alpha}_n$ that attains the minimum on the right-hand side is referred to as the oracle selector. Thus, the oracle selector selects the α that minimizes the dissimilarity with ψ_0 for the given sample P_n. By choosing a grid with width $1/n$, we obtain a grid that is more than fine enough so that no precision is lost. In that case, the $\log K(n)$ is bounded by a constant times $\log n$.

This oracle inequality has been applied to log-likelihood loss $L(\psi) = -\log \psi$ in the case that ψ_0 is a conditional density and squared error loss $L(\psi) = (Y - \psi(W))^2$ in the case that $\psi_0 = E_0(Y \mid W)$ [81], among others.

4.3.2 *Oracle Inequality for Loss Function Indexed by Nuisance Parameter*

Throughout the following theorem we introduce and use the following notation:

$$L^*\left(\hat{\Psi}_k\left(P^0_{n,B_n}\right), \psi_0\right) \equiv L_{\gamma_0}\left(\hat{\Psi}_k\left(P^0_{n,B_n}\right)\right) - L_{\gamma_0}(\psi_0)$$

$$L^{*0}_{n,B_n}\left(\hat{\Psi}_k\left(P^0_{n,B_n}\right),\psi_0\right) \equiv L_{\hat{\Gamma}(P^0_{n,B_n})}\left(\hat{\Psi}_k\left(P^0_{n,B_n}\right)\right) - L_{\hat{\Gamma}(P^0_{n,B_n})}(\psi_0)$$

$$\left(L^{*0}_{n,B_n} - L^*\right)\left(\hat{\Psi}_k\left(P^0_{n,B_n}\right),\psi_0\right) = L^{*0}_{n,B_n}\left(\hat{\Psi}_k\left(P^0_{n,B_n}\right),\psi_0\right) - L^*\left(\hat{\Psi}_k\left(P^0_{n,B_n}\right),\psi_0\right).$$

We also note that the rates $r_1(n), r_2(n)$ as defined in the theorem are determined by the rate at which the nuisance parameter estimate $\hat{\Gamma}(P_n)$ approximates γ_0.

Theorem 1. *Let $\hat{\Psi}_k(P_n)$, $k = 1, \ldots, K(n)$, be a set of given estimators of $\psi_0 = argmin_{\psi \in \mathscr{F}} \int L_{\gamma_0}(O,\psi)\mathrm{d}P_0(O)$. Suppose that $\hat{\Psi}_k(P_n) \in \mathscr{F}$ for all k, with probability 1. Let $k_n = argmin_k E_{B_n} \int L_{\hat{\Gamma}(P^0_{n,B_n})}(\hat{\Psi}_k(P^0_{n,B_n}))\mathrm{d}P^1_{n,B_n}$ be the cross-validation selector, and let $\tilde{k}_{n(1-p)} = argmin_k E_{B_n} \int L_{\gamma_0}(\hat{\Psi}_k(P^0_{n,B_n}))\mathrm{d}P_0$ be the comparable benchmark selector.*

Assumption 1 *A1. The limit γ_0 of the estimator $\gamma_n = \hat{\Gamma}(P_n)$ for $n \to \infty$ is an element of $\Gamma(P_0) \equiv \{\gamma : \psi_0 = \arg\min_\psi P_0 L_\gamma(\psi)\}$.*

A2. There exists a $M_1^* < \infty$ so that

$$\sup_{\psi \in \mathscr{F}} \sup_O L^*(\psi,\psi_0)(O) \leq M_1^*,$$

where the supremum over O is taken over a support of the distribution P_0 of O.

A3. There exists a $M_2 < \infty$ so that for all $\psi \in \mathscr{F}$

$$VAR_{P_0}\left[L^*(\psi,\psi_0)\right] \leq M_2 E_{P_0} L^*(\psi,\psi_0). \tag{4.1}$$

Definition 1. We define the following constants:

$$M_1 = 2M_1^*$$

$$c(M_1,M_2,\delta) = 2(1+\delta)^2\left(\frac{M_1}{3} + \frac{M_2}{\delta}\right)$$

$$a_0 \equiv 2M_1/3$$

$$M_3(n) = 3a_0 + \sqrt{2}\frac{\sqrt{\log(2)}}{\log(K(n))} + \frac{\sqrt{2}}{\sqrt{\log(K(n))}} + b_0 + \int_{b_0}^{\infty} 2K(n)^{1-m(x)}\mathrm{d}x,$$

where b_0 is the smallest constant larger than the solution of $1 - m(x) = 0$ with $m(x) \equiv 0.5\frac{x^2}{1/\log(K(n)) + a_0 x}$. We note that $M_3(n) \downarrow$ in n. We also define the following sequences in n:

$$r_1(n) \equiv \max_{\bar{k} \in \{k_n, \tilde{k}_{n(1-p)}\}} \frac{E\int\left(L^{*0}_{n,B_n} - L^*\right)\left(\hat{\Psi}_{\bar{k}}\left(P^0_{n,B_n}\right),\psi_0\right)\mathrm{d}P_0}{\sqrt{E\int L^*\left(\hat{\Psi}_{\bar{k}}\left(P^0_{n,B_n}\right),\psi_0\right)\mathrm{d}P_0}}$$

$$r_2(n) \equiv E \max_{k \in \{1,\ldots,K(n)\}} \sqrt{\int \left(L^{*0}_{n,B_n} - L^*\right)^2 \left(\hat{\Psi}_k\left(P^0_{n,B_n}\right), \psi_0\right) \mathrm{d}P_0}$$

$$\tilde{r}(n) \equiv \sqrt{E\, d\left(\hat{\Psi}_{\tilde{k}_{n(1-p)}}\left(P^0_{n,B_n}\right), \psi_0\right)}.$$

Finally, for any $\delta > 0$ we define

$$\varepsilon_n(\delta) \equiv (1+2\delta)\tilde{r}^2(n) + 2c(M_1, M_2, \delta)\frac{1+\log(K(n))}{np} + (1+\delta)r_1(n)\tilde{r}(n)$$
$$+ \frac{2M_3(1+\delta)\log(K(n))}{(np)^{0.5}} \max(r_2(n), (np)^{-0.5} I(r_2(n) > 0)).$$

Finite Sample Result

For any $\delta > 0$, we have

$$\sqrt{E\, d\left(\hat{\Psi}_{k_n}\left(P^0_{n,B_n}\right), \psi_0\right)} \le \frac{r_1(n)(1+\delta) + \sqrt{r_1(n)^2(1+\delta)^2 + 4\varepsilon_n(\delta)}}{2}. \quad (4.2)$$

In the special case that γ_0 is known so that $r_1(n) = r_2(n) = 0$, we have that the finite sample result (4.2) reduces to

$$E\, d\left(\hat{\Psi}_{k_n}\left(P^0_{n,B_n}\right), \psi_0\right) = (1+2\delta)E\, d\left(\hat{\Psi}_{\tilde{k}_{n(1-p)}}\left(P^0_{n,B_n}\right), \psi_0\right)$$
$$+ 2c(M_1, M_2, \delta)\frac{1+\log(K(n))}{np}.$$

Asymptotic Implication

For any $\delta > 0$

$$E\, d\left(\hat{\Psi}_{k_n}\left(P^0_{n,B_n}\right), \psi_0\right) \le (1+2\delta)E\, d\left(\hat{\Psi}_{\tilde{k}_{n(1-p)}}\left(P^0_{n,B_n}\right), \psi_0\right) + O(H(n)),$$

where

$$H(n) \equiv \max\left(\frac{\log(K(n))}{np}, \frac{\log(K(n))r_2(n)}{(np)^{0.5}}, r_1^2(n), r_1(n)\tilde{r}(n), r_1(n)^{1.5}\tilde{r}(n)^{0.5},\right.$$
$$\left.\frac{\sqrt{\log(K(n))}r_1(n)}{(np)^{0.5}}, \frac{\sqrt{\log(K(n)}r_2(n)^{0.5}r_1(n)}{(np)^{0.25}}\right).$$

Consequently, we have the following scenarios.

Optimal Rate

If $\max\left(\frac{\log(K(n))}{np}, r_1(n)^2, \log(K(n))r_2(n)^2\right) = O(\tilde{r}(n)^2)$, then $H(n) = O(\tilde{r}(n)^2)$, and thus $E\,d(\hat{\Psi}_{k_n}(P^0_{n,B_n}), \psi_0) = O(\tilde{r}(n)^2)$. If either $\max\left(\frac{\log(K(n))}{np}, r_1(n)^2,\right.$ $\left.\log(K(n))r_2(n)^2\right) = o(\tilde{r}(n)^2)$ or $\max\left(\frac{\log(K(n))^2}{np}, r_1(n)^2, r_2(n)^2\right) = o(\tilde{r}(n)^2)$, then

$$H(n) = o\left(\tilde{r}(n)^2\right).$$

In particular, we note that if $\max(r_1(n)^2, \log(K(n))r_2(n)^2) = o(\tilde{r}(n)^2)$, then

$$H(n) = O\left(\frac{\log(K(n))}{np}\right) + o\left(\tilde{r}^2(n)\right).$$

Asymptotic Optimality

Consequently, under these two possible scenarios under which $H(n) = o\left(\tilde{r}(n)^2\right)$, we have

$$\frac{E\,d\left(\hat{\Psi}_{k_n}\left(P^0_{n,B_n}\right), \psi_0\right)}{E\,d\left(\hat{\Psi}_{\tilde{k}_{n(1-p)}}\left(P^0_{n,B_n}\right), \psi_0\right)} \to 1 \text{ for } n \to \infty. \tag{4.3}$$

Finally, if these two possible scenarios hold with $\tilde{r}(n)^2$ replaced by the random quantity $E_{B_n} d\left(\hat{\Psi}_{\tilde{k}_{n(1-p)}}\left(P^0_{n,B_n}\right), \psi_0\right)$, then

$$\frac{E_{B_n} d\left(\hat{\Psi}_{k_n}\left(P^0_{n,B_n}\right), \psi_0\right)}{E_{B_n} d\left(\hat{\Psi}_{\tilde{k}_{n(1-p)}}\left(P^0_{n,B_n}\right), \psi_0\right)} \to 1 \text{ in probability for } n \to \infty. \tag{4.4}$$

The final convergence in probability statement is a straightforward consequence of our finite sample inequality and the following Lemma 1.

Lemma 1. *Consider a sequence of random variables $Z_1, Z_2, \ldots$, with finite expectation $E|Z_n| = O(g(n))$, for a positive function $g(n)$. Then $Z_n = O_P(g(n))$.*

This lemma is a direct consequence of Markov's inequality. The proof of this theorem is presented in [81].

4.3.3 Asymptotic Equivalence of Cross-Validation Selector with Oracle Procedure

Theorem 1 provides a finite sample bound for the expected value of the loss-based dissimilarity $E_{B_n} d\left(\hat{\Psi}_{k_n}\left(P^0_{n,B_n}\right), P_0\right)$ of the cross-validated selected estimator, and the loss based dissimilarity $E_{B_n} d\left(\hat{\Psi}_{\tilde{k}_{n(1-p)}}\left(P^0_{n,B_n}\right), P_0\right)$ of the oracle-selected estimator. This allows us to compare the performance of the cross-validated selector k_n to the benchmark $\tilde{k}_{n(1-p)}$ in terms of the conditional (true) risks based on $n(1-p)$ training observations. Specifically, the finite sample bound provided by Theorem 1 implies that the two loss-based dissimilarities are asymptotically equivalent as long as the loss-based dissimilarity for the oracle-selected estimator does not converge as fast to zero as $1/\sqrt{n}$ or as fast as the rate at which the nuisance parameter is estimated (i.e. $r_1(n)$ and $r_2(n)$).

However, one would like the cross-validated selector k_n to perform as well as the benchmark selector $\tilde{k}_n$ based on the whole sample of size n, defined as

$$\tilde{k}_n = \arg\min_k d\left(\hat{\Psi}_k(P_n), \psi_0\right),$$

rather than only $n(1-p)$ as above. The following is an immediate corollary of Theorem 1 that addresses this wished optimality. In this corollary, we use the notation $p = p_n$ to emphasize the dependence of the validation set proportion p on n. The corollary proves that, if $p = p_n$ converges slowly enough to zero when the sample size n converges to infinity, then, given the mild condition (4.6) given below, the desired asymptotic optimality of the cross-validation selector k_n follows. The proof of this corollary is straightforward and provided in [81] (and also in subsequent articles).

Corollary 1. *If $p = p_n \to 0$, the conditions of Theorem 1 hold so that for $n \to \infty$*

$$\frac{E\, d\left(\hat{\Psi}_{k_n}\left(P^0_{n,B_n}\right), \psi_0\right)}{E\, d\left(\hat{\Psi}_{\tilde{k}_{n(1-p_n)}}\left(P^0_{n,B_n}\right), \psi_0\right)} \to 1 \text{ as } n \to \infty,$$

and for $n \to \infty$

$$\frac{E\, d\left(\hat{\Psi}_{\tilde{k}_n}(P_n), \psi_0\right)}{E\, d\left(\hat{\Psi}_{\tilde{k}_{n(1-p_n)}}\left(P^0_{n,B_n}\right), \psi_0\right)} \to 1 \text{ as } n \to \infty, \tag{4.5}$$

then

$$\left(\frac{E\, d\left(\hat{\Psi}_{k_n}\left(P^0_{n,B_n}\right), \psi_0\right)}{E\, d\left(\hat{\Psi}_{\tilde{k}_n}(P_n), \psi_0\right)}\right) \to 1 \text{ as } n \to \infty. \tag{4.6}$$

Consider the estimator $\hat{\Psi}(P_n) \equiv \hat{\Psi}_{k_n}(P_n)$. Suppose that for n large enough

$$ER\left(\hat{\Psi}\left(P^0_{n,B_n}\right), P_0\right) \geq ER\left(\hat{\Psi}(P_n), P_0\right),$$

then (4.6) implies the wished asymptotic equivalence result:

$$\frac{E\,d\left(\hat{\Psi}_{k_n}(P_n), \psi_0\right)}{E\,d\left(\hat{\Psi}_{\tilde{k}_n}(P_n), \psi_0\right)} \to 1.$$

In other words, if the estimator $\hat{\Psi}(P_n)$ is capable of learning, then (4.6) implies the wished optimality result. We also note that the condition (4.5) is not more than a very weak regularity condition.

4.4 Examples: Super Learning

In this section, we present a number of examples in which the super learner method can be applied. We provide examples of both linear risk functions and nonlinear risk functions defined by loss functions indexed by a nuisance parameter, noting, as above, that the optimal properties of the super learner in the latter case will depend on the extent to which a good estimator of the nuisance parameter can be obtained. Specifically, for a range of data and target parameters we present an appropriate loss and corresponding risk function and the resulting cross-validation selector. User-supplied input in the form of a library of algorithms and parametric family of weighted combinations of these algorithms results in a set of candidate estimators indexed by α. The corresponding super learner is defined as the cross-validation selector applied to this set of candidates.

The examples presented are by no means exhaustive. As originally highlighted in [81], this template of cross-validation methodology and super learning covers, in particular, the generalization of a cross-validation selection method based on observing a particular full data structure X to any censored data structure $O = \Phi(C, X)$ for a known many to one-mapping Φ and censoring variable C.

4.4.1 Prediction

We observe n i.i.d. observations of $O = (Y, W) \sim P_0$, where Y is an outcome and W is a vector of covariates. Let $\psi_0(W) = E_0(Y \mid W)$ be the parameter of interest. If we define

$$L(\psi)(O) = (Y - \psi(W))^2$$

Table 4.1 Library of prediction algorithms for the corresponding R package

Algorithm	Description	Author
glm	Linear model	[55]
interaction	Polynomial linear model	[55]
randomForest	Random Forest	[13,46]
bagging	Bootstrap aggregation of trees	[11,52]
gam	Generalized additive models	[38,39]
gbm	Gradient boosting	[29,56]
nnet	Neural network	[92]
polymars	Polynomial spline regression	[28,44]
bart	Bayesian additive regression trees	[20,21]
loess	Local polynomial regression	[22]
bayesglm	Bayesian linear model	[33,34]
glmnet	Elastic net	[30,31]
DSA	DSA algorithm	[50,72]
step	Stepwise regression	[92]
ridge	Ridge regression	[92]
svm	Support vector machine	[19,25]

as the squared error loss function, then $\psi_0 = \text{argmin}_\psi E_0 L(\psi)(O)$. Given candidate estimators $\psi_{\alpha,n} = \hat{\Psi}_\alpha(P_n)$, the loss-based dissimilarity is given by

$$d(\psi_{\alpha,n}, \psi_0) = \int (\psi_{\alpha,n}(w) - \psi_0(w))^2 \, \mathrm{d}P_0(w).$$

The cross-validation selector is given by:

$$\alpha_n = \text{argmin}_\alpha E_{B_n} \sum_{i:B_n(i)=1} \left(Y_i - \hat{\Psi}_\alpha\left(P^0_{n,B_n}\right)(W_i)\right)^2.$$

4.4.1.1 Practical Demonstration of Super Learning in Prediction

This section is based on a previously published study as presented in Chap. 3 in [84]. To study the super learner in real data examples, [54] collected a set of publicly available datasets. The sample sizes ranged from 200 to 654 observations and the number of covariates ranged from 3 to 18. All 13 datasets have a continuous outcome and no missing values. The datasets can be found either in public repositories like the UCI data repository or in textbooks, with the corresponding citations listed in the above reference.

Above, we defined the risk function and corresponding cross-validation selector. The super learner template further requires definition of the library of initial algorithms and of the parametric family of weighted combinations of these algorithms. In Table 4.1, we list the algorithms that were included in the library. These algorithms represent a diverse set of basis functions and should allow the super learner to work well in most real settings. For the comparison across all datasets, we kept the library of algorithms the same. When applying the super learner to a specific application, the library can be expanded or modified to include algorithms based on contextual knowledge of the data problem. Our parametric family of algorithms was defined as all convex combinations of the algorithms included in the library. In other words, we used:

$$\hat{\Psi}_\alpha = \sum_{j=1}^{J} \alpha(j)\hat{\Psi}_j,$$

where $\alpha(j) \geq 0$ and $\sum_j \alpha(j) = 1$.

In order to compare the performance of the prediction algorithms across diverse datasets across which the outcome scale differed, we used the mean squared error relative to the mean squared error of a linear model:

$$\text{relMSE}(j) = \frac{\text{MSE}(j)}{\text{MSE}(lm)}, \quad j = 1, \ldots, J, \tag{4.7}$$

for each of the $J = 16$ algorithms in the library. We compared these to the relative MSE for the super learner, as well as for the cross-validation selector applied to the initial J algorithms (rather than all convex combinations of these algorithms). We refer to the latter estimator as the "discrete super learner."

The results for the super learner, the discrete super learner, and each individual algorithm can be found in Fig. 4.1. Each point represents the ten-fold cross-validated relative mean squared error for a single dataset and the plus sign represents the geometric mean across all 13 datasets. The super learner slightly outperforms the discrete super learner, and both outperform any individual algorithm in the library. Among the individual library algorithms the bayesian additive regression trees performs the best, but overfits on one of the datasets with a relative mean squared error of almost 3.0, demonstrating the dangers of reliance on a single algorithm. Further, in many real data applications it is unlikely that one single algorithm contains the true relationship between predictors and outcome. These results demonstrate how the super learner is able to adapt to the true underlying structure across various real data examples. The additional estimation of the combination parameters (α) does not appear to cause an overfit in terms of the risk assessment.

4.4.2 Density Estimation

We observe n i.i.d. observations on $O \sim f_0 \equiv \frac{dP_0}{d\mu}$, where μ is a dominating measure of the data-generating distribution P_0. Let the parameter of interest $\psi_0 = f_0$ be the density itself. If we define as loss function the negative log likelihood,

$$L(\psi) = -\log \psi(O),$$

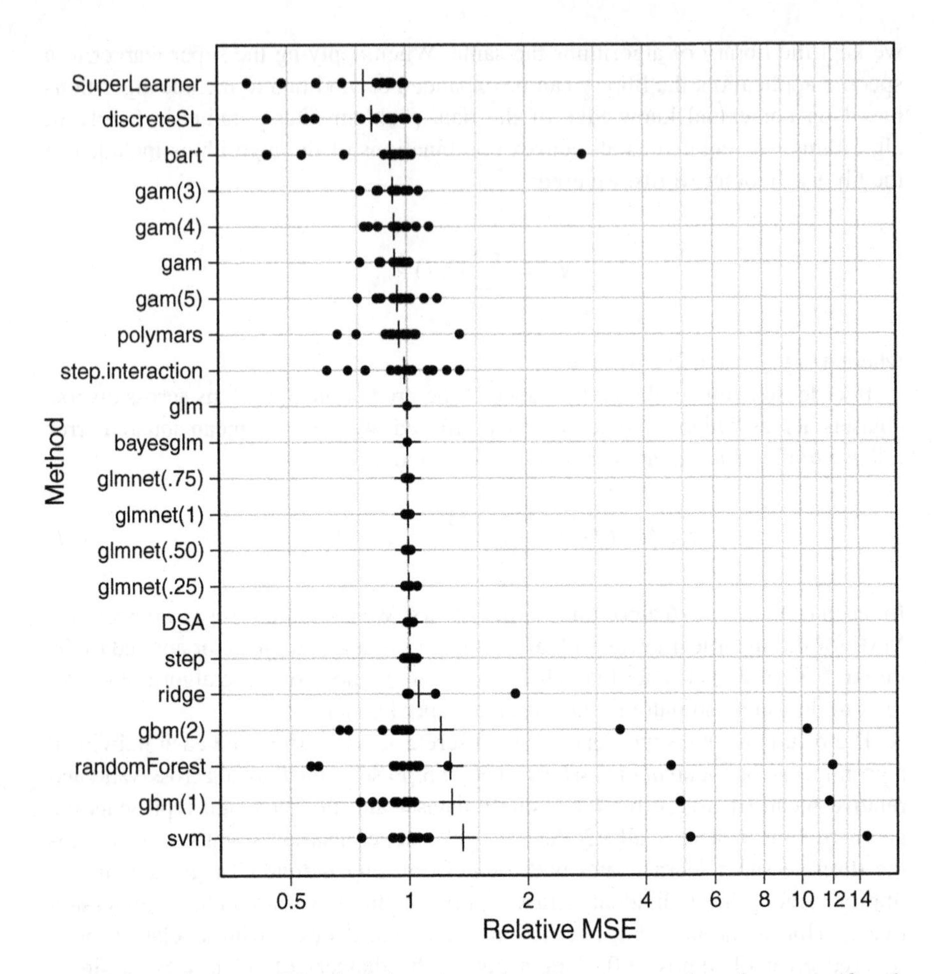

Fig. 4.1 Ten-fold cross-validated relative mean squared error compared to glm across 13 real datasets. Sorted by the geometric mean, denoted with the plus (+) sign

then $f_0 = \text{argmin}_\psi E_0 L(\psi)(O)$. Given candidate density estimators $\psi_{\alpha,n} = \hat{\Psi}_\alpha(P_n)$ of $\psi_0 = f_0$, the loss-based dissimilarity is given by

$$d(\psi_{\alpha,n}, \psi_0) = \int \log\left(\frac{\psi_{\alpha,n}(o)}{\psi_0(o)}\right) \psi_0(o) d\mu(o),$$

that is, $d(\psi_{\alpha,n}, \psi_0)$ is the Kullback–Leibler divergence between $\psi_{\alpha,n}$ and ψ_0. The cross-validation selector is given by

$$\begin{aligned} \alpha_n &= \hat{K}(P_n) \\ &= \text{argmin}_\alpha E_{B_n} \sum_{i:B_n(i)=1} -\log\left(\hat{\Psi}_\alpha\left(P^0_{n,B_n}\right)(O_i)\right). \end{aligned}$$

4.4.3 Multivariate Prediction

Multivariate prediction provides an example of the case of a nonpathwise differentiable parameter indexed by a nuisance parameter. Let $O = (Y = (Y(1), \ldots, Y(l)), W) \sim P_0$, where Y is a multivariate random outcome vector and W a vector of covariates. Let $\psi_0(W) \equiv E_0(Y \mid W) = (E_0(Y(1) \mid W), \ldots, E_0(Y(l) \mid W))$ be the multivariate conditional expectation of Y, given W. For a candidate multivariate predictor $\psi(W)$, we define

$$L_{\gamma_0}(\psi) \equiv (Y - \psi(W))^\top \gamma_0(W)(Y - \psi(W)),$$

where γ_0 is a symmetric $l \times l$-matrix function of W. If γ_0 is a user-supplied known matrix, then it is not a nuisance parameter and we can denote the loss function with $L(\psi)$. However, γ_0 can also denote the desired limit of an estimator of an unknown matrix such as

$$\left[E_0\left(\{Y - E_0(Y \mid W)\}\{Y - E_0(Y|W)\}^\top \mid W\right)\right]^{-1}.$$

In this case, γ_0 denotes a nuisance parameter which needs to be estimated from the data. For any symmetric matrix function $\gamma(W)$, we have

$$\psi_0 = \operatorname{argmin}_\psi E_0 L_\gamma(\psi). \tag{4.8}$$

Given candidate estimators $\psi_{\alpha,n} = \hat{\Psi}_\alpha(P_n)$ of ψ_0 indexed by α, and an estimator $\hat{\Gamma}(P_n)$ (e.g., an estimate of the inverse of the conditional covariance matrix according to a working model such as the independence working model) of γ_0, the cross-validation selector α_n is given by:

$$\begin{aligned}
\alpha_n &= \operatorname{argmin}_\alpha E_{B_n} \int L_{\hat{\Gamma}(P^0_{n,B_n})}\left(\hat{\Psi}_\alpha\left(P^0_{n,B_n}\right)\right) dP^1_{n,B_n}(O) \\
&= \operatorname{argmin}_\alpha E_{B_n} \frac{1}{np} \sum_{i=1}^{n} \Big\{ I(B_n(i) = 1) \\
&\quad (Y(i) - \hat{\Psi}_\alpha\left(P^0_{n,B_n}\right)(W_i))^\top \hat{\Gamma}\left(P^0_{n,B_n}\right)(W_i)(Y(i) - \hat{\Psi}_\alpha\left(P^0_{n,B_n}\right)(W_i)) \Big\}.
\end{aligned}$$

We note that

$$\begin{aligned}
&\left(Y - \hat{\Psi}_\alpha\left(P^0_{n,B_n}\right)(W)\right)^\top \gamma_0(W)\left(Y - \hat{\Psi}_\alpha\left(P^0_{n,B_n}\right)\right)(W)) \\
&= \left\| \gamma_0^{0.5}(W)\left(Y - \hat{\Psi}_\alpha\left(P^0_{n,B_n}\right)(W)\right)\right\|^2,
\end{aligned}$$

where $\| x \| = \sqrt{\sum_{j=1}^{l} x_j^2}$ is the Euclidean norm in $\mathbb{R}^l$, and $\gamma_0^{0.5}$ is the square root of γ_0. This shows that

$$d(\psi_{\alpha,n}, \psi_0) = \int \left\| \gamma_0^{0.5} \left(\psi_{\alpha,n}(W) - \psi_0(W)\right) \right\|^2 \mathrm{d}P_0(W).$$

Application of our general Theorem 1 to this example results in a finite sample result and asymptotic optimality of the cross-validation selector.

4.4.4 Prediction of Survival with Right Censoring

Prediction of survival given a vector of baseline covariates in the presence of right censoring provides a second example of a nonpathwise differentiable parameter indexed by a nuisance parameter. Let $X(t)$, $t \geq 0$, be a time-dependent process, which includes as component $R(t) = I(T \leq t)$, where T is a survival time. Let $X = \bar{X}(T) \equiv \{X(t) : t \leq T\}$ be the full-data structure of interest. Let $W = X(0)$ denote the baseline covariates measured at baseline. The distribution of X will be denoted with F_{X0}. Let C be a right-censoring time so that the observed data structure is given by

$$O = \left(\tilde{T} \equiv \min(T, C)\right), \Delta \equiv I\left(\tilde{T} = T\right), \bar{X}\left(\tilde{T}\right).$$

We will assume that the conditional distribution $G_0(\cdot|X)$ of C, given X, satisfies coarsening at random (CAR, see [83]), that is, for $t < T$,

$$\lambda_C(t \mid X) = m\left(t, \bar{X}(t)\right) \text{ for some measurable function } m,$$

where $\lambda_C(t \mid X)$ denotes the discrete or continuous conditional hazard of C, given X. If $X = (T, W)$ does not include time-dependent covariates, then CAR is equivalent with assuming that C is conditionally independent of T, given W. Under CAR, the density of P_0 factors into a F_{X0}-part and G_0-part [35]. The F_{X0}-part of the density will be denoted with Q_{X0}.

We observe n i.i.d. observations $O_1, \ldots, O_n$ of $O \sim P_0 = P_{F_{X,0},G_0}$. Let $\psi_0(W) = E_0(Y \mid W)$ be the parameter of interest, where $Y \equiv \log(T)$ denotes log-survival time. The corresponding full-data loss function is given by $L(\psi)(X) = (Y - \psi(W))^2$:

$$\psi_0 = \text{argmin}_{\psi} \int L(\psi) \mathrm{d}F_{X0}.$$

Suppose that

$$\bar{G}_0(T \mid X) \equiv P(C > t \mid X)|_{t=T} > \delta > 0, \; F_{X0}\text{-a.e. for some } \delta > 0. \quad (4.9)$$

Then it follows that

$$\psi_0(W) = \text{argmin}_\psi \int L(\psi) \mathrm{d}F_{X0}$$
$$= \text{argmin}_\psi E_{F_{X0}}(Y - \psi(W))^2$$
$$= \text{argmin}_\psi E_{P_0}\left\{L(\psi)\frac{\Delta}{\bar{G}_0(T \mid X)}\right\},$$

which is called the inverse probability of censoring weighted full data loss function [61, 83]. Thus, if we choose

$$L_{G_0}(\psi) = L(\psi)\frac{\Delta}{\bar{G}_0(T \mid X)},$$

then $\psi_0 = \text{argmin}_\psi E_0 L_{G_0}(\psi)$. A more sophisticated loss function is obtained by applying the so called double robust (DR) mapping [83, Chap. 3] to full data function $L(\psi)$:

$$L_{Q_0,G_0}(\psi) = L(\psi)\frac{\Delta}{\bar{G}_0(T \mid X)} + \int E_{Q_{X0},G_0}\left(L_{G_0}(\psi) \mid \bar{X}(u), \tilde{T} \geq u\right) \mathrm{d}M_{G_0}(u), \quad (4.10)$$

where

$$\mathrm{d}M_{G_0}(u) = I(\tilde{T} \in du, \Delta = 0) - I\left(\tilde{T} \geq u\right)\frac{\mathrm{d}G_0(u \mid X)}{\bar{G}_0(u- \mid X)}.$$

Here, $E_{Q_{X0},G_0} = E_{P_0}$. In [83] it is shown that, if G_1 satisfies the condition (4.9), then

$$E_{P_0} L_{Q_1,G_1}(\psi) = E_{F_{X0}} L(\psi) \text{ if either } G_1 = G_0 \text{ or } Q_1 = Q_0.$$

In [83], this identity is referred to as double robustness of the estimating function for full-data parameter $E_0 L(\psi)$ w.r.t. misspecification of Q_0, g_0. Thus, if γ_0 is an element of $\Gamma(P_0) \equiv \{(Q, G) : Q = Q_0 \text{ or } G = G_0\}$, where G ranges over conditional distributions satisfying (4.9), then

$$\psi_0 = \text{argmin}_\psi E_{P_0} L_{\gamma_0}(\psi) \text{ if } \gamma_0 \in \Gamma(P_0).$$

Given candidate estimators $\hat{\Psi}_\alpha(P_n)$, the corresponding distance is given by:

$$d(\psi_{\alpha,n}, \psi_0) = \int L_{\gamma_0}(\psi_{\alpha,n}) - L_{\gamma_0}(\psi_0) \mathrm{d}P_0$$
$$= \int (\psi_{\alpha,n}(w) - \psi_0(w))^2 \, \mathrm{d}P_0(w).$$

To define this set of candidate estimators indexed by α, it remains to define a library of algorithms for estimating ψ_0, together with a parametric family of weighted combinations of these estimators indexed by α. The library of candidate estimators might include, for example, estimators based on parametric Cox proportional hazard models or (log) linear regression models, as well as data adaptive estimators.

For the IPCW-choice of loss function, we have:

$$\alpha_n = \text{argmin}_\alpha E_{B_n} \sum_{i:B_n(i)=1} \left(Y_i - \hat{\Psi}_\alpha\left(P^0_{n,B_n}\right)(W_i)\right)^2 \frac{\Delta_i}{\bar{G}^0_{n,B_n}(T_i \mid W_i)},$$

where $\bar{G}^0_{n,B_n}$ denotes an estimator of the survivor function $\bar{G}_0(\cdot \mid X)$ based on the training sample. We note that this cross-validation selector reduces to the standard cross-validation selector in prediction (Sect. 4.4.1) in the special case that there is no censoring. The asymptotic validity of this selector relies on the consistency of the estimator of the survivor function $\bar{G}_0$. If one uses the DR loss function, then the asymptotic validity of the corresponding selector only relies on the consistency of either $\bar{G}_n$ or of the estimator Q_{Xn} of Q_{X0}, and the assumption that (4.9) is satisfied at the limit of $\bar{G}_n$.

4.4.5 Estimation of Causal Dose–Response Curve

Estimation of the causal dose–response curve provides a second example of the case of a nonpathwise differentiable parameter indexed by nuisance parameter. Let $X = ((Y(a), a \in \mathscr{A}), W) \sim F_{X0}$ be the full data structure of interest, where W denotes baseline covariates and $Y(a)$ denotes the outcome on a subject if the subject would have taken treatment a. Such potential outcomes $Y(a)$ are called counterfactuals (e.g., [69]). Let A be a random variable with conditional probability distribution $g_0(a \mid X) \equiv P(A = a \mid X)$, $a \in \mathscr{A}$, which denotes the treatment the subject actually took; we only observe the potential outcome indexed by the treatment the subject took. Thus, we observe n i.i.d. observations of $O = (W, A, Y \equiv Y(A))$, where Y denotes the observed outcome corresponding with the treatment taken by the subject. We assume that treatment is randomized within strata of W: $g_0(a \mid X) = g_0(a \mid W)$ for all $a \in \mathscr{A}$. We have that the distribution $P_0 = P_{F_{X0},g_0}$ is indexed by the full data distribution F_{X0} and the conditional density g_0 (referred to as the treatment mechanism). Suppose that the parameter of interest is $\psi_0(a, V) = E_{F_{X0}}(Y(a) \mid V)$, where V is a user supplied baseline covariate that can be extracted from W. That is, we want to estimate the multivariate regression of the vector $(Y(a) : a \in \mathscr{A})$ of potential outcomes on V. If we would observe the full-data structure X, then this would be the same problem as covered in the previous multivariate prediction example, and we could use as loss function

$$L(\psi)(X) = \int_{a \in \mathscr{A}} (Y(a) - \psi(a, V))^2 \mathrm{d}\mu(a),$$

for some measure μ so that the conditional distribution of A, given W, dominated by μ. Thus, we have

$$\psi_0 = \mathrm{argmin}_{\psi} E_{F_{X0}} L(\psi).$$

However, in this example, we only observe one of the outcomes $Y(A)$ for each subject so that the loss function $L(\psi)$ is not a function of the observed data structure O. A fundamental objective in the estimating function theory for censored data structures presented in [83] involves mapping an estimating function of a full data structure into an observed data function which has the same expectation as the full data function. In particular, [83] provide an inverse probability of censoring weighted mapping and an optimal (i.e., minimal variance) DR mapping. Consequently, we can choose as loss function the inverse probability of treatment weighted (IPTW) or DR mapping applied to this full data loss function $L(X, \psi)$ [83, Sect. 6.3]. The DR mapping is given by

$$\begin{aligned} L_{\gamma_0}(\psi) = & \frac{(Y - \psi(A, V))^2}{g_0(A \mid W)} - \frac{1}{g_0(A \mid W)} E_0((Y - \psi(A, V))^2 \mid A, W) \\ & + \int_{a \in \mathcal{A}} E_0((Y - \psi(A, V))^2 \mid A = a, W) \mathrm{d}\mu(a). \end{aligned}$$

Here, $\gamma_0 = (g_0, Q_0)$ and $Q_0(A, W) = (E_0(Y \mid A, W), E_0(Y^2 \mid A, W))$. Note that the conditional expectations in this observed data loss function are indeed identified by these first two conditional moments of the conditional distribution of Y, given A, W. It can be verified [83, Sect. 6.3] that for any treatment mechanism g_1 satisfying the so-called experimental treatment assignment assumption (ETA), that is, $\min_{a \in \mathcal{A}} g_1(a \mid W) > 0$ P_0-a.e., we have

$$E_{P_0} L_{Q_1, g_1}(\psi) = E_{F_{X0}} L(\psi) \text{ if either } g_1 = g_0 \text{ or } Q_1 = Q_0.$$

In [83], this identity is referred to as double robustness of the augmented IPCW estimating function for $E_0 L(X, \psi)$ w.r.t. misspecification of Q_0, g_0. Thus, if γ_0 is an element of $\Gamma(P_0) \equiv \{(Q, g) : Q = Q_0 \text{ or } g = g_0\}$, where g ranges over conditional distributions satisfying ETA, then

$$\psi_0 = \mathrm{argmin}_{\psi} E_{P_0} L_{\gamma_0}(\psi).$$

Consequently, for any $\gamma_0 \in \Gamma(P_0)$, we have

$$d(\psi_{\alpha,n}, \psi_0) = \int_{a \in \mathcal{A}} \int (\psi_{\alpha,n}(a, V) - \psi_0(a, V))^2 \mathrm{d}P_0(V) \mathrm{d}\mu(a).$$

Let $\hat{\Psi}_j(P_n)$ be an estimator of ψ_0 based on n i.i.d. observations $O_1, \ldots, O_n$. For example, $\hat{\Psi}_j(P_n)$ is an IPTW estimator, double robust IPTW estimator or targeted maximum likelihood estimator corresponding to a j-specific marginal structural model $E(Y(a) \mid V) = m_j(a, V \mid \beta_j)$ (see [60], and [83]). Now, $\hat{\Psi}_\alpha(P_n)$ could be defined as the α-specific weighted combination over the j-specific algorithms in the library, each of which is itself an estimator of $E(Y(a)|V)$.

Given an estimator $\hat{\Gamma}(P_n)$ of (Q_0, g_0), our cross-validation selector α_n is given by:

$$\alpha_n = \text{argmin}_\alpha E_{B_n} \sum_{i=1}^{n} I(B_n(i) = 1) L_{\hat{\Gamma}(P^0_{n,B_n})} \left(\hat{\Psi}_\alpha\left(P^0_{n,B_n}\right)\right)(O_i).$$

A general data-adaptive estimator of the treatment specific mean based on this cross-validation selector was developed and programmed in [94].

Application of our general Theorem 1 yields a finite sample result and asymptotic optimality for this selector α_n under specific conditions. One of the main conditions is that either g_n is consistent for g_0 or Q_n is consistent for Q_0 at a rate faster than the rate at which ψ_0 is estimated. Informally, in other words, the wished for optimality results are achieved in settings where estimation of the nuisance parameter presents an easier estimation problem than estimation of ψ_0.

Estimation of the causal dose–response curve provides an example of a nonlinear risk function which is linear for a known nuisance parameter. As a result, the above estimator of the risk, based on taking the empirical mean of a plug-in estimator of the loss function (i.e., an estimator of the loss function based on plugging in an estimate of the nuisance parameter γ_0) will provide a super learner with the desired optimality properties in some but not all scenarios. For example, if the treatment mechanism g_0 is known or is correctly modeled by a parametric model, then the cross-validation selector will perform as if it is not indexed by a nuisance parameter, and thus will be asymptotically equivalent with the corresponding oracle selector. However, in other settings nuisance parameter estimation may present a sufficient challenge such that the risk estimator presented here, corresponding to the empirical mean of a plug in estimator of the loss function, will no longer provide a good estimate of the true risk. In particular, if treatment probabilities can be close to zero (i.e., if there are "practical" violations of the experimental treatment assignment assumption) then the empirical estimator of the risk presented here is very sensitive to variations in the estimator of the treatment mechanism g_0, due to the fact that this empirical risk estimator is not a substitution estimator. In such settings, a robust approach to risk estimation is required. We return to this challenge in Sect. 4.7 by using targeted minimum loss-based estimation.

4.5 Targeted Minimum Loss-Based Super Learning

The super learner template discussed up to this point in the chapter has focused on estimation of nonpathwise differentiable parameters. In this section, we turn to the problem of construction of an estimator of a pathwise differentiable parameter.

We are motivated to develop an extension of our initial super learning template by two considerations. First, the optimality result for the cross-validation selector among candidate estimators of the target parameter presented in Sect. 4.3 relies on the assumption that any nuisance parameter (required to estimate the risk) is easy to estimate relative to the target parameter itself, and thus the empirical mean of a plug in estimator of the loss function provides a good estimator of the risk. Second, the optimality theory for the cross-validation selector that formed the principle of super learning applies to candidate estimators that will not achieve the rate $1/\sqrt{n}$-rate of convergence. Since pathwise differentiable parameters can be estimated, in principle, at the (best possible) rate $1/\sqrt{n}$, we can no longer rely on the super learner presented in Sect. 4.2 to provide us with the optimal estimator of a pathwise differentiable parameter.

One wants estimators of a pathwise differentiable parameter to be asymptotically linear and preferably asymptotically efficient according to semiparametric efficiency theory, as developed for the class of regular estimators. Super learning can provide a crucial ingredient to achieve the desired asymptotic linearity by providing a tool for estimating nonpathwise differentiable nuisance parameters for a target parameter that is pathwise differentiable. However, an additional targeted bias reduction step must then be employed. In addition, to make estimators of the target parameter finite sample robust, it is preferable that these estimators respect the global constraints of the model and target parameter. This latter property can be achieved by defining the estimators as substitution/plug-in estimators. Finally, one wants to provide valid estimators of the variance of these estimators so that confidence intervals can be constructed. We review here the method of targeted minimum loss-based learning that was developed for the estimation of pathwise differentiable parameters with these motivations.

4.5.1 General Algorithm

Let O be the observed data structure, and let P_0 be its probability distribution. In addition, let $\mathcal{M}$ be the statistical model for P_0, and let $\Psi : \mathcal{M} \to \mathbb{R}^d$ be a pathwise differentiable d-dimensional parameter. Let $D^*(P)$ be the canonical gradient of the pathwise derivative at $P \in \mathcal{M}$, which is also called the efficient influence curve at P. One observes n i.i.d. copies $O_1, \ldots, O_n$ of O and one wishes to construct an estimator of $\Psi(P_0)$. Suppose that $Q_0 = Q(P_0)$ represents a parameter $Q : \mathcal{M} \to \mathcal{Q}$ so that for some Ψ^1 we have $\Psi(P) = \Psi^1(Q(P))$ for all $P \in \mathcal{M}$. Let $\mathcal{Q} = \{Q(P) : P \in \mathcal{M}\}$ be the parameter space for Q. For notational convenience, we will use notation $\Psi(P)$ and $\Psi(Q)$ interchangeably. We wish to construct a substitution estimator $\Psi(Q_n^*)$ of ψ_0 obtained by substitution of an estimator $Q_n^* \in \mathcal{Q}$ of Q_0 into the parameter mapping Ψ. Let $L(Q)$ be a loss function for Q_0 so that $Q_0 = \arg\min_{Q \in \mathcal{Q}} P_0 L(Q)$. We will allow this loss function to be indexed by a nuisance parameter: $L(Q) = L_{g_0}(Q)$ for some unknown nuisance parameter $g_0 = G(P_0)$. Given an estimator of g_0, one can use loss-based (e.g., super) learning to construct an estimator Q_n^0 of Q_0

However, we are not satisfied with a good estimator of Q_0. Instead, we wish to construct an optimal estimator of $\Psi(Q_0)$, a lower dimensional parameter. We pursue this goal by constructing an updated estimator Q_n^* such that Q_n^* and g_n solve a particular estimating equation $P_n D(Q_n^*, g_n) = 0$ for a user-supplied target-parameter-specific estimating function $D(Q, g)$. The choice of this estimating function D is tailored so that solving this equation implies good properties for the substitution estimator $\Psi(Q_n^*)$ of ψ_0. For example, $D(Q_0, g_0)$ is often defined as the canonical gradient $D^*(Q_0, g_0)$ (i.e., efficient influence curve) of the pathwise derivative of Ψ at P_0. In this case, Q_n^* and g_n will solve the efficient influence curve estimating equation, a property which is known to imply that $\Psi(Q_n^*)$ is asymptotically linear with influence cure equal to the efficient influence curve under appropriate conditions.

For any possible (Q, g), let $\{Q_g(\varepsilon) : \varepsilon\} \subset \mathcal{Q}$ be a submodel with a finite-dimensional parameter ε that contains Q at $\varepsilon = 0$, typically indexed by g, that satisfies the following local condition at $\varepsilon = 0$:

$$\tfrac{\mathrm{d}}{\mathrm{d}\varepsilon} L_g(Q_g(\varepsilon))\big|_{\varepsilon=0} = D(Q, g).$$

The targeted minimum loss based estimator (TMLE) is now defined by the following iterative algorithm. Start with initial estimator Q_n^0, and for $k = 1, \ldots$, define $Q_n^k = Q_{n,g_n}^{k-1}\left(\varepsilon_n^k\right)$, where $\varepsilon_n^k = \arg\min_\varepsilon P_n L_{g_n}\left(Q_{n,g_n}^{k-1}(\varepsilon)\right)$, and stop at step k when $\varepsilon_n^k \approx 0$. If $\varepsilon_n^k = 0$ and it is a local minima at an interior point, then it follows that the final update $Q_n^* = Q_n^k$ solves $0 = P_n D\left(Q_n^*, g_n\right)$. It is also possible to simultaneously update the nuisance parameter g_n in the loss function. The substitution estimator $\Psi(Q_n^*)$ is the targeted minimum-loss-based estimator of ψ_0.

Suppose $\frac{\mathrm{d}}{\mathrm{d}\varepsilon_j} L_g(Q_g(\varepsilon))\Big|_{\varepsilon=0} = D_j(Q, g)$, while $D(Q, g) = \sum_j D_j(Q, g)$. One can also select an ordering for $(\varepsilon_1, \ldots, \varepsilon_J)$ (e.g., starting at ε_J and going backward) and, according to this ordering, iteratively carry out the update step $Q_n^k = Q_{n,g_n}^{k-1}\left(\varepsilon_n^k\right)$, but where ε_n^k is now obtained by minimizing $P_n L_{g_n}(Q_{n,g_n}^{k-1}(\varepsilon))$ only over the next ε-component according to the ordering of the ε-components, setting all other components of ε equal to zero. The next ε-component of the last ε-component in this ordering is defined as the first ε-component in the ordering, so that one keeps circling through all ε-components. At convergence, we have $P_n D_j\left(Q_n^*, g_n\right) = 0$ for all j, and thus, in particular, $P_n D\left(Q_n^*, g_n\right) = 0$.

The asymptotic linearity of $\Psi\left(Q_n^*\right)$ can now be based on the fact that Q_n^* solves the estimating equation corresponding to the estimating function D, and on the statistical properties of nuisance parameter $\left(Q_n^*, g_n\right)$ as an estimator of Q_0, g_0. By selecting a loss function for Q_0, and a fluctuation working model so that the linear span of the derivative of $L_g(Q_g(\varepsilon))$ at $\varepsilon = 0$ includes the components of the efficient influence curve of Ψ at P, one obtains a TMLE that is locally efficient under appropriate conditions.

It remains to choose a loss function for Q_0. One option is to use a standard loss function such as the—log likelihood. However, more targeted choices of loss function are also available. The next subsection discusses one specific alternative choice of loss function that is particularly attractive in that it is able to construct a more targeted initial estimator Q_n^0 of Q_0 in the case that g_0 is easy to estimate relative to Q_0, thereby enhancing efficiency of the resulting TMLE [84]. Other targeted loss functions are presented in [82].

4.5.2 The Squared Efficient Influence Curve Loss for Selecting Initial Estimator or to Select Among Candidate TMLEs

Let $\Psi(Q_0) \in \mathbb{R}^d$ be a d-dimensional parameter of Q_0 which is pathwise differentiable with efficient influence curve $D^*(Q_0, g_0)$ at $P_0 \in \mathcal{M}$, where g_0 is some nuisance parameter. In many cases, the efficient influence curve can be represented as $D^*(\psi_0, Q_0, g_0)$.

Consider the loss function $L_{g_0,\psi_0}(Q) = D^*(Q, g_0)^2$ and assume that D^* satisfies $\arg\min_Q P_0 L_{g_0,\psi_0}(Q) = Q_0$ if the minimum is taken over all $Q \in \mathcal{Q}$ that satisfy $\Psi(Q) = \psi_0$, and for each Q with $\Psi(Q) = \psi_0$, $D^*(Q_0, g_0) = \Pi(D^*(Q, g_0)|T(P_0))$, where $T(P_0)$ is a subspace of $L_0^2(P_0)$ such as the tangent space of the model at P_0. One can view both g_0 as well as ψ_0 as nuisance parameters of this loss function for Q_0. Let $d_{g_0,\psi_0}(Q, Q_0) = P_0\{L_{g_0,\psi_0}(Q) - L_{g_0,\psi_0}(Q_0)\}$ denote the loss-function-based dissimilarity. By the Theorem of Pythagoras in $L_0^2(P_0)$, we have

$$d_{g_0,\psi_0}(Q, Q_0) = P_0 \{D^*(Q, g_0) - D^*(Q_0, g_0)\}^2 .$$

Thus, the loss-based dissimilarity is the $L^2(P_0)$-norm of the efficient influence curve at Q minus the true efficient influence curve at Q_0. This is therefore an excellent loss function for Q_0 since it targets what is needed to efficiently estimate ψ_0. Therefore, we wish to employ a super learner based on this loss function, which can then be incorporated as initial estimator in the TMLE, or we can use its cross-validation selector to select among different candidate targeted maximum likelihood estimators indexed by different initial estimators of Q_0. For that purpose, we have to verify the conditions of Theorem 1.

Assume that the loss function is bounded: $M_1 \equiv \sup_Q \mid L_{g_0,\psi_0}(Q) - L_{g_0,\psi_0}(Q_0) \mid < \infty$. This holds if $D^*(Q, g_0)$ is a uniformly bounded function in O uniformly in Q. In addition, we need that $P_0 \left\{L_{g_0,\psi_0}(Q) - L_{g_0,\psi_0}(Q_0)\right\}^2 \leq M_2 P_0\{L_{g_0,\psi_0}(Q) - L_{g_0,\psi_0}(Q_0)\}$. By the Theorem of Pythagoras, we have $P_0 L_{g_0,\psi_0}(Q) - P_0 L_{g_0,\psi_0}(Q_0) = P_0\{D^*(Q, g_0) - D^*(Q_0, g_0)\}^2$. Thus, to prove the second property of the loss function L_{g_0}, it remains to show that

$P_0\{D^{*2}(Q,g_0) - D^{*2}(Q_0,g_0)\}^2 \leq M_2 P_0\{D^*(Q,g_0) - D^*(Q_0,g_0)\}^2$ for some $M_2 < \infty$. The latter trivially holds for bounded D^*:

$$\begin{aligned} &P_0\{D^{*2}(Q,g_0) - D^{*2}(Q_0,g_0)\}^2 \\ &\quad = P_0\{D^*(Q,g_0) - D^*(Q_0,g_0)\}^2\{D^*(Q,g_0) + D^*(Q_0,g_0)\}^2 \\ &\quad \leq \sup_o \mid \{D^*(Q,g_0) + D^*(Q_0,g_0)\}^2 \mid P_0\{D^*(Q,g_0) - D^*(Q_0,g_0)\}^2, \end{aligned}$$

which completes the proof of second property.

This allows us to apply the oracle inequality for the cross-validation selector as presented in [81] and Theorem 1 above with g_0 treated as known: if the cross-validation selector α_n is defined as a minimizer over a grid with $K(n)$ α-values, then for any $\delta > 0$,

$$\begin{aligned} E d_{g_0,\psi_0}\left(\hat{Q}_{\alpha_n}\left(P^0_{n,B_n}\right), Q_0\right) &\leq (1+2\delta) E \min_{\alpha} E_{B_n} d_{g_0,\psi_0}\left(\hat{Q}_{\alpha}\left(P^0_{n,B_n}\right), Q_0\right) \\ &\quad + C(M_1, M_2, \delta)\frac{\log K(n)}{np}, \end{aligned}$$

where $C(M_1, M_2, \delta)$ is a specified constant. The $\tilde{\alpha}_n$ that attains the minimum on the right-hand side is referred to as the oracle selector that selects the α that minimizes the dissimilarity with Q_0 for the given sample P_n. By choosing a grid with width $1/n$ we obtain a grid that is more than fine enough so that no precision is lost. In that case, the $\log K(n)$ is bounded by a constant times $\log n$. By Theorem 1 we also have a finite sample oracle inequality for the case that g_0, ψ_0 in the loss function L_{g_0,ψ_0} is estimated with (g_n, ψ_n). From this finite sample inequality it follows that, if g_n, ψ_n converges faster to g_0, ψ_0 than $Q^*_{\alpha_n,n}$ converges to Q_0, then the finite sample oracle inequality is asymptotically equivalent with the above one (i.e., the estimation of g_n has an asymptotically negligible effect). For example, if g_0 is known, then one can construct root-n estimators ψ_n so that this would hold.

4.6 Targeted Minimum Loss Based Estimation in Variable Importance Analysis

Suppose one observes n i.i.d. copies of $O = (V, Y)$, where Y is a binary outcome, and V is a high-dimensional covariate vector. One is often interested in assessing the effect of one univariate covariate component of V. Let A be such a variable, and, for the sake of illustration, let us assume A is binary. Let W be the set of adjustment variables contained in V one wishes to adjust for. So (A, W) is a function of V. A particular measure of variable importance of the variable A can now be defined as follows:

$$\psi_0 = E_0(E_0(Y \mid A = 1, W) - E_0(Y \mid A = 0, W)).$$

Other measures can be considered as well. In a variable importance analysis, one would estimate such a target parameter across a large list of variables (i.e., A), with corresponding adjustment sets (i.e., W). One can then carry out multiple testing methods to test all the resulting null hypotheses while controlling a specified type-I error rate. We refer to [4] for a particular application of such a TMLE-based variable importance analysis involving assessment of the effect of mutations in the HIV virus on resistance to a drug.

Let the statistical model $\mathcal{M}$ be nonparametric. We note that $\psi_0 = \Psi(Q_0)$ where $Q_0 = (Q_{W,0}, \bar{Q}_0)$, $Q_{W,0}$ denotes the probability distribution of W under P_0, and $\bar{Q}_0(A, W) = E_0(Y \mid A, W)$. Let g_0 be the conditional distribution of A, given W. The efficient influence curve $D^*(P)$ is given by

$$D^*(P)(O) = H(g)(A, W)\left(Y - \bar{Q}(A, W)\right) + \bar{Q}(1, W) - \bar{Q}(0, W) - \Psi(Q),$$

where $\bar{Q}(a, W) = E_P(Y \mid W, A = a)$, and $H(g)(A, W) = (2A - 1)/g(A \mid W)$.

4.6.1 The TMLE

If Y is binary, then we use the fluctuation working model $\text{logit}\bar{Q}(\varepsilon) = \text{logit}\bar{Q} + \varepsilon H(g)$ and use as loss function for $\bar{Q}$ the log-likelihood for a binary distribution given by $L(\bar{Q}) = -\log\left(\bar{Q}^Y(1 - \bar{Q})^{(1-Y)}\right)$. We can also propose a fluctuation working model for Q_W with score $D_W(Q) = \bar{Q}(1, W) - \bar{Q}(0, W) - \Psi(Q)$ being the appropriate component of the efficient influence curve $D^*(Q, g)$, but since we will use as initial estimator of $Q_{W,0}$ the empirical distribution $Q_{W,n}$, the maximum likelihood estimator of the fluctuation parameter will be zero, so that no updates of $Q_{W,n}$ will occur. Let $\bar{Q}_n^0$ be an initial estimator of $\bar{Q}_0$, and g_n an estimator of g_0. The initial estimator of $\bar{Q}_0$ could be a super learner based on the squared efficient influence curve loss function (as described in Sect. 4.5.2), for example, so that it is fully targeted to fit the efficient influence curve, and thereby maximizes the efficiency of the resulting TMLE if g_n is consistent. If one is not comfortable relying on consistency of g_n, then it is better to fit $\bar{Q}_0$ with a loss function $L_{g_n}(Q)$ that is always valid, even if g_n is inconsistent. For example, one could use the log-likelihood loss, possibly penalized with cross-validated empirical risk of the square of the efficient influence curve divided by n, as in [82]. One now computes $\varepsilon_n = \arg\min_\varepsilon P_n L\left(\bar{Q}_n^0(\varepsilon)\right)$, and one defines the TMLE update as $\bar{Q}_n^1 = \bar{Q}_n^0(\varepsilon_n)$. Further iteration does not result in further updates, so that the TMLE Q_n^* is defined as $Q_n^* = (Q_{W,n}, \bar{Q}_n^1)$. The TMLE of ψ_0 is thus given by $\Psi(Q_n^*) = 1/n \sum_{i=1}^n \{\bar{Q}_n^*(1, W_i) - \bar{Q}_n^*(0, W_i)\}$. If Y is continuous with values in $(0, 1)$, one can use the same loss function $L(\bar{Q})$ and fluctuation function for the conditional mean $\bar{Q}$. If Y is bounded in (a, b), then one can transform the outcome into $Y^* = (Y - a)/(b - a) \in (0, 1)$ and apply this same TMLE. For continuous Y, one might alternatively consider the squared error loss function and the linear fluctuation function; however, such a fluctuation function is not guaranteed to respect known bounds on Y and is thus not generally recommended.

4.6.2 *Influence Curve, Confidence Interval, and p-Value*

Under regularity conditions, $\psi_n^* = \Psi(Q_n^*)$ is asymptotically linear at P_0. In particular, if g_n is a consistent estimator of g_0, then one can use as conservative influence curve $D^*(Q, g_0)$, where Q denotes the possibly misspecified limit of Q_n^*. As a consequence, the variance of $\sqrt{n}(\psi_n^* - \psi_0)$ can then be conservatively estimated as

$$\sigma_n^2 = \frac{1}{n}\sum_{i=1}^{n}\{D^*(Q_n^*, g_n)(O_i)\}^2.$$

One can use as test-statistic for $H_0 : \psi_0 = 0$, $t_n = \sqrt{n}\psi_n^*/\sigma_n \sim_{H_0} N(0,1)$, and asymptotic 0.95-confidence interval $\psi_n^* \pm 1.96\sigma_n/\sqrt{n}$.

If one estimates a whole collection of variable importance measures $(\psi_0(j) : j)$ with the corresponding TMLEs, then the vector-TMLE $(\psi_n^*(j) : j)$ is asymptotically linear with a vector influence curve, so that statistical inference in terms of multiple testing and simultaneous confidence intervals can now proceed based on the multivariate normal limit distribution.

4.7 TMLE of Risk of Candidate Estimator of Nonpathwise Differentiable Parameter

4.7.1 *Motivation, Example, and Overview*

Let $\Psi : \mathcal{M} \to \mathcal{F}$ be a nonpathwise differentiable parameter. Let $R(\psi, P_0)$ be a risk of a candidate $\psi \in \mathcal{F}$ so that $\psi_0 = \arg\min_{\psi\in\mathcal{F}} R(\psi, P_0)$. We assume that for each given ψ, the parameter $P \to R(\psi, P)$ is a pathwise differentiable parameter on the model $\mathcal{M}$.

As a consequence, we can use TMLE to estimate this risk at any ψ. For loss functions with a nuisance parameter, this will generally be a more targeted and robust estimator than the empirical risk using a plug-in estimator of the nuisance parameter: that is, the latter estimator corresponds with an estimating equation-based estimator, and TMLE has fundamental advantages relative to such estimators. In addition, the TMLE is defined for any risk function, not only for risk functions defined by loss-functions indexed by a nuisance parameter. In particular, the TMLE of the risk is not dependent on the particular parameterization chosen for this risk.

We previously used cross-validated empirical means to estimate the risk. We now use a CV-TMLE to estimate the risk, as presented in [96] and the corresponding chapter in [84]. Subsequently, we use this targeted estimator of risk to propose a cross-validation selector defined by the minimizer of the CV-TMLE of the risk for a candidate estimator $\hat{\Psi} : \mathcal{M}_{NP} \to \mathcal{F}$ among all candidate estimators.

We will demonstrate this with one of our previous examples: estimation of the causal dose–response curve (Sect. 4.4.5). Let $O = (W, A, Y)$, where A is a continuous dose of a drug, W baseline covariates, and Y is an outcome of interest. Let $\psi_0(a) = E_0 Y(a) \equiv E_0 E_0(Y \mid A = a, W)$ be the marginal dose–response curve, controlling for confounding by baseline covariates W. Let $\mathcal{M}$ be the nonparametric model and let $\mathcal{F}$ be all functions of a. The parameter $\Psi : \mathcal{M} \to \mathcal{F}$ is not pathwise differentiable.

Let us now develop a risk function for ψ_0. Let μ be a measure that dominates the conditional distribution of A, given W, on the set of possible dose values. In other words, the true conditional distribution G_0 of A, given W, has a density g_0 w.r.t. μ. Typically, μ is the Lebesgue measure on an interval $[a, b]$ chosen so that $P_0(A \in [a, b]) = 1$. We have

$$\psi_0 = \arg\min_{\psi} E_0 \int_a \left\{ \bar{Q}_0(a, W) - \psi(a) \right\}^2 \mathrm{d}\mu(a),$$

where $\bar{Q}_0(a, W) = E_0(Y \mid A = a, W)$. This risk function can be decomposed as

$$E_0 \int_a \left\{ \bar{Q}_0(a, W) - \psi(a) \right\}^2 \mathrm{d}\mu(a) = \int_a E_0 \bar{Q}_0^2(W, a) \mathrm{d}\mu(a) + \int_a \psi^2(a) \mathrm{d}\mu(a)$$
$$-2 \int_a E_0 Y(a) \psi(a) \mathrm{d}\mu(a).$$

The first term does not depend on ψ and can thus be removed without affecting the validity of the risk function. This yields the following relevant risk function for ψ_0:

$$R(\psi, P_0) = \int_a \psi^2(a) \mathrm{d}\mu(a) - 2 \int_a \left\{ E_0 \bar{Q}_0(a, W) \right\} \psi(a) \mathrm{d}\mu(a).$$

In the sequel of this section we will present the TMLE of $R(\psi, P_0)$, the CV-TMLE of $R(\psi, P_0)$, the CV-TMLE of the conditional risk of a candidate estimator $\hat{\Psi}$, the corresponding cross-validation selector. We defer presentation of an oracle inequality for this cross-validation selector for future work.

4.7.2 TMLE of Risk

We start out with presenting the efficient influence curve for $R(\psi, P_0)$, which will provide the ingredient for construction of the TMLE of $R(\psi, P_0)$.

4.7.2.1 The Efficient Influence Curve of the Risk at a Candidate Value

The first term of $R(\psi, P_0)$ is a known value. We have $R(\psi, P_0) = \int_a \psi^2(a)d\mu(a) - 2\theta_0(\psi)$, where $\Theta(\psi, P) \equiv \int_a \{E_P Y(a)\}\psi(a)d\mu(a)$, and $\theta_0(\psi) = \Theta(\psi, P_0)$. If A is discrete so that $d\mu(a) = h(a)$ is a discrete measure with finite support, then the efficient influence curve of this parameter $P \rightarrow \Theta(\psi, P)$ at P_0 is given by

$$D^*(\psi, P_0) = \sum_a D_a^*(P_0)\psi(a)h(a),$$

where

$$D_a^*(P_0) = \frac{I(A = a)}{g_0(A \mid W)}(Y - \bar{Q}_0(A, W)) + \bar{Q}_0(a, W) - E_0 Y(a).$$

This yields the following formula for the efficient influence curve of $\Theta(\psi, P)$ if A is discrete:

$$D^*(\psi, P_0) = \frac{h(A)\psi(A)}{g_0(A \mid W)}\left(Y - \bar{Q}_0(A, W)\right) + \int_a \bar{Q}_0(a, W)\psi(a)d\mu(a) - \theta_0(\psi).$$

This is also the efficient influence curve of $\Theta(\psi, P)$ if A is continuous and $d\mu(a) = h(a)da$ for some function h. It can be straightforwardly verified that this is indeed the efficient influence curve: using the CAR model as in [83] it follows that (1) it is a gradient in the model with the censoring mechanism g_0 known, (2) it is orthogonal to the tangent space of g_0 for the nonparametric model that only assumes CAR, and thereby is an element of the tangent space in the model in which g_0 is known. For this argument, one needs to recall that a gradient that is also an element of the tangent space is a canonical gradient, and the canonical gradient in the model in which g_0 is known equals the canonical gradient in the model that only assumes CAR.

Let $D^*(P_0, R_0(\psi))$ be the efficient influence curve of $R(\psi, P)$ at P_0. We will also use the notation $D^*(Q_0, g_0, R_0(\psi))$ with $Q_0 = \left(Q_{W,0}, \bar{Q}_0\right)$. Above, we showed that

$$\begin{aligned} D^*(P_0, R_0(\psi)) &= -2\frac{h(A)\psi(A)}{g_0(A \mid W)}\left(Y - \bar{Q}_0(A, W)\right) - 2\int_a \bar{Q}_0(a, W)\psi(a)d\mu(a) \\ &+2\theta_0(\psi) = -2\frac{h(A)\psi(A)}{g_0(A \mid W)}\left(Y - \bar{Q}_0(A, W)\right) \\ &-2\int_a \bar{Q}_0(a, W)\psi(a)d\mu(a) - R_0(\psi) + \int \psi^2(a)d\mu(a). \end{aligned}$$

The efficient influence curve is DR: for g satisfying the positivity assumption $\sup_a h(a)/g(a \mid W) < \infty$,

$$P_0 D^*(Q, g, R_0(\psi)) = 0 \text{ if } \bar{Q} = \bar{Q}_0 \text{ or } g = g_0.$$

4.7.2.2 TMLE

We now proceed with developing a TMLE of $R_0(\psi) = R(\psi, P_0)$ for a given ψ. For this we only need to develop the TMLE of $\theta_0(\psi) \equiv \int_a \{E_0 Y(a)\}\psi(a)\mathrm{d}\mu(a)$, since $R(\psi, P_0) = -2\theta_0(\psi) + \psi^2$. Suppose $Y \in [0,1]$. We consider the loss function $-L(\bar{Q}) = Y \log \bar{Q}(A,W) + (1-Y)\log\left(1-\bar{Q}(A,W)\right)$ for $\bar{Q}_0$, and the logistic fluctuation model $\mathrm{Logit}\bar{Q}_g(\varepsilon) = \mathrm{Logit}\bar{Q} + \varepsilon H^*(\psi, g)$, where $H^*(\psi,g)(A,W) = \{h(A)\psi(A)\}/g(A \mid W)$. The generalized score $\frac{\mathrm{d}}{\mathrm{d}\varepsilon}L\left(\bar{Q}_g(\varepsilon)\right)$ at $\varepsilon = 0$ equals $H^*(\psi,g)(Y-\bar{Q})$, the first component of the efficient influence curve of the parameter $R(\psi, P)$. The marginal distribution of W is estimated with the empirical distribution $Q_{W,n}$ of $W_1, \ldots, W_n$. One obtains an initial estimator $\bar{Q}_n^0$ of $\bar{Q}_0$, g_n of g_0, and computes the first step TMLE $\bar{Q}_n^1 = \bar{Q}_{n,g_n}^0(\varepsilon_n)$, where $\varepsilon_n = \arg\min_\varepsilon P_n L\left(\bar{Q}_{n,g_n}^0(\varepsilon)\right)$.

The TMLE of $\theta(\psi, P_0)$ is now given by the plug-in estimator

$$\theta_n^*(\psi) = \int_a \left\{\frac{1}{n}\sum_i \bar{Q}_n^1(a, W_i)\right\} \psi(a)\mathrm{d}\mu(a).$$

This results in the following TMLE of the risk $R(\psi, P_0)$:

$$R_n^*(\psi) = \int_a \psi^2(a)\mathrm{d}\mu(a) - 2\theta_n^*(\psi).$$

Under regularity conditions, if g_0 is known and $g_n = g_0$, this TMLE of $R(\psi, P_0)$ is asymptotically linear with influence curve $D^*(Q, g_0, R_0(\psi))$, where $Q = \left(Q_{W,0}, \bar{Q}\right)$ and $\bar{Q}$ is the possibly misspecified limit of $\bar{Q}_n^1$. In particular, if $\bar{Q} = \bar{Q}_0$, then the TMLE is asymptotically efficient. If g_n is a consistent maximum likelihood-based estimator, then this influence curve $D^*(Q, g_0, R_0(\psi))$ will be conservative.Thus, under these conditions, one can estimate the variance of the TMLE $R_n^*(\psi)$ as follows:

$$\sigma_n^2 = \frac{1}{n^2}\sum_{i=1}^n \left\{D^*\left(Q_n^*, g_n, R_n^*(\psi)\right)(O_i)\right\}^2.$$

This also allows one to construct confidence intervals for the true risk $R(\psi, P_0)$. If one uses adaptive estimators $\hat{\bar{Q}}$ of $\bar{Q}_0$, then it is often appropriate to replace the estimate of the variance by a cross-validated estimator:

$$\sigma_{n,\mathrm{CV}}^2 = \frac{1}{n} E_{B_n} P_{n,B_n}^1 \left\{D^*\left(\hat{Q}^*\left(P_{n,B_n}^0\right), \hat{g}\left(P_{n,B_n}^0\right)\right)\right\}^2.$$

The TMLE $R_n^*(\psi)$ is also DR in the sense that it is consistent if either g_n is consistent for g_0 or the TMLE $\bar{Q}_n^*$ is consistent for $\bar{Q}_0$.

4.7.3 Why Estimate Risk with TMLE Instead of Using the Mean of DR-IPTW Loss?

In Sect. 4.4.5 we presented the empirical estimate of risk $E_0 \int_a \{Y(a) - \psi(a)\}^2 d\mu(a)$ given by $1/n \sum_i L_{g_n,Q_n}(\psi)(O_i)$ based on an augmented-IPTW loss function $L_{g,Q}(\psi)$ obtained by applying the augmented IPTW mapping to the full data loss function $\int_a h(a)(Y(a) - \psi(a))^2 d\mu(a)$. Similarly, we presented the cross-validated risk as a cross-validated mean of the augmented-IPTW loss function. This estimator of risk is also DR and asymptotically efficient. However, just like estimating equation-based estimators in general, this estimator does not respect the global constraints of the model. For example, this empirical estimate of risk could even be negative, even though the risk is known to be positive, and similarly this empirical estimate of risk could be larger than the largest possible value of the risk as implied by the bounds on Y. In contrast, the TMLE of the risk fully respects the bounds enforced by the model because it is a substitution estimator. This enhanced robustness of the TMLE will improve the robustness of the risk-estimators and thereby we expect that this will also improve the practical performance of the resulting cross-validation selector. This argument applies to any nonlinear risk function.

4.7.4 Cross-Validated-TMLE of Risk at Candidate Value

Even though an empirical mean of a loss function (possibly indexed by an estimated nuisance parameter) at a (not too adaptive) candidate estimator is a reasonable estimator of the true conditional risk, the finite sample bias increases in the adaptivity of the candidate estimator. This is the sole motivation for using cross-validated empirical means when estimating the risk of candidate estimators in the definition of the cross-validation selector. The TMLE corresponds with the empirical mean of the loss function type estimator and will thereby also result in an estimate of risk that is biased low for adaptive candidate estimators. [96] present a CV-TMLE, which represents the analogue of the cross-validated empirical mean of a loss function. As a result, the finite sample bias of the CV-TMLE is not sensitive to the adaptivity of the candidate estimator. Indeed, the formal theorem in [96] proves that the CV-TMLE will be asymptotically linear under conditions that allow the candidate estimator and the initial estimator in the TMLE to be arbitrarily adaptive. This makes the CV-TMLE appropriate as an estimator of the risk in the cross-validation selector.

The CV-TMLE of $\theta_0(\psi)$, and thereby $R_0(\psi) = R(\psi, P_0)$ is defined as follows. We refer to [96] and corresponding chapter in [84] for the introduction and detailed understanding of this procedure.

Suppose $Y \in [0, 1]$. We consider the loss function $-L\left(\bar{Q}\right) = Y \log \bar{Q}(A, W) + (1-Y) \log\left(1 - \bar{Q}(A, W)\right)$ for $\bar{Q}_0$, and the logistic fluctuation model $\text{Logit}\bar{Q}_g(\varepsilon) = \text{Logit}\bar{Q} + \varepsilon H^*(\psi, g)$, where $H^*(\psi, g)(A, W) = \{h(A)\psi(A)\}/g(A \mid W)$. Let $\hat{\bar{Q}}$, $\hat{g}$ be initial estimators of $\bar{Q}_0$ and g_0. Let $\hat{Q}_W(P_n)$ be the empirical distribution of $W_1, \ldots, W_n$.

Let B_n be a cross-validation scheme. One defines

$$\varepsilon_n^{\text{CV}} = \arg\min_{\varepsilon} E_{B_n} P_{n,B_n}^1 L\left(\hat{\bar{Q}}_{\hat{g}(P_{n,B_n}^0)}\left(P_{n,B_n}^0\right)(\varepsilon)\right).$$

For each B_n, one now defines the update $\bar{Q}_{n,B_n}^* \equiv \hat{\bar{Q}}\left(P_{n,B_n}^0\right)\left(\varepsilon_n^{\text{CV}}\right)$. The CV-TMLE of $\theta_0(\psi)$ is now given by the plug-in estimator

$$\theta_{n,\text{CV}}^*(\psi) = \int_a \left\{E_{B_n} P_{n,B_n}^1 \bar{Q}_{n,B_n}^*(a, \cdot)\right\} \psi(a) \mathrm{d}\mu(a).$$

This results in the following CV-TMLE of the risk $R(\psi, P_0)$:

$$R_{n,\text{CV}}^*(\psi) = \int_a \psi^2(a) \mathrm{d}\mu(a) - 2\theta_{n,\text{CV}}^*(\psi).$$

Under weaker regularity conditions than needed for the regular TMLE of $R_0(\psi)$, this CV-TMLE is asymptotically linear with the same influence curve as the TMLE. Specifically, the CV-TMLE avoids any empirical process conditions on the initial estimator so that arbitrary adaptive estimators of $\bar{Q}_0$ are allowed, as shown in [84, 96].

4.7.5 Cross-Validated-TMLE of Risk of Candidate Estimator

Consider a candidate estimator $\hat{\Psi} : \mathcal{M}_{\text{NP}} \to \mathcal{F}$. The optimal goal would be to estimate the risk $R\left(\hat{\Psi}(P_n), P_0\right)$, treating $\hat{\Psi}(P_n)$ as a given value. We could estimate this with the CV-TMLE above of $R_0(\psi)$ by setting $\psi = \hat{\Psi}(P_n)$. A concern of such an estimator is that the cross-validation selector $\varepsilon_n^{\text{CV}}$ in the CV-TMLE does not satisfy an oracle inequality, due to $\hat{\Psi}(P_n)$ being random. In addition, related to this, the analysis of this estimator of risk will now involve empirical process conditions on an estimated efficient influence curve D^*, contrary to the CV-TMLE presented above for a fixed ψ. Therefore, we apply the CV-TMLE above, but where ψ is replaced by $\hat{\Psi}\left(P_{n,B_n}^0\right)$ in the estimator $\varepsilon_n^{\text{CV}}$, so that the cross-validation selector $\varepsilon_{n,\text{CV}}$ still satisfies the oracle inequality, and the analysis of the CV-TMLE can follow the proof as given in [96], which avoids empirical process conditions.

In this manner, it can be shown that our CV-TMLE of risk is still an asymptotically linear estimator of the conditional risk $R_n\left(\hat{\Psi}, P_0\right) \equiv E_{B_n} R\left(\hat{\Psi}\left(P^0_{n,B_n}\right), P_0\right)$ under minimal conditions.

The proposed CV-TMLE of the conditional risk $R_n\left(\hat{\Psi}, P_0\right)$ of $\hat{\Psi}$ is defined as follows. Suppose $Y \in [0,1]$. As before, we consider the loss function $-L(\bar{Q}) = Y \log \bar{Q}(A,W) + (1-Y)\log\left(1-\bar{Q}(A,W)\right)$ for $\bar{Q}_0$, the logistic fluctuation model $\text{Logit}\bar{Q}_g(\varepsilon) = \text{Logit}\bar{Q} + \varepsilon H^*(\psi, g)$, where $H^*(\psi, g)(A,W) = \{h(A)\psi(A)\}/g(A|W)$. Let $\hat{\bar{Q}}$, $\hat{g}$ be initial estimators of $\bar{Q}_0$ and g_0. The estimator $\hat{Q}_W$ is the empirical distribution. For each B_n, let

$$\hat{\bar{Q}}\left(P^0_{n,B_n}\right)(\varepsilon) = \hat{\bar{Q}}\left(P^0_{n,B_n}\right) + \varepsilon H\left(\hat{\Psi}\left(P^0_{n,B_n}\right), \hat{g}\left(P^0_{n,B_n}\right)\right).$$

Thus, the ψ in the clever covariate of the TMLE is now replaced by the candidate estimator based on the training sample. With this modification, the CV-TMLE is now defined as in previous subsection. Thus

$$\varepsilon^{\text{CV}}_n = \arg\min_{\varepsilon} E_{B_n} P^1_{n,B_n} L\left(\hat{\bar{Q}}\left(P^0_{n,B_n}\right)(\varepsilon)\right).$$

For each B_n, one defines the update $\bar{Q}^*_{n,B_n} \equiv \hat{\bar{Q}}\left(P^0_{n,B_n}\right)\left(\varepsilon^{\text{CV}}_n\right)$. This results in the plug-in estimator

$$\theta^*_{n,\text{CV}}\left(\hat{\Psi}\right) = E_{B_n} \int_a \{P^1_{n,B_n} \bar{Q}^*_{n,B_n}(a,\cdot)\} \hat{\Psi}\left(P^0_{n,B_n}\right)(a) d\mu(a),$$

the following CV-TMLE of the conditional risk $R_n\left(\hat{\Psi}, P_0\right)$:

$$R^*_{n,\text{CV}}\left(\hat{\Psi}\right) = E_{B_n} \int_a \hat{\Psi}\left(P^0_{n,B_n}\right)^2(a) d\mu(a) - 2\theta^*_{n,\text{CV}}\left(\hat{\Psi}\right).$$

4.7.6 Cross-Validation Selector Based on CV-TMLE of Risk of Candidate Estimator

Let $R^*_n\left(\hat{\Psi}\right)$ be the CV-TMLE presented above of $R(\hat{\Psi}, P_0, P_n) \equiv E_{B_n} R(\hat{\Psi}(P^0_{n,B_n}), P_0)$. Given a collection of candidate estimators $\hat{\Psi}_k$, $k = 1, \ldots, K$, the cross-validation selector is given by $k_n = \arg\min_k R^*_n\left(\hat{\Psi}_k\right)$. Let $\tilde{k}_n = \arg\min_k E_{B_n} R(\hat{\Psi}_k(P^0_{n,B_n}), P_0)$ be the comparable oracle selector. In future work, we will present an oracle inequality comparing k_n with $\tilde{k}_n$, analogue to the previously presented oracle inequalities for loss functions indexed by a nuisance parameter.

4.8 Summary

In this chapter, we presented three very general templates for estimator selection in semiparametric statistical models. Specifically, we first introduced the method of super learning for nonpathwise differentiable parameters, in which cross-validation was used to select among weighted combinations of some user-supplied family of initial algorithms for estimating the target parameter ψ_0. Specifically, the final estimator (super learner) was defined as the weighted combination with the lowest estimated risk, where risk was estimated as the cross-validated empirical mean of some loss function. Finite sample oracle inequalities were presented that demonstrated the asymptotic optimality of the resulting super learner for nonpathwise differentiable parameters and linear risk functions, including the case of risk functions indexed by a known nuisance parameter. We also discussed and provided examples of cases in which the nuisance parameter was not known, but could be estimated at a rate faster than the target parameter, resulting in approximately linear risk functions and justifying application of the super learner template.

We next extended this method to build targeted estimators of pathwise differentiable parameters using TMLE. Specifically, super learning was used to provide an initial estimator of a nonpathwise differentiable nuisance parameter Q_0, some function of which provided a substitution estimator of the target parameter ($\psi_0 = \Psi(Q_0)$ for some Ψ). In recognition of the fact that the initial estimator of Q_0 was optimized for a higher dimensional parameter than the target $\Psi(Q_0)$, the initial fit was then updated with a targeted bias reduction step designed to result, whenever possible, in an asymptotically efficient and maximally robust estimator of the target parameter. We further introduced one possible choice of targeted loss function that was designed to provide an initial estimator of Q_0 with properties expected to improve performance of the final estimator of the target parameter .

Finally, we returned to nonpathwise differentiable parameters and presented novel results that generalize super learning to the case of nonlinear risk functions. Specifically, we proposed a (cross-validated) TMLE of the risk itself. The resulting super learner was defined as the weighted combination of algorithms for estimating the target parameter that minimizes this improved risk estimate. This novel approach has considerable theoretical appeal, full exploration of which, together with demonstrations using real and simulated data, we leave as a topic for future work.

4.9 Bibliographical Remarks

Unified loss-function-based cross-validation was presented and developed in [81], including the finite sample oracle inequality, the asymptotic equivalence of the cross-validation selector, and the oracle selector. This general theory was applied and further worked out in [26,43,72,86,87,91]. A finite sample result for the single-split cross-validation selector was originally presented for the squared error loss function was in [37]. This result was generalized in [81] and [26] to handle general cross-validation schemes and a general class of loss functions.

Loss-based super learning was presented in [88]. Super learner represents a generalization of the stacking algorithm in the classification context [45,95] and adapted to the regression context by [10]. The name super learner was introduced because of its theoretical optimality property implied by the finite sample oracle inequality, as presented in [81]. The relationship between stacking and the model-mix algorithm of [76] and the predictive sample-reuse method of [32] is discussed in [45]. For some recent literature on ensemble learners, we refer to [16–18, 23, 24, 42, 78]. For some applications of super learning, including prediction with right-censored survival data, we refer to chapters 15 and 16 in [53, 84].

Different types of cross-validation have been proposed in the literature, such as V-fold cross-validation, bootstrap cross-validation, Monte Carlo cross-validation, and leave-one-out cross-validation [2, 8, 9, 12, 14, 15, 27, 37, 40, 57, 76, 77]. We refer to [51] for simulations demonstrating the good practical performance of likelihood-based cross-validation relative to other selection methods in the context of mixture models, such as Akaike's information criterion [1,7], Bayesian Information criterion [71], minimum description length [58], and informational complexity [6].

Reference [40] provides a comprehensive book on machine-learning algorithms and related topics, including stepwise selection procedures, ridge regression, LASSO, principal component regression, least angle regression, nearest neighbor methods, random forests, support vector machines, neural networks, classification methods, kernel smoothing methods, and ensemble learning. All these algorithms could be included in the super learner to build a powerful library and thereby super learner.

IPTW estimation is presented and discussed in detail in [41, 59]. Augmented IPTW is originally developed in [61]. Further development on estimating equation methodology and double robustnness is presented in [60, 62, 63, 83]. For a detailed bibliography on locally efficient estimating equation methodology, we refer to Chap. 1 in [83].

For the original paper on TMLE, we refer to [85]. Subsequent papers on TMLE in observational and experimental studies include [3, 4, 36, 47–49, 53, 64–68, 73, 75, 79, 82, 93]. For a general comprehensive book on this topic, which includes most of these applications on TMLE and many more, we refer to [84].

An original example of a particular type of TMLE (based on a DR parametric regression model) for estimation of a causal effect of a point-treatment intervention was presented in [70] and we refer to [67] for a detailed review of this earlier literature and its relation to TMLE. References [80] and [74] (see also [84]) present a closed form TMLE, based on the log-likelihood loss function, for estimation of a causal effect of a multiple time point intervention on an outcome of interest (including survival outcomes that are subject to right-censoring) based on general longitudinal data structures.

We refer to [5] for a text on efficiency theory for semiparametric models. In addition, [89] provides a thorough presentation on asymptotic statistics, and for empirical process theory, we refer to [90].

References

1. H. Akaike. Information theory and an extension of the maximum likelihood principle. In B.N. Petrov and F. Csaki, editors, *Second International Symposium on Information Theory*, Budapest, 1973. Academiai Kiado.
2. C. Ambroise and G.J. McLachlan. Selection bias in gene extraction on the basis of microarray gene-expression data. *Proc Natl Acad Sci*, 99(10):6562–6566, 2002.
3. O. Bembom and M.J. van der Laan. A practical illustration of the importance of realistic individualized treatment rules in causal inference. *Electron J Stat*, 1:574–596, 2007.
4. O. Bembom, M.L. Petersen, S.-Y. Rhee, W.J. Fessel, S.E. Sinisi, R.W. Shafer, and M.J. van der Laan. Biomarker discovery using targeted maximum likelihood estimation: application to the treatment of antiretroviral resistant HIV infection. *Stat Med*, 28:152–172, 2009.
5. P.J. Bickel, C.A.J. Klaassen, Y. Ritov, and J. Wellner. *Efficient and adaptive estimation for semiparametric models*. Springer, Berlin/Heidelberg/New York, 1997.
6. H. Bozdogan. Choosing the number of component clusters in the mixture model using a new informational complexity criterion of the inverse fisher information matrix. In O. Opitz, B. Lausen, and R. Klar, editors, *Information and classification*. Springer, Berlin/Heidelberg/New York, 1993.
7. H. Bozdogan. Akaike's information criterion and recent developments in information complexity. *J Math Psychol*, 44:62–91, 2000.
8. L. Breiman. Heuristics of instability and stabilization in model selection. *Ann Stat*, 24(6): 2350–2383, 1996a.
9. L. Breiman. Out-of-bag estimation. Technical Report, Department of Statistics, University of California, Berkeley, 1996b.
10. L. Breiman. Stacked regressions. *Mach Learn*, 24:49–64, 1996c.
11. L. Breiman. Bagging predictors. *Mach Learn*, 24(2):123–140, 1996d.
12. L. Breiman. Arcing classifiers. *Ann Stat*, 26:801–824, 1998.
13. L. Breiman. Random forests. *Mach Learn*, 45:5–32, 2001.
14. L. Breiman and P. Spector. Submodel selection and evaluation in regression. The X random case. *Int Stat Rev*, 60:291–319, 1992.
15. L. Breiman, J.H. Friedman, R. Olshen, and C.J. Stone. *Classification and regression trees*. Chapman & Hall, Boca Raton, 1984.
16. F. Bunea, A.B. Tsybakov, and M.H. Wegkamp. Aggregation and sparsity via L1 penalized least squares. In G. Lugosi and H.-U. Simon, editors, *COLT*, volume 4005 of *Lecture Notes in Computer Science*, Berlin/Heidelberg/New York, 2006. Springer.
17. F. Bunea, A.B. Tsybakov, and M.H. Wegkamp. Aggregation for gaussian regression. *Ann Stat*, 35(4):1674–1697, 2007a.
18. F. Bunea, A.B. Tsybakov, and M.H. Wegkamp. Sparse density estimation with L1 penalties. In N.H. Bshouty and C. Gentile, editors, *COLT*, volume 4539 of *Lecture Notes in Computer Science*, Berlin/Heidelberg/New York, 2007b. Springer.
19. C.-C. Chang and C.-J. Lin. *LIBSVM: a library for support vector machines*, 2001. URL http://www.csie.ntu.edu.tw/~cjlin/libsvm.
20. H.A. Chipman and R.E. McCulloch. *BayesTree: Bayesian methods for tree-based models*, 2009. URL http://CRAN.R-project.org/package=BayesTree. R package version 0.3-1.
21. H.A. Chipman, E.I. George, and R.E. McCulloch. BART: Bayesian additive regression trees. *Ann Appl Stat*, 4(1):266–298, 2010.
22. W.S. Cleveland, E. Groose, and W.M. Shyu. Local regression models. In J.M. Chambers and T.J. Hastie, editors, *Statistical models in S*. Chapman & Hall, Boca Raton, 1992.
23. A.S. Dalalyan and A.B. Tsybakov. Aggregation by exponential weighting and sharp oracle inequalities. In N.H. Bshouty and C. Gentile, editors, *COLT*, volume 4539 of *Lecture Notes in Computer Science*, Berlin/Heidelberg/New York, 2007. Springer.
24. A.S. Dalalyan and A.B. Tsybakov. Aggregation by exponential weighting, sharp pac-Bayesian bounds and sparsity. *Mach Learn*, 72(1–2):39–61, 2008.

25. E. Dimitriadou, K. Hornik, F. Leisch, D. Meyer, and A. Weingessel. *e1071: misc functions of the Department of Statistics (e1071)*, 2009. URL http://CRAN.R-project.org/package=e1071. R package version 1.5-22.
26. S. Dudoit and M.J. van der Laan. Asymptotics of cross-validated risk estimation in estimator selection and performance assessment. *Stat Methodol*, 2(2):131–154, 2005.
27. B. Efron and R. J. Tibshirani. *An Introduction to the bootstrap*. Chapman & Hall, Boca Raton, 1993.
28. J.H. Friedman. Multivariate adaptive regression splines. *Ann Stat*, 19(1):1–141, 1991.
29. J.H. Friedman. Greedy function approximation: a gradient boosting machine. *Ann Stat*, 29:1189–1232, 2001.
30. J.H. Friedman, T.J. Hastie, and R.J. Tibshirani. Regularization paths for generalized linear models via coordinate descent. *J Stat Softw*, 33(1), 2010a.
31. J.H. Friedman, T.J. Hastie, and R.J. Tibshirani. *glmnet: lasso and elastic-net regularized generalized linear models*, 2010b. URL http://CRAN.R-project.org/package=glmnet. R package version 1.1–5.
32. S. Geisser. The predictive sample reuse method with applications. *J Am Stat Assoc*, 70(350):320–328, 1975.
33. A. Gelman, A. Jakulin, M.G. Pittau, and Y.-S. Su. A weakly informative default prior distribution for logistic and other regression models. *Ann Appl Stat*, 2(3):1360–1383, 2009.
34. A. Gelman, Y.-S. Su, M. Yajima, J. Hill, M.G. Pittau, J. Kerman, and T. Zheng. *arm: data analysis using regression and multilevel/hierarchical models*, 2010. URL http://CRAN.R-project.org/package=arm. R package version 1.3-02.
35. R.D. Gill, M.J. van der Laan, and J.M. Robins. Coarsening at random: characterizations, conjectures and counter-examples. In D.Y. Lin and T.R. Fleming, editors, *Proceedings of the First Seattle Symposium in Biostatistics*, pages 255–94, New York, 1997. Springer Verlag.
36. S. Gruber and M.J. van der Laan. An application of collaborative targeted maximum likelihood estimation in causal inference and genomics. *Int J Biostat*, 6(1), 2010.
37. L. Györfi, M. Kohler, A. Krzyżak, and H. Walk. *A distribution-free theory of nonparametric regression*. Springer, Berlin/Heidelberg/New York, 2002.
38. T.J. Hastie. Generalized additive models. In J.M. Chambers and T.J. Hastie, editors, *Statistical models in S*. Chapman & Hall, Boca Raton, 1992.
39. T.J. Hastie and R.J. Tibshirani. *Generalized additive models*. Chapman & Hall, Boca Raton, 1990.
40. T.J. Hastie, R.J. Tibshirani, and J.H. Friedman. *The elements of statistical learning: data mining, inference, and prediction*. Springer, Berlin/Heidelberg/New York, 2001.
41. M.A. Hernan, B. Brumback, and J.M. Robins. Marginal structural models to estimate the causal effect of zidovudine on the survival of HIV-positive men. *Epidemiol*, 11(5):561–570, 2000.
42. A. Juditsky, A.V. Nazin, A.B. Tsybakov, and N. Vayatis. Generalization error bounds for aggregation by mirror descent with averaging. In *NIPS*, 2005.
43. S. Keleş, M.J. van der Laan, and S. Dudoit. Asymptotically optimal model selection method for regression on censored outcomes. Technical Report 124, Division of Biostatistics, University of California, Berkeley, 2002.
44. C. Kooperberg. *polspline: Polynomial spline routines*, 2009. URL http://CRAN.R-project.org/package=polspline. R package version 1.1.4.
45. M. LeBlanc and R.J. Tibshirani. Combining estimates in regression and classification. *J Am Stat Assoc*, 91:1641–1650, 1996.
46. A. Liaw and M. Wiener. Classification and regression by randomforest. *R News*, 2(3):18–22, 2002. URL http://CRAN.R-project.org/package=randomForest.
47. K.L. Moore and M.J. van der Laan. Application of time-to-event methods in the assessment of safety in clinical trials. In Karl E. Peace, editor, *Design, summarization, analysis & interpretation of clinical trials with time-to-event endpoints*, Boca Raton, 2009a. Chapman & Hall.
48. K.L. Moore and M.J. van der Laan. Covariate adjustment in randomized trials with binary outcomes: targeted maximum likelihood estimation. *Stat Med*, 28(1):39–64, 2009b.

49. K.L. Moore and M.J. van der Laan. Increasing power in randomized trials with right censored outcomes through covariate adjustment. *J Biopharm Stat*, 19(6):1099–1131, 2009c.
50. R. Neugebauer and J. Bullard. *DSA: Data-adaptive estimation with cross-validation and the D/S/A algorithm*, 2009. URL http://www.stat.berkeley.edu/~laan/Software/. R package version 3.1.3.
51. M. Pavlic and M.J. van der Laan. Fitting of mixtures with unspecified number of components using cross validation distance estimate. *Comput Stat Data An*, 41:413–428, 2003.
52. A. Peters and T. Hothorn. *ipred: Improved Predictors*, 2009. URL http://CRAN.R-project.org/package=ipred. R package version 0.8-8.
53. E.C. Polley and M.J. van der Laan. Predicting optimal treatment assignment based on prognostic factors in cancer patients. In K.E. Peace, editor, *Design, summarization, analysis & interpretation of clinical trials with time-to-event endpoints*, Boca Raton, 2009. Chapman & Hall.
54. E.C. Polley and M.J. van der Laan. Super learner in prediction. Technical Report 266, Division of Biostatistics, University of California, Berkeley, 2010.
55. R Development Core Team. *R: a language and environment for statistical computing*. R Foundation for Statistical Computing, Vienna, 2010. URL http://www.R-project.org.
56. G. Ridgeway. *gbm: generalized boosted regression models*, 2007. R package version 1.6-3.
57. B.D. Ripley. *Pattern recognition and neural networks*. Cambridge, New York, 1996.
58. J. Rissanen. Modelling by shortest data description. *Automatica*, 14:465–471, 1978.
59. J.M. Robins. Marginal structural models versus structural nested models as tools for causal inference. In *Statistical models in epidemiology: the environment and clinical trials*. Springer, Berlin/Heidelberg/New York, 1999.
60. J.M. Robins. Robust estimation in sequentially ignorable missing data and causal inference models. In *Proceedings of the American Statistical Association on Bayesian Statistical Science* 1999. pp. 6–10. 2000.
61. J.M. Robins and A. Rotnitzky. Recovery of information and adjustment for dependent censoring using surrogate markers. *AIDS Epidemiology – Methodological issues*, eds. N. Jewell, K. Dietz, V. Farewell, Boston, MA: Bikhäuser. pp. 297–331 (*includes errata sheet*). 1992.
62. J.M. Robins and A. Rotnitzky. Comment on the Bickel and Kwon article, "Inference for semiparametric models: some questions and an answer". *Stat Sinica*, 11(4):920–936, 2001.
63. J.M. Robins, A. Rotnitzky, and M.J. van der Laan. Comment on "On profile likelihood". *J Am Stat Assoc*, 450:431–435, 2000.
64. S. Rose and M.J. van der Laan. Simple optimal weighting of cases and controls in case-control studies. *Int J Biostat*, 4(1):Article 19, 2008.
65. S. Rose and M.J. van der Laan. Why match? Investigating matched case-control study designs with causal effect estimation. *Int J Biostat*, 5(1):Article 1, 2009.
66. S. Rose and M.J. van der Laan. A targeted maximum likelihood estimator for two-stage designs. *Int J Biostat*, 7(17), 2011.
67. M. Rosenblum and M.J. van der Laan. Targeted maximum likelihood estimation of the parameter of a marginal structural model. *Int J Biostat*, 6(2):Article 19, 2010.
68. M. Rosenblum, S.G. Deeks, M.J. van der Laan, and D.R. Bangsberg. The risk of virologic failure decreases with duration of HIV suppression, at greater than 50% adherence to antiretroviral therapy. *PLoS ONE*, 4(9): e7196.doi:10.1371/journal.pone.0007196, 2009.
69. Donald B. Rubin. Bayesian inference for causal effects: the role of randomization. *Ann Stat*, 6:34–58, 1978.
70. D.O. Scharfstein, A. Rotnitzky, and J.M. Robins. Adjusting for nonignorable drop-out using semiparametric nonresponse models, (with discussion and rejoinder). *J Am Stat Assoc*, 94:1096–1120 (1121–1146), 1999.
71. G. Schwartz. Estimating the dimension of a model. *Ann Stat*, 6:461–464, 1978.
72. S.E. Sinisi and M.J. van der Laan. Deletion/Substitution/Addition algorithm in learning with applications in genomics. *Stat Appl Genet Mol*, 3(1), 2004. Article 18.
73. O.M. Stitelman and M.J. van der Laan. Collaborative targeted maximum likelihood for time-to-event data. *Int J Biostat*, 6(1):Article 21, 2010.

74. O.M. Stitelman and M.J. van der Laan. Targeted maximum likelihood estimation of time-to-event parameters with time-dependent covariates. Technical Report, Division of Biostatistics, University of California, Berkeley, 2011a.
75. O.M. Stitelman and M.J. van der Laan. Targeted maximum likelihood estimation of effect modification parameters in survival analysis. *Int J Biostat*, 7(1), 2011b.
76. M. Stone. Cross-validatory choice and assessment of statistical predictions. *J R Stat Soc Ser B*, 36(2):111–147, 1974.
77. M. Stone. Asymptotics for and against cross-validation. *Biometrika*, 64(1):29–35, 1977.
78. A.B. Tsybakov. Optimal rates of aggregation. In B. Schölkopf and M.K. Warmuth, editors, *COLT*, volume 2777 of *Lecture Notes in Computer Science*, Berlin/Heidelberg/New York, 2003. Springer.
79. M.J. van der Laan. Estimation based on case-control designs with known prevalance probability. *Int J Biostat*, 4(1):Article 17, 2008.
80. M.J. van der Laan. Targeted maximum likelihood based causal inference: Part I. *Int J Biostat*, 6(2):Article 2, 2010.
81. M.J. van der Laan and S. Dudoit. Unified cross-validation methodology for selection among estimators and a general cross-validated adaptive epsilon-net estimator: finite sample oracle inequalities and examples. Technical Report 130, Division of Biostatistics, University of California, Berkeley, 2003.
82. M.J. van der Laan and S. Gruber. Collaborative double robust penalized targeted maximum likelihood estimation. *Int J Biostat*, 6(1), 2010.
83. M.J. van der Laan and J.M. Robins. *Unified methods for censored longitudinal data and causality*. Springer, Berlin/Heidelberg/New York, 2003.
84. M.J. van der Laan and S. Rose. *Targeted Learning: Causal Inference for Observational and Experimental Data*. Springer, Berlin/Heidelberg/New York, 2011.
85. M.J. van der Laan and Daniel B. Rubin. Targeted maximum likelihood learning. *Int J Biostat*, 2(1):Article 11, 2006.
86. M.J. van der Laan, S. Dudoit, and S. Keleş. Asymptotic optimality of likelihood-based cross-validation. *Stat Appl Genet Mol*, 3(1):Article 4, 2004.
87. M.J. van der Laan, S. Dudoit, and A.W. van der Vaart. The cross-validated adaptive epsilon-net estimator. *Stat Decis*, 24(3):373–395, 2006.
88. M.J. van der Laan, E.C. Polley, and A.E. Hubbard. Super learner. *Stat Appl Genet Mol*, 6(1):Article 25, 2007.
89. A.W. van der Vaart. *Asymptotic statistics*. Cambridge, New York, 1998.
90. A.W. van der Vaart and J.A. Wellner. *Weak convergence and empirical processes*. Springer, Berlin/Heidelberg/New York, 1996.
91. A.W. van der Vaart, S. Dudoit, and M.J. van der Laan. Oracle inequalities for multi-fold cross-validation. *Stat Decis*, 24(3):351–371, 2006.
92. W.N. Venables and B.D. Ripley. *Modern applied statistics with S*. Springer, Berlin/Heidelberg/New York, 4th edition, 2002.
93. H. Wang, S. Rose, and M.J. van der Laan. Finding quantitative trait loci genes with collaborative targeted maximum likelihood learning. *Stat Prob Lett*, published online 11 Nov (doi: 10.1016/j.spl.2010.11.001), 2010.
94. Y. Wang, O. Bembom, and M.J. van der Laan. Data adaptive estimation of the treatment specific mean. *J Stat Plan Infer*, 137(6):1871–1877, 2007.
95. D. H. Wolpert. Stacked generalization. *Neural Networks*, 5:241–259, 1992.
96. W. Zheng and M.J. van der Laan. Asymptotic theory for cross-validated targeted maximum likelihood estimation. Technical Report 273, Division of Biostatistics, University of California, Berkeley, 2010.

Chapter 5
Random Forests

Adele Cutler, D. Richard Cutler, and John R. Stevens

5.1 Introduction

Random Forests were introduced by Leo Breiman [6] who was inspired by earlier work by Amit and Geman [2]. Although not obvious from the description in [6], Random Forests are an extension of Breiman's bagging idea [5] and were developed as a competitor to boosting. Random Forests can be used for either a categorical response variable, referred to in [6] as "classification," or a continuous response, referred to as "regression." Similarly, the predictor variables can be either categorical or continuous.

From a computational standpoint, Random Forests are appealing because they

- naturally handle both regression and (multiclass) classification;
- are relatively fast to train and to predict;
- depend only on one or two tuning parameters;
- have a built in estimate of generalization error;
- can be used directly for high-dimensional problems;
- can easily be implemented in parallel.

Statistically, Random Forests are appealing because of the additional features they provide, such as

- measures of variable importance;
- differential class weighting;
- missing value imputation;
- visualization;
- outlier detection;
- unsupervised learning.

A. Cutler (✉) • D.R. Cutler • J.R. Stevens
Department of Mathematics and Statistics, Utah State University, Logan, UT 84322-3900, USA
e-mail: adele.cutler@usu.edu; richard.cutler@usu.edu; John.R.Stevens@usu.edu

C. Zhang and Y. Ma (eds.), *Ensemble Machine Learning: Methods and Applications*,
DOI 10.1007/978-1-4419-9326-7_5, © Springer Science+Business Media, LLC 2012

This chapter gives an introduction to the Random Forest method for classification and regression, including a brief description of the types of classification and regression trees used in the Random Forests algorithm. The chapter describes how out-of-bag data are used not only to give a fast estimate of generalization error but also to estimate variable importance. A discussion of some important practical issues such as tuning the algorithm and weighting classes to deal with unequal sample sizes is also included. Methods for finding Random Forest proximities and using them to give illuminating plots as well as imputing missing values are presented. Finally, references to extensions of the Random Forest method are given.

5.2 The Random Forest Algorithm

As the name suggests, a Random Forest is a tree-based ensemble with each tree depending on a collection of random variables. More formally, for a p-dimensional random vector $X = (X_1, \ldots, X_p)^T$ representing the real-valued input or predictor variables and a random variable Y representing the real-valued response, we assume an unknown joint distribution $P_{XY}(X, Y)$. The goal is to find a prediction function $f(X)$ for predicting Y. The prediction function is determined by a loss function $L(Y, f(X))$ and defined to minimize the expected value of the loss

$$E_{XY}(L(Y, f(X))) \tag{5.1}$$

where the subscripts denote expectation with respect to the joint distribution of X and Y.

Intuitively, $L(Y, f(X))$ is a measure of how close $f(X)$ is to Y; it penalizes values of $f(X)$ that are a long way from Y. Typical choices of L are *squared error loss* $L(Y, f(X)) = (Y - f(X))^2$ for regression and *zero-one loss* for classification:

$$L(Y, f(X)) = I(Y \neq f(X)) = \begin{cases} 0 \text{ if } Y = f(X) \\ 1 \text{ otherwise.} \end{cases} \tag{5.2}$$

It turns out (see, for example, [10, Sect. 2.4]) that minimizing $E_{XY}(L(Y, f(X)))$ for squared error loss gives the conditional expectation

$$f(x) = E(Y|X = x) \tag{5.3}$$

otherwise known as the *regression function*. In the classification situation, if the set of possible values of Y is denoted by $\mathcal{Y}$, minimizing $E_{XY}(L(Y, f(X)))$ for zero one loss gives

$$f(x) = \arg\max_{y \in \mathcal{Y}} P(Y = y|X = x), \tag{5.4}$$

otherwise known as the *Bayes rule*.

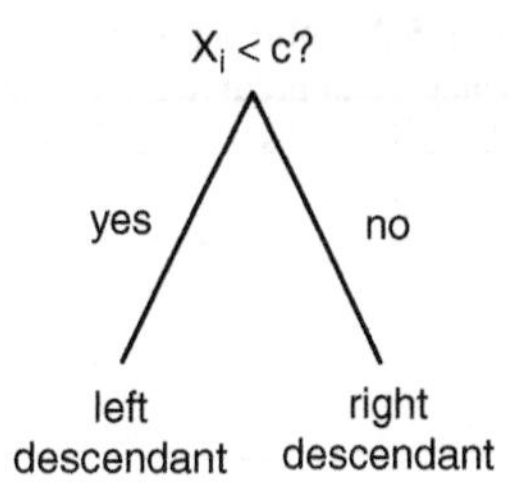

Fig. 5.1 Splitting on a continuous predictor variable X_i, using split point c

Ensembles construct f in terms of a collection of so-called "base learners" $h_1(x), \ldots, h_J(x)$ and these base learners are combined to give the "ensemble predictor" $f(x)$. In regression, the base learners are averaged

$$f(x) = \frac{1}{J} \sum_{j=1}^{J} h_j(x), \tag{5.5}$$

while in classification, $f(x)$ is the most frequently predicted class ("voting")

$$f(x) = \arg\max_{y \in \mathscr{Y}} \sum_{j=1}^{J} I(y = h_j(x)). \tag{5.6}$$

In Random Forests the jth base learner is a tree denoted $h_j(X, \Theta_j)$, where Θ_j is a collection of random variables and the Θ_j's are independent for $j = 1, \ldots, J$. Although the definition of a Random Forest is very general, they are almost invariably implemented in the specific way described in Subsect. 5.2.2. To understand the Random Forest algorithm, it is important to have a fundamental knowledge of the type of trees used as base learners.

5.2.1 Introduction to Classification and Regression Trees

The trees used in Random Forests are based on the binary recursive partitioning trees in the monograph [4] and also described in [10, 14, 26]. These trees (Algorithm 1) partition the predictor space using a sequence of binary partitions ("splits") on individual variables. The "root" node of the tree comprises the entire predictor space. The nodes that are not split are called "terminal nodes" and form the final partition of the predictor space. Each nonterminal node splits into two descendant nodes, one on the left and one on the right, according to the value of one of the predictor variables. For a continuous predictor variable, a split is determined by a split point; points for which the predictor is smaller than the split point go to the left, the rest go to the right (see Fig. 5.1).

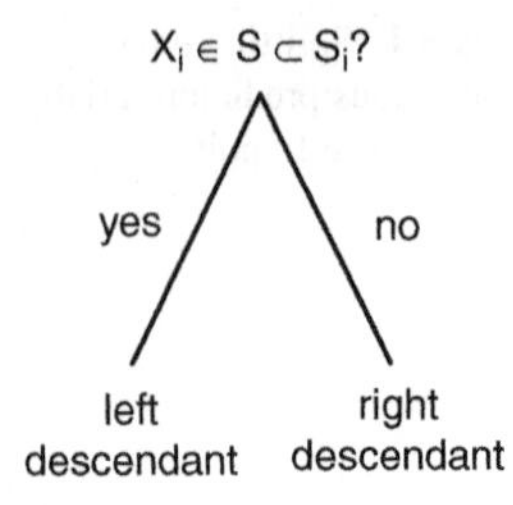

Fig. 5.2 Splitting on a categorical predictor variable X_i, using subset $S \subset S_i$

A categorical predictor variable X_i takes values from a finite set of categories $S_i = \{s_{i,1}, \ldots, s_{i,m}\}$. A split sends a subset of these categories $S \subset S_i$ to the left and the remaining categories to the right (see Fig. 5.2).

The particular split a tree uses to partition a node into its two descendants is chosen by considering every possible split on every predictor variable and choosing the "best" according to some criterion. In the regression context, if the response values at the node are $y_1, \ldots, y_n$, a typical splitting criterion is the mean squared residual at the node

$$Q = \frac{1}{n} \sum_{i=1}^{n} (y_i - \bar{y})^2, \tag{5.7}$$

where $\bar{y} = \frac{1}{n} \sum_{i=1}^{n} y_i$ is the predicted value at the node (the average of the response values). In the classification context where there are K classes denoted $1, \ldots, K$, a typical splitting criterion is the Gini index

$$Q = \sum_{k \neq k'}^{K} \hat{p}_k \hat{p}_{k'}, \tag{5.8}$$

where $\hat{p}_k$ is the proportion of class k observations in the node:

$$\hat{p}_k = \frac{1}{n} \sum_{i=1}^{n} I(y_i = k). \tag{5.9}$$

The splitting criterion gives a measure of "goodness of fit" (regression) or "purity" (classification) for a node, with large values representing poor fit (regression) or an impure node (classification). A candidate split creates two descendant nodes, one on the left and one on the right. Denoting the splitting criteria for the two candidate descendants as Q_{L} and Q_{R} and their sample sizes by n_{L} and n_{R}, the split is chosen to minimize $Q_{\mathrm{split}} = n_{\mathrm{L}} Q_{\mathrm{L}} + n_{\mathrm{R}} Q_{\mathrm{R}}$.

For a continuous predictor variable, finding the best possible split entails sorting the values of the predictor and considering splits between every distinct pair of consecutive values. Typically the midpoint of the interval is used, although any value in the interval would suffice. The values of Q_{L}, Q_{R} and hence Q_{split} are computed for each of these possible split points, usually using a fast update algorithm. For a categorical predictor variable, Q_{L}, Q_{R}, and Q_{split} are computed for all possible ways of choosing a subset of categories to go to each descendant node.

Algorithm 1 Binary Recursive Partitioning

Let $\mathscr{D} = \{(x_1, y_1), \ldots, (x_N, y_N)\}$ denote the training data, with $x_i = (x_{i,1}, \ldots, x_{i,p})^T$.

1. Start with all observations $(x_1, y_1), \ldots, (x_N, y_N)$ in a single node.
2. Repeat the following steps recursively for each unsplit node until the stopping criterion is met:

    a. Find the best binary split among all binary splits on all p predictors.
    b. Split the node into two descendant nodes using the best split (Step 2a).

3. For prediction at x, pass x down the tree until it lands in a terminal node. Let k denote the terminal node and let $y_{k_1}, \ldots, y_{k_n}$ denote the response values of the training data in node k. Predicted values of the response variable are given by:

- $\hat{h}(x) = \bar{y}_k = \frac{1}{n} \sum_{i=1}^{n} y_{k_i}$ for regression
- $\hat{h}(x) = \arg\max_y \sum_{i=1}^{n} I(y_{k_i} = y)$ for classification, where $I(y_{k_i} = y) = 1$ if $y_{k_i} = y$ and 0 otherwise.

Once a split has been selected, the data are partitioned into the two descendant nodes and each of these nodes is treated in the same way as the original node. The procedure continues recursively until a stopping criterion is met. For example, the procedure may stop when all unsplit nodes contain fewer than some fixed number of cases. When the stopping criterion is met, unsplit nodes are called "terminal nodes." A predicted value is obtained for all observations in the terminal nodes by averaging the response for regression problems or computing the most frequent class for classification problems. To predict at a new point, its set of predictor values are used to pass the point down the tree until it falls into a terminal node and the prediction for the terminal node is used as the prediction for the new point.

Often, trees are deliberately grown larger than necessary and "pruned" back to prevent overfitting [4]. Although pruning is very important to prevent overfitting for stand-alone trees, it is not used in Random Forests, so it will not be described here, but the interested reader is referred to [4] or [14].

5.2.1.1 Example 1: Prostate Cancer Data

To illustrate regression trees, data from the prostate cancer study of [23], also studied in [10] is used. The response variable is the level of prostate-specific antigen (lpsa). The predictor variables are log cancer volume (lcavol), log prostate weight (lweight), age, log of the amount of benign prostatic hyperplasia (lbph), seminal vesicle invasion (svi), log of capsular penetration (lcp), Gleason score (gleason), and percentage of Gleason scores 4 or 5 (pgg45). A regression tree was fit to two of

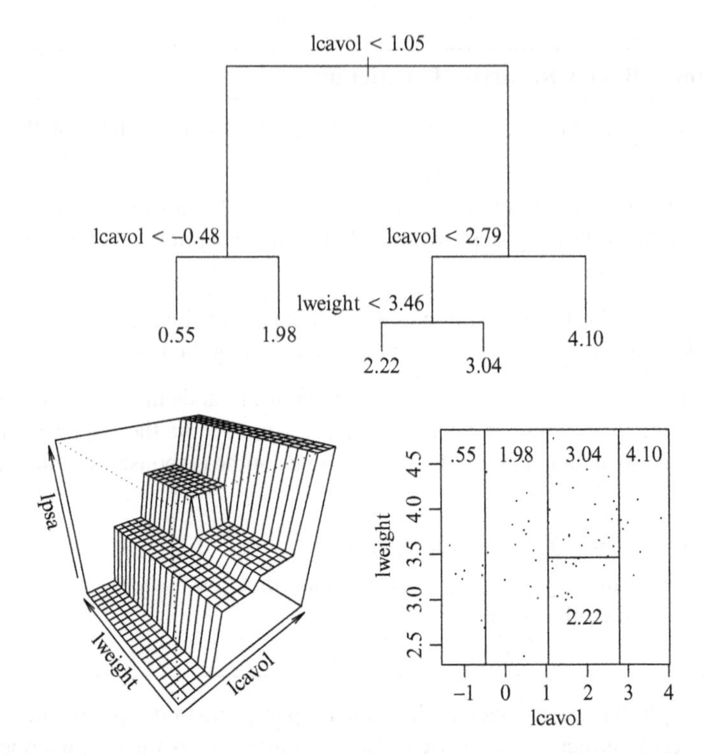

Fig. 5.3 Regression tree for two-dimensional prostate cancer data (Example 1). The *top panel* shows the tree diagram, the *bottom left* contains a perspective plot of the fitted regression surface, the *bottom right* shows the partitioning of the predictor space

the predictor variables, namely, lcavol and lweight. The top panel of Fig. 5.3 shows the regression tree. At each node, cases that satisfy the inequality go to the left, while ones that do not satisfy the inequality go to the right. Each terminal node results in a single predicted value, namely the average value of the response for the observations falling into the node. At the bottom left, Fig. 5.3 shows a perspective plot of the piecewise linear regression surface corresponding to the regression tree in the top panel. On the bottom right, Fig. 5.3 shows the partitioning of the predictor space. For continuous predictors such as these, the splits are parallel to the coordinate axes and the predictor space is divided into (hyper-) rectangles, each with a single predicted value. Each of the five rectangles corresponds to one of the terminal nodes in the tree.

Trees are popular for a wide range of problems, in part because trees can model complex interactions. The rank-based nature of the splits makes trees robust to outliers and insensitive to monotone transformations of the predictor variables. A summary of the characteristics that make trees popular, even for low-dimensional problems, is [10, Sect. 10.7] that trees:

- can model interactions;
- naturally handle both regression and (multiclass) classification;
- naturally handle both continuous and categorical predictor variables;
- handle missing values in the predictor variables;
- are robust to outliers in the predictor variables;
- are insensitive to monotone transformations of the predictor variables;
- scale well for large sample sizes;
- deal well with irrelevant predictor variables.

Neither support vector machines nor neural networks rate highly on any of the above characteristics [10, Sect. 10.7]. On the downside, regression trees have sharp jumps in the predictions at the edges of the nodes. Also they

- are not good at capturing relationships involving linear combinations of predictor variables;
- are known to be unstable in the sense that if the data are perturbed slightly, the tree can change substantially;
- are not as accurate as some of the more recently developed methods.

Trees enjoy a mixed reception when it comes to interpretability. Tree diagrams are easily understood, but interpretation can be difficult because adjacent or nearby rectangles can appear in quite distant parts of the tree. A less obvious problem occurs when two or more predictor variables are highly correlated within a node. Such variables are called surrogates, and lead to similar splits of the node. However, they make interpretation more difficult because *different* surrogates may be selected for splits at this and descendant nodes. If there are only a few predictor variables, good software can help keep track of surrogates, but in very high-dimensional examples the task becomes much more difficult and it may be impossible to extract a coherent story from the tree diagram.

Perhaps the single largest drawback of trees is that they are not as accurate as more recently developed methods. However, they are the building blocks of many ensemble methods including Random Forests.

5.2.2 Random Forest Definition

As mentioned earlier in this section, a Random Forest uses trees $h_j(X, \Theta_j)$ as base learners. For training data $\mathscr{D} = \{(x_1, y_1), \ldots, (x_N, y_N)\}$, where $x_i = (x_{i,1}, \ldots, x_{i,p})^T$ denotes the p predictors and y_i denotes the response, and a particular realization θ_j of Θ_j, the fitted tree is denoted $\hat{h}_j(x, \theta_j, \mathscr{D})$. While this is the original formulation from Breiman [6], in practice the random component θ_j is not considered explicitly but is implicitly used to inject randomness in two ways. First, as with bagging, each tree is fit to an independent bootstrap sample from the original data. The randomization involved in bootstrap sampling gives one part of θ_j. Second, when splitting a node, the best split is found over a randomly selected

Algorithm 2 Random Forests

Let $\mathscr{D} = \{(x_1, y_1), \dots, (x_N, y_N)\}$ denote the training data, with $x_i = (x_{i,1}, \dots, x_{i,p})^T$. For $j = 1$ to J:

1. Take a bootstrap sample $\mathscr{D}_j$ of size N from $\mathscr{D}$.
2. Using the bootstrap sample $\mathscr{D}_j$ as the training data, fit a tree using binary recursive partitioning (Subsect. 5.2.1):
   a. Start with all observations in a single node.
   b. Repeat the following steps recursively for each unsplit node until the stopping criterion is met:
      (i) Select m predictors at random from the p available predictors.
      (ii) Find the best binary split among all binary splits on the m predictors from Step (i).
      (iii) Split the node into two descendant nodes using the split from Step (ii).

To make a prediction at a new point x,

- $\hat{f}(x) = \frac{1}{J}\sum_{j=1}^{J} \hat{h}_j(x)$ for regression
- $\hat{f}(x) = \arg\max_y \sum_{j=1}^{J} I\left(\hat{h}_j(x) = y\right)$ for classification

where $\hat{h}_j(x)$ is the prediction of the response variable at x using the jth tree (Algorithm 1).

subset of m predictor variables instead of all p predictors, independently at each node. The randomization used to sample the predictors gives the remaining part of θ_j.

The trees are grown without pruning. Initially, Breiman [6] suggested growing them until the terminal nodes were pure (classification) or until there were fewer than a prespecified number of data points in each terminal node (regression). More recently [21] suggests controlling the maximum number of terminal nodes.

The resulting trees are combined by unweighted voting if the response is categorical (classification) or unweighted averaging if the response is continuous (regression), as described in Algorithm 2.

5.2.3 Using Out-of-Bag Data

When a bootstrap sample is taken from the data, some observations do not make it into the bootstrap sample. These are called "out-of-bag data," and are extremely useful for estimating generalization error and variable importance (see Sect. 5.3).

Algorithm 3 Out-of-Bag Predictions

Let $\mathscr{D}_j$ denote the jth bootstrap sample and $\hat{h}_j(x)$ denote the prediction at x from the jth tree, for $j = 1, \ldots, J$. For $i = 1$ to N:

1. Let $\mathscr{J}_i = \{j : (x_i, y_i) \notin \mathscr{D}_j\}$ and let J_i be the cardinality of $\mathscr{J}_i$ (Algorithm 2).
2. Define the out-of-bag prediction at x_i to be
   - $\hat{f}_{\text{oob}}(x_i) = \frac{1}{J_i} \sum_{j \in \mathscr{J}_i} \hat{h}_j(x_i)$ for regression
   - $\hat{f}_{\text{oob}}(x_i) = \arg\max_y \sum_{j \in \mathscr{J}_i} I\left(\hat{h}_j(x_i) = y\right)$ for classification

   where $\hat{h}_j(x_i)$ is the prediction of the response variable at x_i using the jth tree (Algorithm 1).

To estimate generalization error, first note that if the trees are large, predictions naively obtained using all the trees will be overly optimistic if used to predict the response variable for observations that were in the training set $\mathscr{D}$. For this reason, prediction of the response variable for observations that were in the training set is only done using trees for which the observation is out-of-bag. These predictions are called out-of-bag predictions (Algorithm 3).

For regression with squared error loss, generalization error is typically estimated using the out-of-bag mean squared error (MSE):

$$\text{MSE}_{\text{oob}} = \frac{1}{N} \sum_{i=1}^{N} \left(y_i - \hat{f}_{\text{oob}}(x_i)\right)^2 \tag{5.10}$$

where $\hat{f}_{\text{oob}}(x_i)$ is the out-of-bag prediction for observation i.

For classification with zero one loss, generalization error rate is estimated using the out-of-bag error rate:

$$E_{\text{oob}} = \frac{1}{N} \sum_{i=1}^{N} I\left(y_i \neq \hat{f}_{\text{oob}}(x_i)\right). \tag{5.11}$$

A common misconception is that the out-of-bag error rate is obtained by computing the out-of-bag error rate for each tree, and averaging these error rates to give the out-of-bag error rate for the forest. Instead, we use the error rate of the out-of-bag predictions. This allows us to obtain a classwise error rate for each class, and an out-of-bag "confusion matrix" by cross-tabulating y_i and $\hat{f}_{\text{oob}}(x_i)$.

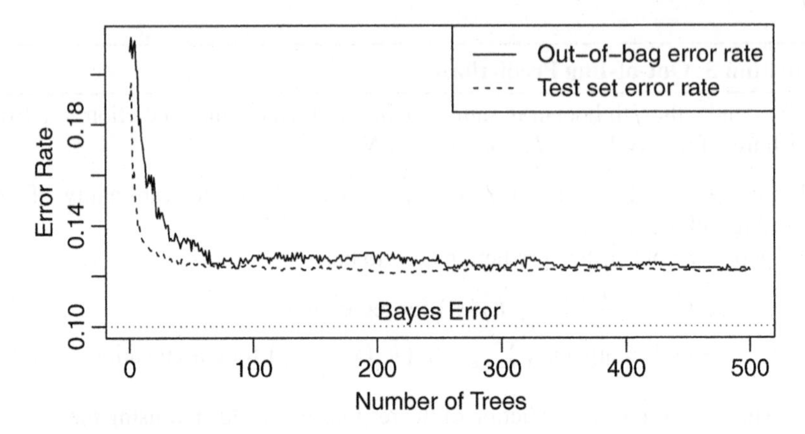

Fig. 5.4 Out-of-bag and test set error rate for Mease and Wyner data (Example 2)

5.2.3.1 Example 2 Mease and Wyner Data

To illustrate the use of the out-of-bag data, we consider a simulation model used by Mease and Wyner [17]. For input variables $X_1, \ldots, X_p$ independently taken from the standard uniform distribution $U[0, 1]$, the response variable $Y \in \{0, 1\}$ is generated using

$$P(Y = 1|X_1, \ldots, X_p) = \begin{cases} q & \text{if } \sum_{l=1}^{L} X_l \leq L/2 \\ 1 - q & \text{otherwise.} \end{cases}$$

For $q < 0.5$, the Bayes' rule classifies an observation $x_1, \ldots, x_p$ into class 0 if $\sum_{l=1}^{L} x_l \leq L/2$ and into class 1 otherwise. For $q > 0.5$, the class labels are reversed, and in both cases the Bayes' error is q. In this way, the first L predictors are important and the remaining $p - L$ predictors are noise. Using $L = p = 2$ and $q = 0.1$, Fig. 5.4 shows the out-of-bag error estimate and the test set error estimate for a training set of size $N = 1000$ and a test set of size 10,000 as the number of trees increases from $J = 1$ to $J = 500$. The out-of-bag error rate tracks the test set error rate quite closely in this Fig. 5.4. We chose a case for which the out-of-bag error rate and test set error rate were quite similar. Other runs showed the out-of-bag error rate to be somewhat higher or somewhat lower than the test set error rate. Table 5.1 shows the out-of-bag confusion matrix and test set confusion matrix for the run shown in Fig. 5.4. Note that the out-of-bag confusion matrix is obtained using the out-of-bag prediction for each observation in the training set against the nominal class.

Table 5.1 Out-of-bag and test set confusion matrices for Mease and Wyner data (Example 2)

Out-of-bag confusion matrix

	Predicted		
	Class 0	Class 1	Total
Nominal Class 0	417	58	475
Nominal Class 1	64	461	525
Total	481	519	1000

Test set confusion matrix

	Predicted		
	Class 0	Class 1	Total
Nominal Class 0	4409	626	5035
Nominal Class 1	590	4375	4965
Total	4999	5001	10000

5.2.4 Tuning

Although Random Forests have the reputation of working quite well right out of the box, there are three parameters that may be tuned to give improved accuracy for particular situations:

- m, the number of randomly selected predictor variables chosen at each node
- J, the number of trees in the forest
- *tree size*, as measured by the smallest node size for splitting or the maximum number of terminal nodes.

The only one of these parameters to which Random Forests is somewhat sensitive appears to be m. In classification, the standard default is $m = \sqrt{M}$, where M is the total number of predictors. In regression, the default is $m = N/3$, where N is the sample size. If tuning is necessary, m can be chosen using the out-of-bag error rate, but then this no longer gives an unbiased estimate of generalization error. However, typically Random Forests are not very sensitive to m, so fine-tuning is not required and overfitting effects due to choice of m should be relatively small, as demonstrated by [9].

For many ensemble methods, generalization error initially decreases as J increases, but at some point J becomes too large and overfitting sets in, with an associated increase in generalization error. This is not the case with Random Forests. For small values of J, the out-of-bag estimate can be unstable and inaccurate. However, as J increases Breiman showed [6] that the generalization error for Random Forests converges almost surely to a limit. In practice, this means J can be chosen as large as desired, without fear of increasing the generalization error. The only real concern with J is that it not be too small, and usually the out-of-bag error rate can be used to decide when J is large enough that the estimated generalization error has stabilized. Often a plot such as that shown in Fig. 5.4 is used to decide whether or not J is large enough.

Breiman's original work [6] recommends growing very large trees. In a recent paper by Segal and Xiao [21], the authors give a classification example for which a forest of large trees overfits and suggest this was not observed in Breiman's original work because the benchmark data sets all came from the University of California at Irvine (UCI) repository and happen to share properties that make large trees nearly optimal. In problems for which large trees overfit, users can tune using either the number of nodes or the smallest nodesize. Out-of-bag error rates can be used to choose the tuning parameter, understanding that such use will lead to a bias in the estimated generalization error.

5.2.5 *Weighting*

Unbalanced data sets, where some classes are much smaller than others, present a challenge to many classifiers. A naive classifier will work on getting the large classes right, while allowing a high-error rate for the small classes. Random Forests has an effective method for weighting the classes to give balanced results in unbalanced data (`www.math.usu.edu/~adele/forests`). One reason to do this is that the important predictor variables may be different when the method is forced to pay greater attention to a small class. Even in the balanced case, the weights can be adjusted to give lower error rates to decisions that have a high-misclassification cost. For example, it is often more serious to incorrectly conclude that someone is healthy than it would be to incorrectly conclude that someone is ill. Example 3 in Subsect. 5.3.1 illustrates the effect of different weights on permutation variable importance.

5.3 Variable Importance

Measures of the importance of the predictor variables are useful for variable selection and for interpreting the fitted forest. While it is standard in many applications to run a principal components analysis (PCA) to reduce dimensionality before fitting a classifier or regression predictor, it is possible that the principal components do not capture the important information for the prediction problem. In this case, it may be preferable to obtain variable importance directly from the algorithm and then re-fit using only the most important predictors.

5.3.1 *Permutation Importance*

Random Forests use an unusual but intuitive measure of variable importance. To measure the importance of variable k, the following procedure is performed for each tree. First, the out-of-bag observations are passed down the tree and the predicted values are computed. Next, the values of variable k are randomly permuted in the out-of-bag data, keeping all the other predictor variables fixed. These modified out-of-bag data are passed down the tree and the predicted values are computed. This process gives two sets of out-of-bag predictions for each observation: one set obtained from real data, the other set from variable-k-permuted data.

Algorithm 4 Permutation Variable Importance

To find the importance of variable k, for $k = 1$ to p:

1. (Find $\hat{y}_{i,j}$) For $i = 1$ to N:

   a. Let $\mathcal{J}_i = \{j : (x_i, y_i) \notin \mathcal{D}_j\}$ and let J_i be the cardinality of $\mathcal{J}_i$ (Algorithm 2).
   b. Let $\hat{y}_{i,j} = \hat{h}_j(x_i)$ for all $j \in \mathcal{J}_i$.

2. (Find $\hat{y}^{\star}_{i,j}$) For $j = 1$ to J:

   a. Let $\mathcal{D}_j$ be the jth bootstrap sample (Algorithm 2).
   b. Let $\mathcal{F}_j = \{i : (x_i, y_i) \notin \mathcal{D}_j\}$.
   c. Randomly permute the value of variable k for the data points $\{x_i : i \in \mathcal{F}_j\}$ to give $\mathcal{P}_j = \{x_i{}^{\star} : i \in \mathcal{F}_j\}$.
   d. Let $\hat{y}^{\star}_{i,j} = \hat{h}_j(x_i{}^{\star})$ for all $i \in \mathcal{F}_j$.

3. For $i = 1$ to N:

   - For classification: $\text{Imp}_i = \frac{1}{J_i}\sum_{j \in \mathcal{J}_i} I\left(y_i \neq \hat{y}^{\star}_{i,j}\right) - \frac{1}{J_i}\sum_{j \in \mathcal{J}_i} I(y_i \neq \hat{y}_{i,j})$.
   - For regression: $\text{Imp}_i = \frac{1}{J_i}\sum_{j \in \mathcal{J}_i}\left(y_i - \hat{y}^{\star}_{i,j}\right)^2 - \frac{1}{J_i}\sum_{j \in \mathcal{J}_i}(y_i - \hat{y}_{i,j})^2$.

For classification, the difference between the error rate of the predictions obtained from permuted data and those obtained using permuted data gives a measure of variable importance for the observation. The same procedure is used for regression, but using MSE instead of error rates. For classification, classwise variable importance is computed by averaging over observations from the same class. Overall variable importance is computed by averaging over all the observations.

Algorithm 4 gives the importance of a particular variable, denoted by k in the algorithm description, on the predictions for a particular observation, denoted by i. The values can be used as measures of *local* variable importance, or they can be averaged over all observations to give measures of overall importance of the variable. The largest values are generally plotted (Fig. 5.5).

Intuitively, the permutation-based importance of variable k is an estimate of the how much the prediction error or MSE on a test set would increase if the value of variable k were randomly permuted in the test set. In this sense, it is similar to the coefficient-based measures of importance used in methods such as linear regression or logistic regression—they measure how much the prediction would change if the value of the predictor increased by one unit, keeping everything else the same. Quite a different measure is obtained, for both Random Forests and classical methods, if variable k is removed and the model is refit, because in this case predictors that are correlated with the one of interest can give a similar fit and make the variable appear unimportant. In contrast, if an important predictor variable is correlated with other predictor variables, Random Forests sometimes splits on one and sometimes on another, due to the random choice of predictors at each node. Therefore, Random

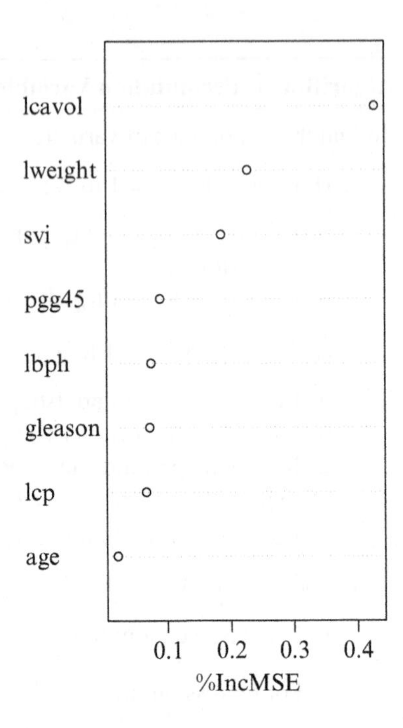

Fig. 5.5 Permutation variable importance, prostate data (Example 1)

Forests permutation importance tends to identify all of the correlated predictors as important if any one of them is important.

One attractive feature of all tree-based methods is their ability to capture complex interactions between predictors. If Random Forests captures such an interaction, the variables involved are likely to show up as "important" because randomly permuting one of them destroys the predictive power of the interaction.

To illustrate the behavior of Random Forest permutation importance, a regression forest was fit to the prostate data (Example 1). A permutation importance plot is given in Fig. 5.5, showing that the three most important variables are lcavol, lweight, and svi. Interestingly, these are the same three variables chosen by lasso (see [10, Fig. 3.10]).

5.3.1.1 Example 3 Normal Mixture

To illustrate the behavior of Random Forest variable importance when classes are weighted differently, consider a bivariate normal mixture of three classes

$$\pi_1 N(\mu_1, I) + \pi_2 N(\mu_2, I) + \pi_3 N(\mu_3, I),$$

where $N(\mu, I)$ denotes the bivariate normal density with mean μ and covariance matrix the identity. Generating $N = 300$ observations from such a mixture with $\mu_1 = (0,0)^T, \mu_2 = (0,3)^T, \mu_3 = (3,3)^T$, and $\pi_1 = 0.4, \pi_2 = 0.4, \pi_3 = 0.2$ gave

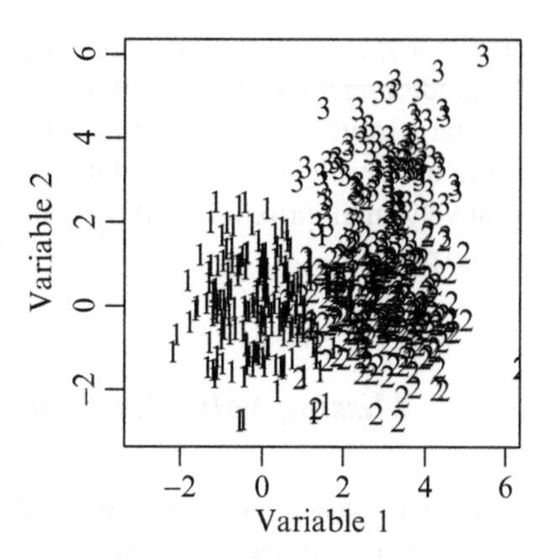

Fig. 5.6 Bivariate normal mixture of three classes (Example 3)

Table 5.2 Impact of class weights on error rates and permutation importance

	Class 1	Class 2	Class 3	Class 1	Class 2	Class 3
Weights	1/3	1/3	1/3	1/7	1/7	5/7
Classwise error rate (percent)	8.2	13.0	8.9	8.7	17.7	5.0
Overall error rate (percent)	10.4			11.8		
Permutation importance (variable 1)	32.7			73.8		
Permutation importance (variable 2)	22.3			49.1		

the data shown in Fig. 5.6. Fitting Random Forests using $J = 500$ trees and $m = 1$ for two different weighting schemes gave the results in Table 5.2. Equal weighting gives the lowest overall error rate. Increasing the weight on class 3, the smallest class, reduces the class 3 error rate from 8.9% to 5.0% and increases the error rates of the other two classes, giving an overall increase in error rate from 10.4% to 11.8%. More interestingly, equal weighting ranks variable 1 as more important than variable 2, while increasing the weight on class 3 reverses the ranking.

5.4 Proximities

Random Forests proximities are used for missing value imputation, and visualization.

5.4.1 Definition

The proximity between two observations is the proportion of the time that they end up in the same terminal node, where the proportion is taken over the trees in the forest. If two observations are always in the same terminal node, their proximity will be 1. If they are never in the same terminal node, their proximity will be 0. The proximity between two observations is a measure of how close together they are in predictor space, but it automatically gives more weight to predictors that are useful for predicting the response. Observations that are very far apart in Euclidean space may have quite a large proximity if they only differ on weak or irrelevant predictors, while observations that are relatively close together in Euclidean space may have relatively small proximities if they differ on predictors that are crucial for predicting the response.

5.4.2 *Missing Value Imputation*

Random Forests imputes missing values using the proximities described above. The procedure is iterative: an initial forest is built using median imputation, proximities are calculated, and new imputations are obtained by a proximity-weighted average for a continuous predictor or a proximity-weighted vote for a categorical predictor. A new forest is built, giving new proximities and imputations. Usually five or six iterations are sufficient to give stable imputations. Although no formal analysis has been done, the fact that the method uses proximity-based nearest neighbors suggests that it will be valid if values of the predictors are missing at random.

5.4.3 *Visualization*

From a statistical perspective, one of the difficult aspects of high-dimensional data analysis is that it is not obvious how to get a good "feel" for the data. Are there interesting patterns or structures, such as subgroups within the known classes? Are there outliers? In a multiclass situation, are some of the groups separated while others overlap? Random Forests provide a way to look at the data to give some insight into these questions. This is done by computing proximities, deriving a distance matrix, and performing classical multidimensional scaling (MDS) to obtain two- or three-dimensional plots. Each point on such a plot represents one of the observations and the distances between the points reproduce, as closely as possible, the proximity-based distances. Such a plot can be used to pick out subgroups of cases that almost always stay together in the trees, or outliers that are almost always alone in a terminal node.

5.4.3.1 Example 4 Microarray Data

To illustrate the potential usefulness of visualization using the proximity matrix, we consider the prostate cancer microarray data [22]. These data have 6033 gene expression values for 102 arrays (50 normal samples and 52 tumor samples). We used the normalization described by Dettling [8]. Figure 5.7 (left) shows the first two dimensions of the MDS plot based on the Random Forest proximity matrix.

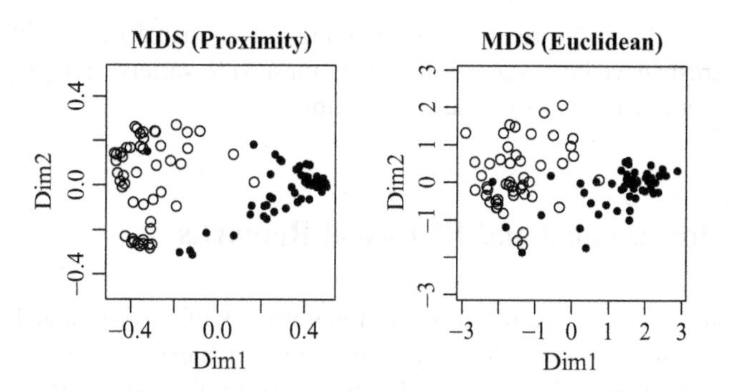

Fig. 5.7 MDS plot from the Random Forests proximities (*left*) and from Euclidean distance (*right*) for Example 4. Solid circles represent cancer cases, open circles represent controls

A natural question at this point is whether it would be just as good to use MDS on a conventional distance, such as Euclidean distance or one of the other distances commonly used in cluster analysis. This can certainly be done, but one of the difficulties is that a conventional distance can be dominated by noisy and uninformative predictors that may drown out the effects of the important predictors. This behavior can be seen in Fig. 5.7, which presents the MDS plot derived from the proximity matrix and the MDS plot derived from Euclidean distance for the microarray data in Example 4. The proximity plot reveals much more structure than the plot based on Euclidean distances, including an outlier that could be of interest to the investigators.

5.5 Software

Commercial software for Random Forests is available from www.salford-systems.com. The R package is `randomForest` [15] and this, along with R [19], is available from the CRAN website `www.cran.r-project.org`. Open source FORTRAN software for Random Forests is available from `www.math.usu.edu/~adele/forests`.

5.6 Summary

Random Forests are a multipurpose tool, applicable to both regression and classification problems, including multiclass classification. They give an internal estimate of generalization error so cross-validation is unnecessary. They can be tuned, but often work quite well with default tuning parameters. Variable importance measures are available, which can be used for variable selection. Random Forests produce proximities, which can be used to impute missing values. Proximities can also provide a wealth of information by enabling novel visualizations of the data. Random Forests have been successfully used for a wide variety of applications and enjoy considerable popularity in several disciplines.

5.7 Bibliographical and Historical Remarks

The Random Forest algorithm was the last major work of Leo Breiman [6].

Theoretical developments have been difficult to achieve. In the original paper, Breiman [6] suggested that Random Forests work by reducing correlation, while keeping the variance relatively small. Lin and Jeon [16] show that RF behaves like a nearest neighbor classifier with an adaptive metric. More recently, Biau et al. address consistency [3].

Several extensions have been published, for example [9] developed a variable selection procedure, [18] introduced quantile regression forests, and [12, 13] considered forests for survival analysis. More recently, [21] extends Random Forests for multivariate responses. Amaratunga et al. [1] suggest an extension to very high-dimensional data.

Applications of Random Forests are numerous and only a few can be mentioned here. Statnikov [24] compares random forests and support vector machines for microarray-based cancer classification. Schroff et al. [20] used Random Forests for image segmentation. Chen et al. [7] use Random Forests to identify genetic interactions, while Goldstein et al. [11, 25] apply Random Forests to SNP-based genomewide association data.

References

1. Amaratunga, D., Cabrera, J., Lee, Y.-S.: Enriched random forests. Bioinformatics **24** (18) pp. 2010–2014 (2008).
2. Amit, Y., Geman, D.: Shape quantization and recognition with randomized trees. Neural Computation **9**(7) pp. 1545–1588 (1997).
3. Biau, G., Devroye, L., Lugosi, G.: Consistency of Random Forests and Other Averaging Classifiers. Journal of Machine Learning Research **9** pp. 2039–2057 (2008).
4. Breiman, L., Friedman, J., Olshen, R., Stone, C.: Classification and Regression Trees. Wadsworth, New York (1984).
5. Breiman, L.: Bagging Predictors. Machine Learning **24** (2) pp. 123–140 (2001).
6. Breiman, L.: Random Forests. Machine Learning **45** (1) pp. 5–32 (2001).
7. Chen, X., Liu, C.-T., Zhang, M., Zhang, H.: A forest-based approach to identifying gene and genegene interactions. Proc Natl Acad Sci USA **104** (49) pp. 19199–19203 (2007).

8. Dettling, M.: BagBoosting for Tumor Classification with Gene Expression Data. Bioinformatics **20** (18) pp. 3583–3593 (2004).
9. Diaz-Uriarte, R., Alvarez de Andres, S.: Gene Selection and Classification of Microarray Data Using Random Forest. BMC Bioinformatics **7** (1) 3 (2006).
10. Hastie, T., Tibshirani, R., Friedman, J.: The Elements of Statistical Learning: Data Mining, Inference, and Prediction, Second Edition. Springer Series in Statistics, Springer, New York (2009).
11. Goldstein, B., Hubbard, A., Cutler, A. Barcellos, L.: An application of Random Forests to a genome-wide association dataset: Methodological considerations & new findings. BMC Genetics **11** (1) 49 (2010).
12. Hothorn, T., Bühlmann, P., Dudoit, S., Molinaro, A., Van Der Laan, M.: Survival Ensembles. Biostatistics **7** (3) pp. 355–373 (2006).
13. Ishwaran, H., Kogalur, U.B., Blackstone, E.H., Lauer, M.S.: Random survival forests. Annals of Applied Statistics **2** (3) pp. 841–860 (2008).
14. Izenman, A.: Modern Multivariate Statistical Techniques. Springer Texts in Statistics, Springer, New York (2008).
15. Liaw, A., Wiener, M.: Classification and Regression by randomForest. R News **2** (3) pp. 18–22 (2002).
16. Lin, Y., Jeon, Y.: Random Forests and Adaptive Nearest Neighbors. Journal of the American Statistical Association **101** (474) pp. 578–590 (2006).
17. Mease, D., Wyner, A.: Evidence Contrary to the Statistical View of Boosting. Journal of Machine Learning Research **9** pp. 131–156 (2008).
18. Meinshausen, N.: Quantile Regression Forests. Journal of Machine Learning Research **7** pp. 983–999 (2006).
19. R Development Core Team: R: A Language and Environment for Statistical Computing. R Foundation for Statistical Computing, Vienna, Austria (2011). http://www.R-project.org.
20. Schroff, F., Criminisi, A., Zisserman, A.: Object Class Segmentation using Random Forests. Proceedings of the British Machine Vision Conference 2008, British Machine Vision Association, **1** (2008).
21. Segal, M., Xiao, Y.: Multivariate Random Forests. Wiley Interdisciplinary Reviews: Data Mining and Knowledge Discovery **1** (1) pp. 80–87 (2011).
22. Singh D., Febbo P.G., Ross K., Jackson D.G., Manola J., Ladd C., Tamayo P., Renshaw A.A., D'Amico A.V., Richie J.P., Lander E.S., Loda M., Kantoff P.W., Golub T.R., Sellers W.R.: Gene expression correlates of clinical prostate cancer behavior. Cancer Cell **1** (2) pp. 203–209 (2002).
23. Stamey, T., Kabalin, J., McNeal J., Johnstone I., Freiha F., Redwine E., Yang N.: Prostate specific antigen in the diagnosis and treatment of adenocarcinoma of the prostate. II. Radical prostatectomy treated patients. Journal of Urology **16** pp. 1076–1083 (1989).
24. Statnikov, A., Wang, L., Aliferis, C.: A comprehensive comparison of random forests and support vector machines for microarray-based cancer classification. BMC Bioinformatics **9** (1) 319 (2008).
25. Wang, M., Chen, X., Zhang, H.: Maximal conditional chi-square importance in random forests. **26** (6): pp. 831–837 (2010).
26. Zhang, H., Singer, B.H.: Recursive Partitioning and Applications, Second Edition. Springer Series in Statistics, Springer, New York (2010).

Chapter 6
Ensemble Learning by Negative Correlation Learning

Huanhuan Chen, Anthony G. Cohn, and Xin Yao

6.1 Introduction

This chapter investigates a specific ensemble learning approach by negative correlation learning (NCL) [21–23]. NCL is an ensemble learning algorithm which considers the cooperation and interaction among the ensemble members. NCL introduces a correlation penalty term into the cost function of each individual learner so that each learner minimizes its mean-square-error (MSE) error together with the correlation with other ensemble members.

The chapter describes the traditional algorithms of NCL and their robust implementation, regularized negative correlation learning (RNCL) [8]. This chapter also treats ensemble learning as an optimization problem to balance the trade-off among accuracy, negative correlation, and regularization by a multi-objective optimization algorithm [9]. The numerical results demonstrate the superiority of negatively correlated ensembles. In general, negatively correlated ensembles can be viewed as a *framework*, rather than an algorithm itself, meaning several other learning techniques could make use of it.

H. Chen (✉) • X. Yao
CERCIA, School of Computer Science, University of Birmingham, Birmingham B15 2TT, UK
e-mail: H.Chen@cs.bham.ac.uk; X.Yao@cs.bham.ac.uk

A.G. Cohn
School of Computing, University of Leeds, Leeds LS2 9JT, UK
e-mail: A.G.Cohn@leeds.ac.uk

C. Zhang and Y. Ma (eds.), *Ensemble Machine Learning: Methods and Applications*,
DOI 10.1007/978-1-4419-9326-7_6, © Springer Science+Business Media, LLC 2012

6.2 Negatively Correlated Ensemble

6.2.1 *Ensemble Learning by Negative Correlation*

NCL introduces a correlation penalty term into the error function of each individual learner in the ensemble so that all the learners can be trained simultaneously on the same training data set [20].

Given a training set $(\mathbf{x}_n, y_n)_{n=1}^N$, NCL combines M individual learners $f_i(\mathbf{x})$ to constitute the ensemble.

$$f_{\text{ens}}(\mathbf{x}_n) = \frac{1}{M}\sum_{i=1}^{M} f_i(\mathbf{x}_n).$$

To train an individual learner f_i, the cost function e_i of f_i is defined by

$$e_i = \sum_{n=1}^{N}(f_i(\mathbf{x}_n) - y_n)^2 + \lambda p_i, \tag{6.1}$$

where λ is a weighting parameter on the penalty term p_i:

$$\begin{aligned} p_i &= \sum_{n=1}^{N}\left\{(f_i(\mathbf{x}_n) - f_{\text{ens}}(\mathbf{x}_n))\sum_{j\neq i}(f_j(\mathbf{x}_n) - f_{\text{ens}}(\mathbf{x}_n))\right\} \\ &= -\sum_{n=1}^{N}(f_i(\mathbf{x}_n) - f_{\text{ens}}(\mathbf{x}_n))^2. \end{aligned} \tag{6.2}$$

The first term on the right-hand side of (6.1) is the empirical training error of f_i. The second term p_i is a correlation penalty function. The purpose of minimizing p_i is to negatively correlate each learner's error with errors for the rest of the ensemble. The λ parameter controls the trade-off between the training error term and the penalty term. With $\lambda = 0$, we would have an ensemble that is exactly equivalent to training a set of networks independently of one another. If λ is increased, more and more emphasis would be placed on minimizing the correlation.

The parameter λ is crucial for the generalization performance. The λ parameter should be neither negative nor too large. Negative parameters will lead the ensemble to be positively correlated, large positive values will cause the Hessian matrix $\mathbf{H} = \frac{\partial^2 e_i}{\partial w_i \partial w_j}$ to be nonpositive definite (nPD). Although the state of the Hessian during training does not directly relate to the generalization error, the non-PD Hessian during the training will cause weight divergence since there will be no minimum to converge to. In the following, we will derive the conditions under which the Hessian will be non-PD [3].

Assume the individual estimator in NCL is a linear combination of nonlinear functions ϕ:

$$f_i(\mathbf{x}_n) = \sum_{k=1}^{K} w_{ki}\phi_{ki}(\mathbf{x}_n).$$

Examples of estimators in this class are Multi-Layer Perceptions using linear output nodes, Polynomial Neural Networks, and Radial Basis Functions (RBFs). When the Hessian matrix is positive definite, all elements on the leading diagonal are positively valued; therefore, if any element on that diagonal is zero or less, the entire matrix cannot be positive definite. The diagonal elements in the Hessian matrix can be written as

$$\frac{\partial^2 e_i}{\partial w_{i,j}^2} = 2\sum_{n=1}^{N} \phi_{i,j}^2(\mathbf{x}_n) - 2\lambda\left(1 - \frac{1}{M}\right)^2 \sum_{n=1}^{N} \phi_{i,j}^2(\mathbf{x}_n), \quad (6.3)$$

where $w_{i,j}$ is the jth weight in the output layer of the ith learner. In the case of RBF networks, $\phi_{i,j}^2$ is the squared output of the jth basis function in the ith network.

If this element, (6.3), is nonpositive, the entire Hessian matrix is guaranteed to be nPD. Therefore, we would like the following inequality to hold.

$$\frac{\partial^2 e_i}{w_{i,j}^2} > 0 \Rightarrow \lambda < \left(\frac{M}{M-1}\right)^2.$$

Therefore, the bound for the parameter is $\lambda \in \left[0, \left(\frac{M}{M-1}\right)^2\right)$. When λ is varied beyond this upper bound, the Hessian matrix will be nPD.

6.2.2 *Negative Correlation Learning Algorithm*

The training algorithm of NCL can be achieved by gradient descent. The detailed NCL algorithm is given below. We will report some experiments in Subsect. 6.3.4.

Note that NCL is a framework and can be used for any learner aiming to minimize the mean-square-error (MSE). In this chapter, we take RBF networks as an example of individual ensemble members.

The output of RBF network is computed as a linear combination of n_i basis functions

$$f_i(\mathbf{x}) = \sum_{k=1}^{n_i} w_k\phi_k(\mathbf{x}) = \Phi^T\mathbf{w}_i,$$

Algorithm 1 Negative Correlation Learning [21,22]

1: **Input:** Training Set $(\mathbf{x}_n, y_n)_{n=1}^N$, the number of predictors required M, the number of iterations iter, and the parameter λ.
2: **Output:** the trained ensemble function f_{ens}
3: **for** $j = 1$ to iter **do**
4: Calculate $f_{\text{ens}}(\mathbf{x}_n) = \frac{1}{M}\sum_{i=1}^M f_i(\mathbf{x}_n)$
5: **for** $i = 1$ to M **do**
6: for each weight $w_{i,j}$ in learner i, perform a desired number of gradient descent updates
7: $\frac{\partial e_i}{\partial w_{i,j}} = 2\sum_{n=1}^N (f_i(\mathbf{x}_n) - y_n)\frac{\partial f_i(\mathbf{x}_n)}{\partial w_{i,j}} - 2\lambda \sum_{n=1}^N (f_i(\mathbf{x}_n) - f_{\text{ens}}(\mathbf{x}_n))\left(1 - \frac{1}{M}\right)\frac{\partial f_i(\mathbf{x}_n)}{\partial w_{i,j}}$
8: **end for**
9: **end for**

where $\mathbf{w}_i = (w_1, \ldots, w_{n_i})^T$ denotes the weight vector in the output layer and $\Phi = (\phi_1, \ldots, \phi_{n_i})$ is the vector of basis functions. The Gaussian basis functions ϕ_k are defined as

$$\phi_k(\mathbf{x}) = \exp\left(\frac{\|\mathbf{x} - \rho_k\|^2}{2\sigma_k^2}\right),$$

where ρ_k and σ_k denote the center and width of the Gaussian, respectively. The training of the RBF network is divided into two steps. In the first step, the means ρ_k are initialized with randomly selected data points from the training set and the variances σ_k are determined as the Euclidean distance between ρ_k and the closest $\rho_i (i \neq k, i \in \{1, \ldots, n_i\})$. Then in the second step we perform a gradient descent in the regularized error function (weight decay):

$$\min e_i = \sum_{n=1}^N (f_i(\mathbf{x}_n) - y_n)^2 - \lambda \sum_{n=1}^N (f_i(\mathbf{x}_n) - f_{\text{ens}}(\mathbf{x}_n))^2.$$

The derivative with respect to w_k is:

$$\frac{\partial e_i}{\partial w_k} = 2\sum_{n=1}^N (f_i(\mathbf{x}_n) - y_n)\frac{\partial f_i(\mathbf{x}_n)}{\partial w_k} - 2\lambda \sum_{n=1}^N (f_i(\mathbf{x}_n) - f_{\text{ens}}(\mathbf{x}_n))\left(1 - \frac{1}{M}\right)\frac{\partial f_i(\mathbf{x}_n)}{\partial w_k}.$$

In order to fine-tune the centers and widths, we simultaneously adjust the output weights and the RBF centers and variances. Taking the derivative with respect to RBF means ρ_k and variances σ_k^2 we obtain

$$\frac{\partial e_i}{\partial \rho_k} = 2\sum_{n=1}^N (f(\mathbf{x}_n) - y_n)\frac{\partial f_i(\mathbf{x}_n)}{\partial \rho_k} - 2\lambda \sum_{n=1}^N (f_i(\mathbf{x}_n) - f_{\text{ens}}(\mathbf{x}_n))\left(1 - \frac{1}{M}\right)\frac{\partial f_i(\mathbf{x}_n)}{\partial \rho_k},$$

with $\frac{\partial f_i(\mathbf{x}_n)}{\partial \rho_k} = w_k \frac{\mathbf{x}_n - \rho_k}{\sigma_k^2} \phi_k(\mathbf{x}_n)$ and

$$\frac{\partial e_i}{\partial \sigma_k} = 2\sum_{n=1}^{N}(f(\mathbf{x}_n) - y_n)\frac{\partial f_i(\mathbf{x}_n)}{\partial \sigma_k} - 2\lambda \sum_{n=1}^{N}(f_i(\mathbf{x}_n) - f_{\text{ens}}(\mathbf{x}_n))\left(1 - \frac{1}{M}\right)\frac{\partial f_i(\mathbf{x}_n)}{\partial \sigma_k},$$

with $\frac{\partial f_i(\mathbf{x}_n)}{\partial \sigma_k} = w_k \frac{\|\mathbf{x} - \rho_k\|^2}{\sigma_k^3} \phi_k(\mathbf{x}_n)$. These three derivatives are employed in the minimization of (6.1) by a scaled conjugate gradient (SCG) descent [28].

The algorithm can be summarized in Algorithm 1.

6.3 Regularized Negatively Correlated Ensemble

In the previous section, we have introduced the negative correlation learning algorithm. According to the formulation of NCL, it seems that the correlation term in the cost function acts as the regularization term. However, we observe that NCL is prone to overfitting the noise in the training set by training the ensemble as a single estimator and only minimizing the MSE without regularization [18]. Therefore, regularization should be used to address the overfitting problem of NCL. This section will introduce the RNCL and its Bayesian interpretation.

6.3.1 Negative Correlation with Overfitting

Based on the individual error function, i.e., (6.1), the cost function of the ensemble can be obtained by averaging these learners' errors e_i. With $\lambda = 1$, the average error E of all the networks' e_i is obtained as follows:

$$\begin{aligned} E &= \frac{1}{M}\sum_{i=1}^{M} e_i = \frac{1}{M}\sum_{n=1}^{N}\sum_{i=1}^{M}\left\{(f_i(\mathbf{x}_n) - y_n)^2 - (f_i(\mathbf{x}_n) - f_{\text{ens}}(\mathbf{x}_n))^2\right\} \\ &= \sum_{n=1}^{N}(f_{\text{ens}}(\mathbf{x}_n) - y_n)^2. \end{aligned} \tag{6.4}$$

According to (6.4), NCL is equivalent to training a single estimator $f_{\text{ens}}(\mathbf{x}_n)$ instead of training each individual network separately. It is observed that NCL only minimizes the empirical training MSE error $\sum_{n=1}^{N}(f_{\text{ens}}(\mathbf{x}_n) - y_n)^2$ but does not regularize the complexity of the ensemble.

In this case, NCL only reduces the empirical MSE of the ensemble, but it pays less attention to regularizing the complexity of the ensemble and NCL is prone to overfitting the noise in the training set. Similarly, setting a zero or small positive λ corresponds to independently training these estimators without regularization and in this case, NCL is prone to overfitting as well.

NCL can use the penalty coefficient to explicitly alter the emphasis on the individual MSE and correlation portions of the ensemble and thus alleviate the overfitting problem to some extent. However, NCL could not totally overcome the overfitting problem by tuning this parameter without regularization, especially when dealing with data with nontrivial noise, which will be evidenced by the empirical work in this chapter.

6.3.2 *Overfitting Management by Regularization*

In order to improve the generalization ability of NCL, in this subsection we propose RNCL [8]. Following the traditional strategy to avoid overfitting, a regularization term is incorporated into the ensemble error function:

$$E_{\text{ens}} = \frac{1}{M}\sum_{n=1}^{N}\sum_{i=1}^{M}\left\{(f_i(\mathbf{x}_n) - y_n)^2 - (f_i(\mathbf{x}_n) - f_{\text{ens}}(\mathbf{x}_n))^2\right\} + \sum_{i=1}^{M}\alpha_i \mathbf{w}_i^T \mathbf{w}_i, \quad (6.5)$$

where $\mathbf{w}_i = [w_{i,1}, \ldots, w_{i,n_i}]^T$ is the weight vector of network i and n_i is the total number of weights in network i.

This regularization term $\sum_{i=1}^{M}\alpha_i \mathbf{w}_i^T \mathbf{w}_i$ is the weight decay [17] term for the entire ensemble. In order to train each neural network with its regularization, we decompose the regularization term to M parts, each part for a network. The error function for network i can be obtained as follows:

$$e_i = \frac{1}{M}\sum_{n=1}^{N}(f_i(\mathbf{x}_n) - y_n)^2 - \frac{1}{M}\sum_{n=1}^{N}(f_i(\mathbf{x}_n) - f_{\text{ens}}(\mathbf{x}_n))^2 + \alpha_i \mathbf{w}_i^T \mathbf{w}_i. \quad (6.6)$$

Comparing this error function with the cost function of NCL, (6.1), RNCL imposes a regularization term on every individual neural network and it optimizes the regularization parameter α_i instead of the correlation parameter λ.

RNCL is implemented by the SCG [28] algorithm. According to (6.6), the minimization of the error function of the ensemble is achieved by minimizing the error functions of each individual network. The algorithm can be summarized in Algorithm 2.

Algorithm 2 Regularized Negatively Correlated Ensemble [8]

1: **Input:** Training Set $(\mathbf{x}_n, y_n)_{n=1}^N$, the number of predictors required M, the number of iterations iter, and the initial values of parameter $\alpha_i, i = 1, \ldots, M$.
2: **Output:** the trained ensemble function f_{ens}
3: **for** $j = 1$ to iter **do**
4: Calculate $f_{\text{ens}}(\mathbf{x}_n) = \frac{1}{M}\sum_{i=1}^M f_i(\mathbf{x}_n)$
5: **for** $i = 1$ to M **do**
6: for each weight $w_{i,j}$ in learner i, perform a desired number of gradient descent updates
7: $\frac{\partial e_i}{\partial w_{i,j}} = \frac{2}{M}\sum_{n=1}^N (f_i(\mathbf{x}_n) - y_n)\frac{\partial f_i(\mathbf{x}_n)}{\partial w_{i,j}} - \frac{2}{M}\sum_{n=1}^N (f_i(\mathbf{x}_n) - f_{\text{ens}}(\mathbf{x}_n))\left(1 - \frac{1}{M}\right)\frac{\partial f_i(\mathbf{x}_n)}{\partial w_{i,j}} + 2\alpha_i w_{i,j}$
8: **end for**
9: Parameter Optimization by Bayesian inference.
10: **end for**

6.3.3 *Regularized Parameter Optimization by Bayesian Inference*

This subsection describes the probabilistic interpretation of RNCL, the function of the regularization term and how to optimize these parameters by Bayesian inference [8].

6.3.3.1 Bayesian Interpretation

Given the training set $D = \{\mathbf{x}_n, y_n\}_{n=1}^N$, we follow the standard probabilistic formulation and assume that the targets are sampled from the model with additive noise:

$$y_n = f_{\text{ens}}(\mathbf{x}_n) + e_n = \frac{1}{M}\sum_{i=1}^M f_i(\mathbf{x}_n) + e_n,$$

where e_n is an independent sample from some noise process which is further assumed to be mean-zero Gaussian with variance β^{-1}.

According to the Bayesian theorem, given the hyperparameters $\boldsymbol{\mu} = [\mu_1, \mu_2, \ldots, \mu_M]$[1] and β, the weigh vector $\mathbf{w} = \left[\mathbf{w}_1^T, \ldots, \mathbf{w}_M^T\right]^T$ can be obtained by maximizing the posterior $P(\mathbf{w}|D)$.

$$P(\mathbf{w}|D) = \frac{P(D|\mathbf{w},\beta)P(\mathbf{w}|\boldsymbol{\mu})}{P(D|\boldsymbol{\mu},\beta)}, \tag{6.7}$$

where the probability $P(D|\boldsymbol{\mu}, \beta)$ is a normalization factor, which is independent of $\mathbf{w}$.

[1] $\mu_i, i = 1, 2, \ldots M$, is the inverse variance of the Gaussian distribution of weights for network i.

The weight vector of each network $\mathbf{w}_i$ is assumed to have a Gaussian distribution with zero mean and variance μ_i^{-1}. The prior of the weight vector $\mathbf{w}$ is obtained as follows.

$$P(\mathbf{w}|\boldsymbol{\mu}) = \prod_{i=1}^{M} \left(\frac{\mu_i}{2\pi}\right)^{n_i/2} \exp\left(-\frac{1}{2}\mu_i \mathbf{w}_i^T \mathbf{w}_i\right), \tag{6.8}$$

where n_i is the total number of weights in network i.

The traditional Bayesian methods [1, 24, 29] often use an isotropic Gaussian prior over weights $\mathbf{w}$ where the covariance matrix is an identity matrix multiplied by a parameter, which means these weights in the learner share the same prior. RNCL extends this by imposing different regularization parameters for different networks in the ensemble. The prior of RNCL becomes a block-isotropic Gaussian prior whose covariance matrix is diagonal matrix with M different values. That is, each network has its own different prior.

Since noise e_n follows a Gaussian distribution with zero mean and variance β^{-1}, the likelihood $P(D|\mathbf{w},\beta)$ can be written as

$$P(D|\mathbf{w},\beta) = \prod_{n=1}^{N} \left(\frac{\beta}{2\pi}\right)^{1/2} \exp\left(-\frac{\beta}{2}e_n^2\right). \tag{6.9}$$

We omit all constants and normalization factor, and apply Bayesian rules:

$$P(\mathbf{w}|D) \propto \exp\left(-\frac{\beta}{2}\sum_{n=1}^{N} e_n^2\right) \cdot \exp\left(-\sum_{i=1}^{M} \frac{\mu_i}{2}\mathbf{w}_i^T \mathbf{w}_i\right). \tag{6.10}$$

Taking the negative logarithm, the maximum of the posterior model parameters $\mathbf{w}$ is obtained as the solution to the following optimization problem:

$$\min J_1(\mathbf{w}) = \frac{1}{2}\beta \sum_{n=1}^{N} e_n^2 + \frac{1}{2}\sum_{i=1}^{M} \mu_i \mathbf{w}_i^T \mathbf{w}_i. \tag{6.11}$$

The error function J_1 is made up of two terms. The first, $\frac{1}{2}\beta \sum_{n=1}^{N} e_n^2$, is the sum of the empirical training errors. The second, $\frac{1}{2}\sum_{i=1}^{M} \mu_i \mathbf{w}_i^T \mathbf{w}_i$, is the regularization term, measuring the amount of square of weights.

Comparing (6.11) with (6.5), RNCL is equivalent to maximization of the posterior under Bayesian framework. The likelihood $P(D|\mathbf{w},\beta)$ corresponds to the empirical training error term and the prior over weight vector $P(\mathbf{w}|\boldsymbol{\mu})$ corresponds to the regularization term. The regularization term penalizes large weights, causing the weights to converge to smaller absolute values than they otherwise would.

Based on the above analysis, RNCL is an application of a Bayesian framework in an ensemble system. Instead of simultaneously optimizing the weigh vector of ensemble, RNCL manages to train the entire ensemble by decomposing the job into a set of subtasks, which significantly reduces computational complexity.

In order to obtain the posterior of the weight vector $\mathbf{w}$, the Taylor expansion of $J_1(\mathbf{w})$ is employed at point $\mathbf{w}_{\text{MP}}$

$$J_1(\mathbf{w}) = J_1(\mathbf{w}_{\text{MP}}) + \frac{1}{2}(\mathbf{w} - \mathbf{w}_{\text{MP}})^T A(\mathbf{w} - \mathbf{w}_{\text{MP}}), \tag{6.12}$$

where $\mathbf{w}_{\text{MP}}$ is the most probable weight vector, and A is the Hessian matrix of $J_1(\mathbf{w})$:

$$A = \nabla\nabla J_1 = \nabla\nabla\left(\sum_{i=1}^{M} \frac{\mu_i}{2}\mathbf{w}_i^T \mathbf{w}_i + \frac{\beta}{2}\sum_{n=1}^{N} e_n^2\right) = \text{diag}(\Lambda) + \beta\nabla\nabla\left(\frac{1}{2}\sum_{n=1}^{N} e_n^2\right), \tag{6.13}$$

where $\Lambda = \left(\mu_1^{(1)}, \ldots \mu_1^{(n_1)}, \mu_2^{(1)}, \ldots \mu_2^{(n_2)}, \ldots, \mu_M^{(1)}, \ldots \mu_M^{(n_M)}\right)^T$ and the superscript indicates the number of repetitions of μ_i.

The integral can be computed as below:

$$\int \exp(-J_1(\mathbf{w}))\text{d}\mathbf{w} = \int \exp(-J_1(\mathbf{w}_{\text{MP}}) - \frac{1}{2}(\mathbf{w} - \mathbf{w}_{\text{MP}})^T A(\mathbf{w} - \mathbf{w}_{\text{MP}}))\text{d}\mathbf{w}$$
$$= \exp(-J_1(\mathbf{w}_{\text{MP}})) \cdot (2\pi)^{W/2} \det A^{-\frac{1}{2}}.$$

Based on these equations, the approximated posterior of $\mathbf{w}$ is obtained as follows:

$$P(\mathbf{w}|D) = \frac{\exp(-J_1(\mathbf{w}))}{\int \exp(-J_1(\mathbf{w}))\text{d}\mathbf{w}} = \frac{\exp\left(-\frac{1}{2}(\mathbf{w} - \mathbf{w}_{\text{MP}})^T A(\mathbf{w} - \mathbf{w}_{\text{MP}})\right)}{(2\pi)^{W/2} \det A^{-\frac{1}{2}}}. \tag{6.14}$$

6.3.3.2 Inference of Regularization Parameters

In order to find the most probable values of $\boldsymbol{\mu}$ and β, we need to maximize the posterior of $P(\boldsymbol{\mu}, \beta|D)$.

According to the Bayesian rule, the posteriors of $\boldsymbol{\mu}$ and β are obtained by:

$$P(\boldsymbol{\mu}, \beta|D) = \frac{P(D \mid \boldsymbol{\mu}, \beta)P(\boldsymbol{\mu}, \beta)}{P(D)} \propto P(D \mid \boldsymbol{\mu}, \beta), \tag{6.15}$$

where flat priors are assumed on the hyperparameters $\boldsymbol{\mu}$ and β. According to (6.8), (6.9), and (6.14), the marginal likelihood can be obtained in the following way [13]:

$$P(D|\boldsymbol{\mu}, \beta) = \frac{P(D|\mathbf{w}, \beta)P(\mathbf{w}|\boldsymbol{\mu})}{P(\mathbf{w}|D)}$$
$$\approx \frac{(2\pi)^{W/2}|A|^{-\frac{1}{2}} \prod_{i=1}^{M} \left(\frac{\mu_i}{2\pi}\right)^{n_i/2} \left(\frac{\beta}{2\pi}\right)^{N/2} \exp\left(-J_1(\mathbf{w})\right)}{\exp(-\frac{1}{2}(\mathbf{w} - \mathbf{w}_{\text{MP}})^T A(\mathbf{w} - \mathbf{w}_{\text{MP}}))}.$$

By using the Gaussian approximation $J_1(\mathbf{w}) \approx J_1(\mathbf{w}_{\mathrm{MP}}) + \frac{1}{2}(\mathbf{w} - \mathbf{w}_{\mathrm{MP}})^T A(\mathbf{w} - \mathbf{w}_{\mathrm{MP}})$ and the relation $W = \sum n_i$, where W is the total number of weights in the ensemble,

$$\begin{aligned} P(D|\boldsymbol{\mu}, \beta) &\approx (2\pi)^{W/2}|A|^{-\frac{1}{2}} \prod_{i=1}^{M} \left(\frac{\mu_i}{2\pi}\right)^{n_i/2} \left(\frac{\beta}{2\pi}\right)^{N/2} \exp\left(-J_1(\mathbf{w}_{\mathrm{MP}})\right) \\ &\approx \left(\frac{1}{2\pi}\right)^{N/2} \sqrt{\frac{\prod_{i=1}^{M} \mu_i^{n_i} \beta^N}{\det A}} \exp(-J_1(\mathbf{w}_{\mathrm{MP}})) \\ &\propto \sqrt{\frac{\prod_{i=1}^{M} \mu_i^{n_i} \beta^N}{\det A}} \exp(-J_1(\mathbf{w}_{\mathrm{MP}})). \end{aligned} \tag{6.16}$$

In order to maximize the probability $P(D|\boldsymbol{\mu}, \beta)$, a negative logarithm is applied:

$$J_2 = \frac{1}{2}\sum_{i=1}^{M} \mu_i \mathbf{w}_{i,\mathrm{MP}}^T \mathbf{w}_{i,\mathrm{MP}} + \frac{1}{2}\beta \sum_{n=1}^{N} e_{n,\mathrm{MP}}^2 - \frac{1}{2}\sum_{i=1}^{M} n_i \log \mu_i - \frac{1}{2} N \log \beta + \frac{1}{2} \log \det A, \tag{6.17}$$

where the subscript MP indicates the most probable values.

The update rule for $\alpha_i = \mu_i/\beta$ can be obtained from the derivation of J_2. In order to apply a partial derivative to J_2, we need to apply partial derivative to $\log \det A$.

Since $\det A = \prod_{j=1}^{W} (\beta\lambda_j + \Lambda_j)$, where λ_j is the eigenvalue of the Hessian matrix $\nabla\nabla\left(\frac{1}{2}\sum_{n=1}^{N} e_n^2\right)$ and W is the number of weights in the ensemble:

$$\frac{\partial}{\partial \mu_i} \log \det A = \frac{\partial}{\partial \mu_i} \log\left(\prod_{j=1}^{W} (\beta\lambda_j + \Lambda_j)\right) = \sum_{j \in n_i} \frac{1}{\beta\lambda_j + \mu_i},$$

$$\frac{\partial}{\partial \beta} \log \det A = \frac{\partial}{\partial \beta} \log\left(\prod_{j=1}^{W} (\beta\lambda_j + \Lambda_j)\right) = \sum_{j} \frac{\lambda_j}{\beta\lambda_j + \Lambda_j},$$

where $j \in n_i$ indicates the range $\left[\sum_{t=1}^{i-1} n_t + 1, \ldots, \sum_{t=1}^{i} n_t\right]$.

The gradient of $\log P(D|\mu, \beta)$ toward μ_i and β are:

$$\frac{\partial J_2}{\partial \mu_i} = \frac{1}{2}\mathbf{w}_{i,\mathrm{MP}}^T \mathbf{w}_{i,\mathrm{MP}} - \frac{1}{2}\frac{n_i}{\mu_i} + \frac{1}{2}\sum_{j \in n_i} \frac{1}{\beta\lambda_j + \mu_i},$$

$$\frac{\partial J_2}{\partial \beta} = \frac{1}{2}\sum_{n=1}^{N} e_{n,\mathrm{MP}}^2 - \frac{N}{2\beta} + \frac{1}{2}\sum_{j} \frac{\lambda_j}{\beta\lambda_j + \Lambda_j}.$$

Setting the gradient to zero, the optimal μ_i and β can be obtained:

$$\mu_i^{\text{new}} = \frac{1}{\mathbf{w}_{i,\text{MP}}^T \mathbf{w}_{i,\text{MP}}} \left(n_i - \sum_{j \in n_i} \frac{\mu_i}{\beta\lambda_j + \mu_i} \right), \tag{6.18}$$

$$\beta^{\text{new}} = \frac{1}{\sum_{n=1}^{N} e_{n,\text{MP}}^2} \left(N - \sum_{j=1}^{W} \frac{\beta\lambda_j}{\beta\lambda_j + \Lambda_j} \right). \tag{6.19}$$

Combining (6.18) and (6.19) and the relation $\alpha_i = \mu_i/\beta$, we obtain the following equation:

$$\beta \left[\sum_{i=1}^{M} \alpha_i \mathbf{w}_{i,\text{MP}}^T \mathbf{w}_{i,\text{MP}} + \sum_{n=1}^{N} e_{n,\text{MP}}^2 \right] = N. \tag{6.20}$$

In the following, we reformulate the optimization problem, (6.17), in μ_i and β into a scalar optimization problem in $\alpha_i = \mu_i/\beta$. Therefore, we firstly replace that optimization problem by an optimization problem in β and α_i by the relation $\mu_i = \beta\alpha_i$. As (6.20) also holds in the scalar optimization, we search for the optimum only along this curve in the (α_i and β) space.

By elimination of β from (6.20), the minimization problem from J_2 is obtained in a straightforward way:

$$J_3 = \sum_{j=1}^{W} \log\left(1 + \frac{\lambda_j}{\hat{\alpha}_j}\right) + N \log\left(\sum_{i=1}^{M} \alpha_i \mathbf{w}_{i,\text{MP}}^T \mathbf{w}_{i,\text{MP}} + \sum_{n=1}^{N} e_{n,\text{MP}}^2 \right),$$

where $\hat{\alpha}_j = \Lambda_j/\beta$ and $\Lambda = \left(\mu_1^{(1)}, \ldots \mu_1^{(n_1)}, \mu_2^{(1)}, \ldots \mu_2^{(n_2)}, \ldots, \mu_M^{(1)}, \ldots \mu_M^{(n_M)}\right)^T$.

Setting $\frac{\partial J_3}{\partial \alpha_i} = 0$, the update rule $\alpha_i^{\text{new}} = \mu_i/\beta$ can be obtained as follows:

$$\alpha_i^{\text{new}} = \frac{\sum_{n=1}^{N} e_{n,\text{MP}}^2 \left(n_i - \sum_{j \in n_i} \frac{\alpha_i}{\lambda_j + \alpha_i} \right)}{\mathbf{w}_{i,\text{MP}}^T \mathbf{w}_{i,\text{MP}} \left(N - \sum_{j=1}^{W} \frac{\lambda_j}{\lambda_j + \hat{\alpha}_j} \right)}, \tag{6.21}$$

where $j \in n_i$ indicates the range $\left[\sum_{t=1}^{i-1} n_t + 1, \ldots, \sum_{t=1}^{i} n_t\right]$.

The RNCL learning is conducted in an iterative manner. In each iteration, the ensemble is firstly trained by the SCG algorithm with the previous regularization parameters α_i, followed by the estimation of new most probable α_i^{new} values by (6.21), and then we incorporate the new α_i^{new} in the ensemble. The learning algorithm repeats the process, until some suitable convergence criteria have been satisfied.

6.3.4 Computational Examples

In this subsection, RBF networks are used as the individual ensemble members. The number of hidden nodes is randomly selected but restricted in the range 3–12. The initial centers, widths for individual NNs are randomly selected. The details of the specification including the derivations of error with respect to centers and widths of a RBF network are presented in Subsect. 6.2.2.

We employ the SCG algorithm to train NCL and RNCL. Twenty five RBF networks are employed in the ensemble. The input attributes of data sets are scaled to mean zero and unit variance as the preprocessing procedure.

As the first experiment, we compare RNCL, $NCL_{\lambda=1}$, and NCL_{CV} on the sinc data set. Figure 6.1a and b show the output of RNCL and NCL_{CV} on the sinc function with different noise levels. In the noise-free case, both RNCL and NCL_{CV} perfectly approximate the actual function, although there is a little misfit for NCL_{CV} near the tail. When the noise level increases, NCL_{CV}, although it selects the parameter by cross validation, overfits the noise in the training set, while RNCL is more robust to noise than NCL, as indicated by Fig. 6.1b.

In order to explore the behavior of RNCL, NCL_{CV}, and $NCL_{\lambda=1}$ with different noise levels, we add mean zero and different levels of Gaussian noise to sinc. Figure 6.1c shows the average results of 100 runs.

For the sinc data set, when the noise level (variance) is small, RNCL and NCL_{CV} perform similarly and their performances are better than $NCL_{\lambda=1}$. When the noise level becomes greater, MSE of RNCL increases slower than that of NCL_{CV} and $NCL_{\lambda=1}$ and NCL_{CV} performs a little better than $NCL_{\lambda=1}$

In Fig. 6.2, we have illustrated the mean of regularization parameters α obtained in RNCL versus different noise levels and the parameter λ selected by NCL_{CV}.

Figure 6.2a reports the mean α value[2] obtained in RNCL vs. different noise levels on sinc and Friedman data. The results are based on 100 runs. When the noise level increases, the learned model will become more complex to fit the data and in this situation, large regularization is preferred to control the complexity in the model. Bayesian parameter selection in RNCL does reflect this tendency when the noise level increases.

Figure 6.2b reports the selected λ parameter in NCL_{CV} and the performance of RNCL. It is observed that NCL_{CV} could not beat RNCL even if it uses the optimal correlation parameter λ. Figures 6.1 and 6.2 confirm that NCL_{CV} could not overcome the overfitting problem by only tuning the λ parameter for regression problems.

In the following, we demonstrate the application of RNCL on classification problems. Firstly, we apply RNCL and NCL_{CV} to two synthetic data sets in two dimensions in order to illustrate graphically the decision boundary.

---

[2]Since we optimize α_i for each individual networks in the ensemble, in this figure we only show the mean α_i value.

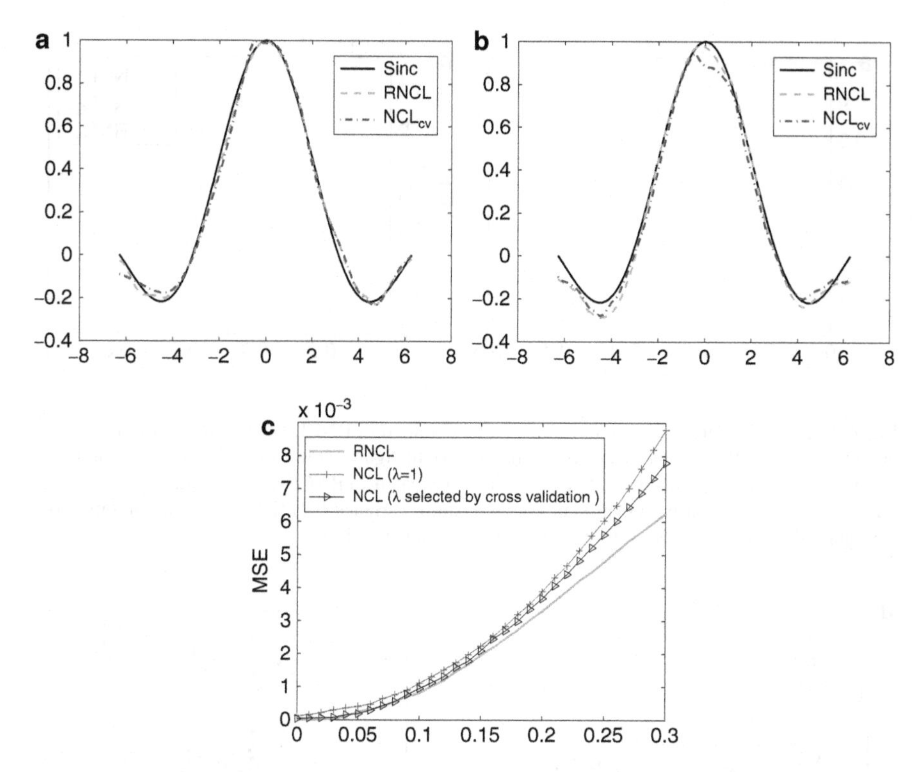

Fig. 6.1 Comparison of RNCL and NCL on sinc data sets. In (**a**) and (**b**), the lines in green (dashed), red (dotted), and black (solid) are obtained by RNCL, NCL_{CV}, and the noise-free function, respectively. Part figure (**c**) shows mean squared error (MSE) of RNCL (green solid), $NCL_{\lambda=1}$ (blue crossed), and NCL_{CV} (black triangled) on sinc with different noise levels. (**a**) Sinc free of noise. (**b**) Sinc with Gaussian noise (mean 0, variance 0.2). (**c**) Sinc with different noise levels (For better interpretation of the figure, the reader is referred to the web version of this article.)

These two data sets are (1) *synth* is generated from mixtures of two Gaussians by [30], and (2) *Bumpy* comes from two equal Gaussians but being rotated by 90°, quadratic boundaries are required.

In Fig. 6.3, we present a comparison of RNCL and NCL_{CV}. In the case of *Synth*, RNCL disregards the outliers in the training points and produces a smooth boundary, while NCL_{CV} generates a corner in the decision boundary due to several outliers. In the case of *Bumpy*, the noise level is greater because of these overlapping points. NCL_{CV} does not generalize very well and produces a little twisty boundary. RNCL generates a quadratic boundary as expected.

In order to check the behaviors of RNCL and NCL_{CV} on noisy classification problems, we conduct similar noise experiments as the regression problems. In the experiments, we select one data set, Gaussian. Gaussian is a synthetic two class

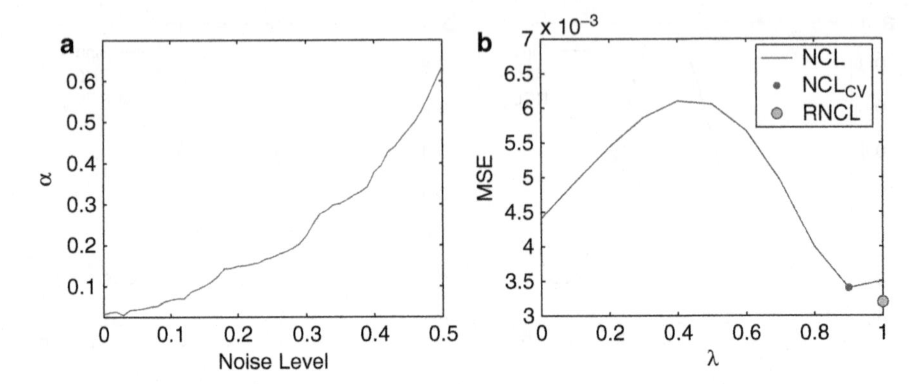

Fig. 6.2 The left figure reports the mean α value obtained in RNCL versus different noise levels on sinc data. Results are based on 100 runs. The right figure shows the selection of λ and the performance of RNCL, on sinc (0.2 noise level). (**a**) Mean α in RNCL on sinc with different noise levels. (**b**) Sinc function with Gaussian noise (mean 0, variance 0.2) (For better interpretation of the figure, the reader is referred to the web version of this article.)

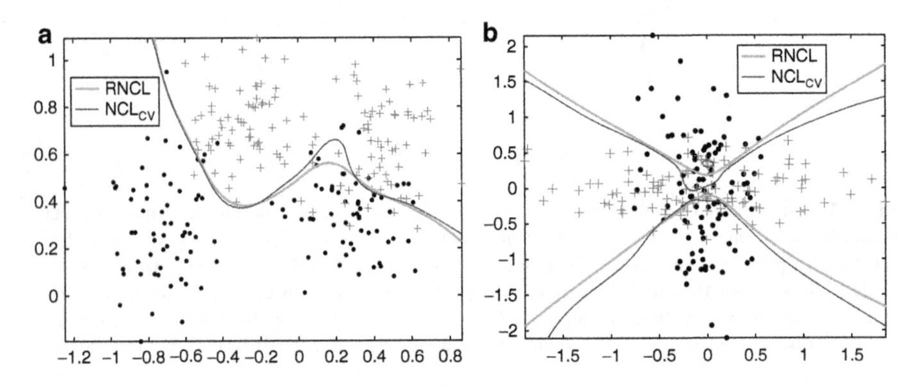

Fig. 6.3 Comparison of RNCL and NCL_{CV} on two synthetic classification data sets. Two classes are shown as crosses and dots. The separating lines were obtained by projecting test data over a grid. The lines in wide and thin were obtained by RNCL and NCL_{CV}, respectively. (**a**) Synth. (**b**) Bumpy (For better interpretation of the figure, the reader is referred to the web version of this article.)

two-dimensional data set which is sampled from a mixture of four Gaussians. Each class is associated with two of the Gaussians so that the optimal decision boundary is nonlinear.

To change the noise level, we randomly select different percentages of data points and reverse their labels. We run the algorithms 100 times and report the average results in Fig. 6.4. Figure 6.4a visualizes the decision boundaries of RNCL and NCL_{CV} with 20% noise points.

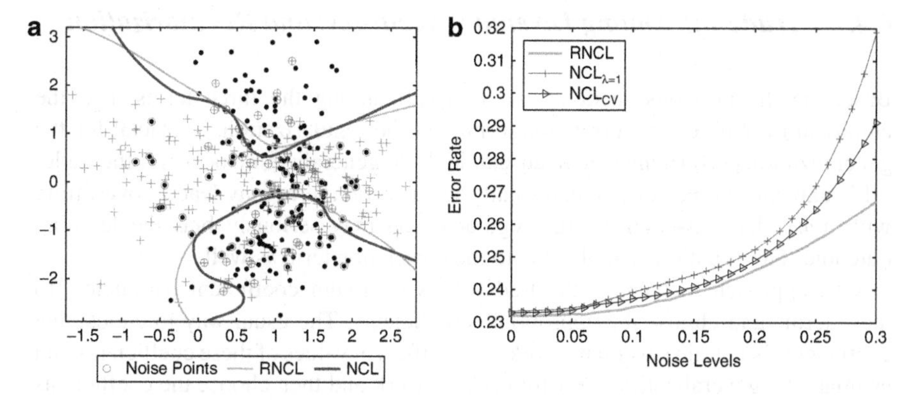

Fig. 6.4 Comparison of RNCL and NCL_{CV} on one classification data set. Two classes are shown as crosses and dots. The separating lines were obtained by projecting test data over a grid. In (**a**), the decision boundaries in green (solid) and blue (dark) are obtained by RNCL and NCL_{CV}, respectively. The randomly-selected noise points are marked with a circle. Part figure (**b**) shows the classification error of RNCL (green solid), $NCL_{\lambda=1}$ (blue crossed) and NCL_{CV} (black triangled) versus noise levels on Gaussian data set. The results are based on 100 runs. (**a**) Gaussian with 20% noise. (**b**) Gaussian with different noise levels (For better interpretation of the figure, the reader is referred to the web version of this article in color.)

Although the noise level is high, RNCL produces a smooth boundary. NCL_{CV} does not generalize well. We also plot the curve, Fig. 6.4b, of classification error versus noise level for the Gaussian data set. In this figure, RNCL is a little better in the beginning, but as the noise level increases, RNCL significantly outperforms NCL_{CV} and $NCL_{\lambda=1}$.

6.4 Multi-objective Ensemble Learning

Most ensemble learning algorithms train the base learners independently or sequentially, so the advantages of interaction and cooperation among the base learner are not exploited. NCL has shown that ensemble methods benefits from considering the cooperation among the base learners. This approach opens a new research area where the design and training of the base learners can be interdependent. Although NCL performs well for a broad range of practical applications by considering the cooperation in the ensemble, it is not regularized, which leads to overfitting, and the weighting coefficient, which controls the trade-off between empirical error and correlation, needs to be tuned.

In order to address these problems, the previous section introduced a regularization term to the error function. This section will describe a multi-objective approach to balance the error, correlation, and regularization terms in the error function [9].

6.4.1 Trade-off Among Diversity, Accuracy, and Regularization

Based on the previous sections, the trade-off among the three terms, i.e., the empirical training error, correlation term, and the regularization, is crucial for the generalization performance of an ensemble. Poor generalization occurs if the trade-off is unbalanced. For example, a small regularization term may lead to overfitting with noise data sets and a large regularization may seriously bias the learning outcome. The situation is applicable to the correlation term as well.

One approach to balance the trade-off is to assign coefficient parameters to these terms and choose the appropriate coefficients. The usual way to choose the coefficients is to train several networks with different values of these coefficients and estimate the generalization error for each network and then choose the coefficients that minimize the estimated generalization error.

Evolutionary multi-objective algorithms are well suited to search the optimal trade-off among different objectives by parallelizing the searching using a population of networks and biasing toward the Pareto front and at the same time maintaining population diversity to obtain as many diverse solutions as possible [4]. These properties are especially important in ensemble design.

This chapter introduces multi-objective regularized negative correlation learning (MRNCL) algorithm, which implements the RNCL algorithm by an evolutionary multi-objective algorithm. MRNCL [9] involves minimization of the three terms: empirical training error term, correlation penalty term, and the regularization term. MRNCL algorithm not only addresses the issues concerned with NCL but also provides the following advantages: (1) Being a multi-objective algorithm, the approach is able to produce a diverse ensemble. Some individuals are good at minimizing the training error; some pay more attention to cooperation and the others manage to control the complexity. (2) The parameters of individual network can be effectively obtained in the evolutionary multi-objective algorithm. (3) Due to the regularization term in MRNCL, the obtained ensemble is regularized and is more robust with respect to noise. (4) There is no need to weigh the different objectives by optimizing the coefficient parameters.

According to (6.6), MRNCL defines the following three objectives.

- Objective of Performance $\sum_{n=1}^{N}(f_i(\mathbf{x}_n) - y_n)^2$

  This objective measures the empirical mean square error based on the training set.
- Objective of Correlation $-\sum_{n=1}^{N}(f_i(\mathbf{x}_n) - f_{\text{ens}}(\mathbf{x}_n))^2$

  This correlation term measures the amount of variability among the ensemble members and this term can also be treated as the diversity measure [19]. From both theoretical and experimental results it has been shown that, if the individual networks in an ensemble are unbiased, the most effective combination of them occurs when the errors of the individual networks are negatively correlated. This objective encourages individual networks to negatively correlate their errors and thus helps to generate a diverse ensemble.

- Objective of Regularization $\mathbf{w}_i^T \mathbf{w}_i = \sum_j w_j^2$
  Based on the regularization theory [32], the weight decay term [17] is employed to punish large weights. The weight decay term causes the weights to converge to smaller absolute values than they otherwise would. The regularization term helps the generalization ability of a neural network because large weights can hurt generalization in two different ways: (a) excessively large weights leading to hidden nodes can cause the output function to be too rough, possibly with near discontinuities. Excessively large weights leading to output nodes can cause wild outputs far beyond the range of the data if the output activation function is not bounded to the same range as the data. (b) Large weights can cause excessive variance of the output [12]. The regularization term is beneficial to NCL since large weights are usually connected with near linear dependence among groups of nodes in the network, and NCL would seem to potentiate the appearance of large weights in the ensemble.

6.4.2 *Trade-off Optimization by Multi-objective Learning*

To optimize the trade-off among these three terms, the evolutionary neural network approach is used. A RBF network is used as the component network for this purpose. The structure of RBF networks is similar to that described in Subsect. 6.2.2.

We use an RBF network as the base learner because of the following advantages. (1) Once the centers and the widths of the basis functions have been fixed, the optimal output weights $\mathbf{w}$ can be efficiently computed in a closed form, which means the performance mostly depends on the selection of basis functions. (2) It is reasonable to define crossover and mutation operators in structural-evolving RBF network by tuning these basis functions.

Based on the above reasons, the crossover operator and mutation operator for RBF networks are described as follows.

- Crossover Operator
  As the performance of a RBF network mostly depends on the basis functions, i.e., the centers and the widths, the crossover operator is defined to exchange the basis functions of two RBF networks. Many crossover techniques exist in the literature, such as one-point crossover, two-point crossover, and "cut and splice" crossover. In a RBF network ensemble, as different networks may have different numbers of basis functions, the "cut and splice" approach has been adopted by randomly choosing separate crossover points for two RBF networks and swap their basis functions beyond those points.
- Mutation Operator
  This algorithm defines two structural mutation operators for RBF networks.

1. Deleting one basis function. Randomly select one basis function and delete it.

2. Adding one basis function. The center of the new basis function is determined by a randomly selected data point from the training set. Then, the width of the basis function is chosen as the minimal distance from other centers in this RBF network.

As the crossover and mutation operations may not generate the optimal combination of basis functions, afterwards, we simultaneously adjust the output weights, the RBF centers, and widths. This procedure is also called parametric mutation [36], which only modifies the parameters of the network without modifying its topology. This parametric mutation is performed for a few iterations (in our experiments, only one SCG update is employed).

In this chapter, nondominated sorting with fitness sharing [31] and rank-based fitness assignment were used. Nondominated sorting is based on layers of Pareto front, which ranks the individuals in the population. The diversity of population is maintained by a niching method.

The nondominated sorting algorithm consists of two stages: One is to obtain the nondominated fronts of different layers and every individual of these fronts is assigned an equal dummy fitness. The algorithm used for obtaining the nondominated set of solutions compares the individuals pairwise and marks these individuals, which are dominated by at least one member of the population, as dominated. The second is that the members of every front share their fitness [11] with the constraint that none of the members of a front gets a higher fitness than any of the members of the previous front.

Since the dummy fitness assigned by nondominated sorting is raw, sometimes the range of the dummy fitness is too large, leading to the situation that some networks reproduce too rapidly, taking over the population too quickly, and preventing the evolutionary algorithm from searching other areas of the solution space. Fitness scaling is used to map an arbitrary fitness range into an appropriate range.

The algorithm employs rank-based fitness assignment to reassign the fitness to the networks because rank-based fitness assignment behaves in a more robust manner than proportional fitness assignment. In the rank-based fitness assignment, the population is sorted according to the raw fitness values. The fitness assigned to each individual depends only on its position in the individual's ranking and not on the actual raw fitness value.

We use a linear rank-based fitness assignment, where the fitness value for an individual is calculated as:

$$\text{fitness}(\text{Pos}) = 2 - \text{SP} + 2(\text{SP} - 1)\frac{\text{Pos} - 1}{M - 1}, \tag{6.22}$$

where M is the number of individuals in the population. Pos is the position of an individual in this population (least fit individual has Pos $= 1$, the fittest individual Pos $= M$) and SP is the selective pressure. Linear ranking allows values of the parameter SP in $[1.0, 2.0]$. Our algorithm adopts 1.5 as the selective pressure.

Algorithm 3 Multi-objective Regularized Negatively Correlated Ensemble [9]

1: **Input:** Training Set $(\mathbf{x}_n, y_n)_{n=1}^N$, the number of population M, the number of iterations iter.
2: **Output:** the trained ensemble function f_{ens}
3: Generate an initial RBF network population.[3]
4: Train the initial RBF network population and recode the three objective values of each network.
5: Apply nondominated sorting with rank-based fitness assignment algorithm to obtain the rank-based fitness.
6: **for** $i = 1$ to iter **do**
7: Perform a desired number of crossover operations.[4]
8: Perform a desired number of mutation operations.
9: Apply nondominated sorting algorithm and obtain the rank-based fitness for the new population.
10: **end for**
11: Combine these classifiers to form the ensemble.

The details about MRNCL are summarized in Algorithm 3.

Note that in the crossover and mutation operations, the comparison of the child network with the parent network is conducted as follows.

1. Evaluate the three objective values of the child network.
2. Include the child network into the population, then apply nondominant sorting with a fitness sharing algorithm to obtain the raw fitness values[5] of the child network and the parent network.
3. Compare the raw fitness values and keep the better one.

To determine the time to stop evolution, we selected three threshold values ($t_1 = t_2 = t_3 = 10^{-3}$ in this paper) and compare the thresholds with the differences between the old minimal objective values with the new minimal objective values. If all the differences are lower than the thresholds, the algorithm will be terminated. The maximal number of generations is 200.

[3]To generate an initial RBF network population: Generate an initial population of M RBF Networks, the number of hidden nodes K for each network is specified randomly restricted by the maximal number of hidden nodes. The centers μ_k are initialized with randomly selected data points from the training set and the width σ_k are determined as the Euclidian distance between μ_k and the closest $\mu_j (j \neq k, j \in \{1, \dots, K\})$.

[4]Choose parents based on roulette wheel selection algorithm and perform crossover. Then perform a few number of updates for weights, centers, and widths. Compare the children with parents and keep the better ones.

[5]The raw fitness values depend on their ranked layers (fronts) in the population. If they are in the same layer (front), e.g., they are both nondominant solutions, the one in the less-crowded area will receive greater fitness according to the fitness sharing algorithm.

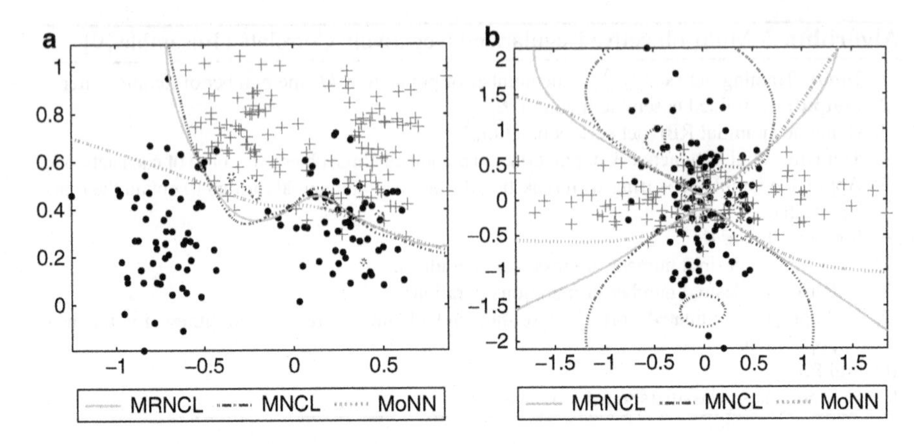

Fig. 6.5 Comparison of MRNCL, MNCL, and MoNN on two synthetic classification data sets. Two classes are shown as crosses and dots. The separating lines were obtained by projecting test data over a grid. (**a**) Synth. (**b**) Bumpy

6.4.3 Computational Examples

In the experiments, RBF networks are used as the individual classifiers. The number of hidden nodes is randomly selected but restricted in the range of 5–15. The parameters in the evolutionary algorithm are set to: the population size M (100), the number of crossovers in one generation 20, the number of mutations in one generation 10, the number of generations (200), the parameter of fitness sharing σ_{share} (0.2). These parameters are chosen after some preliminary experiments. They are not meant to be optimal.

In order to compare our algorithm with previous work on multi-objective ensemble learning, we have obtained the source code from Dr. Yaochu Jin and used the same parameters as their algorithm in [16]. This algorithm evolves multilayer perception (MLP) using two objectives (training error and regularization, i.e., number of connections in MLP) and we name the algorithm as multi-objective neural network (MoNN) in this section.

In the experiments, we restrict the minimal hidden nodes of RBF networks as three in MRNCL and multi-objective negative correlation learning (MNCL) to discourage improperly simple networks.

As the first experiment, we demonstrate the results of MRNCL on two synthetic data sets in two dimensions in order to illustrate graphically the decision boundaries.

These two data sets are synth and bumpy, as described in Subsect. 6.3.4.

In Fig. 6.5, we present a comparison of MRNCL, MNCL, and MoNN. We observe that MRNCL gives more accurate results in these two cases. In the cases of *Synth* and *Bumpy*, MRNCL produces smooth boundaries and disregards the outliers in the training sets. In the case of *Synth*, MoNN tries to use a *near–linear* boundary to separate the *nonlinear* data set consisted of four Gaussians. The generated model is overregularized and thus degrades performance. In the case of *Bumpy*, although

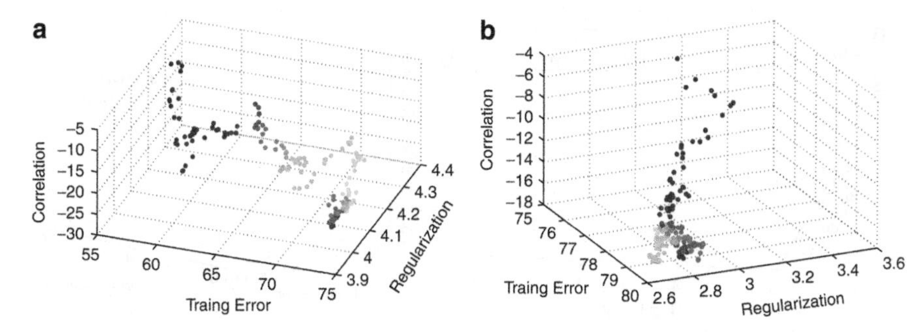

Fig. 6.6 Illustration of the mean value of these three objectives in different generations. The generation starts from the dark points to the light points. The gray scale indicates generations. (**a**) Synth. (**b**) Bumpy

the decision boundary of MoNN is smooth, it does not generate an appropriate boundary. (The optimal boundary is a quadratic one.) Since the noise level is large because of these overlapping points in the case of *Bumpy*, MNCL does not generalize well and produces the twisty boundary. In the case of *Synth*, MNCL concentrates on several outliers and generates a corner in the boundary.

Figure 6.6 illustrates the mean values of these three objectives in different generations. According to these figures, MRNCL algorithm tries to minimize the three objectives. However, the empirical training error is negatively correlated with the correlation term. Instead of minimizing the three objectives simultaneously, MRNCL seeks to find a good balance between the two negatively correlated terms (reduce one will increase another), i.e., training error and correlation term, and MRNCL always minimizes the third objective, the regularization term, in the evolutionary algorithm.

The 3D view of the last population is illustrated in Fig. 6.7. The negative correlation between the empirical error term and the correlation term[6] was confirmed by these figures. The final population distributes a good trade-off between these three objectives for all the data sets. According to this figure, we also notice that almost 80–90% of the solutions in the last generation are nondominated solutions.

Instead of combining all individual members in the population, ensemble selection and pruning algorithms can be used to generate compact yet powerful ensembles. Chen et al. [6] have proposed a probabilistic ensemble pruning algorithm using expectation propagation which can get an estimate of the leave-one-out (LOO) error. The LOO error is used together with Bayesian evidence for ensemble pruning.

---

[6]Negative correlation was used to indicate the correlation between on individual's error with the error of the rest of the ensemble. By minimizing the correlation term, i.e., $-\sum_{n=1}^{N}(f_i(\mathbf{x}_n) - f_{\text{ens}}(\mathbf{x}_n))^2$, the individual in the population will be more diverse, i.e., the term $\sum_{n=1}^{N}(f_i(\mathbf{x}_n) - f_{\text{ens}}(\mathbf{x}_n))^2$ increases. Therefore, the average training error term $\sum_{n=1}^{N}(f_i(\mathbf{x}_n) - y_n)^2$ will increase.

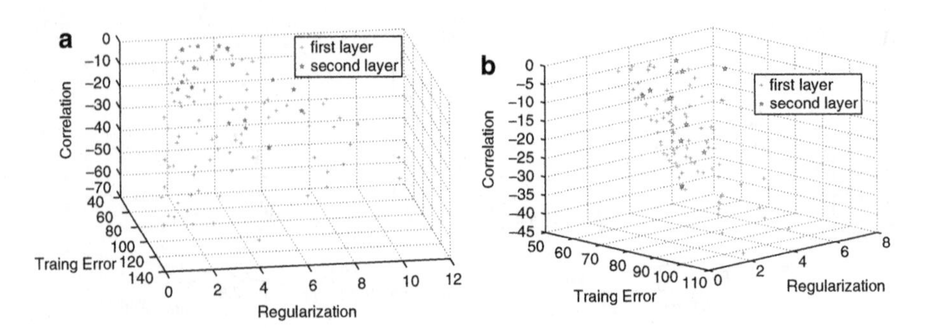

Fig. 6.7 Three-dimensional view of the last population with three objectives: training error, regularization, and correlation for two synthetic classification data sets. (**a**) Synth. (**b**) Bumpy

6.5 Summary

This chapter introduces NCL and demonstrates that NCL is prone to overfitting the noise. To overcome the shortcomings of NCL, RNCL was proposed, which incorporates an additional regularization term into NCL. Moreover, the Bayesian interpretation of RNCL was given and an algorithm to optimize the regularization parameter by Bayesian inference was presented.

This chapter also investigates RNCL from a multi-objective optimization point of view. The resulting algorithm can effectively search the best trade-off among three terms, i.e., empirical error, error correlation and regularization. To effectively evolve these networks, the crossover and mutation operators are defined to vary the structure of RBF networks. The nondominated sorting algorithm with fitness sharing and linear rank-based fitness assignment were employed to promote diversity in this algorithm.

The numerical results and visualization on some data sets have demonstrated that regularization is an important factor in ensemble construction, especially when the noise is nontrivial in data sets.

Compared with RNCL by gradient descent with Bayesian inference, the multi-objective implementation often achieves a little better performance by considering an additional weighting coefficient of the correlation term. The potential advantages of the multi-objective approach include: It enables us to observe the interaction and trade-off among different objectives; and it enables us to add or remove an objective easily without changing the overall algorithm. However, the better performance comes with the price, i.e., more computational time to train MRNCL.

In practice, whether to use RNCL or MRNCL depends on the application and users' specification. If users would like to observe the interaction and trade-off among different objectives and easily modify the code without changing the overall algorithm, MRNCL is more appropriate. If they pay more attention to the computational resource and prefer the explicit combination of coefficients, gradient descent-based RNCL with Bayesian inference is a better choice.

6.6 Bibliographical and Historical Remarks

Negative correlation learning [21, 22] is a successful neural network ensemble learning algorithm that has been researched for decades. In 2001, McKay et al. [25] presented an alternative anti-correlation measure, root-quartic negative correlation learning (RTQRT-NCL), and used the anti-correlation in training neural network ensembles. The empirical results showed significant improvements for both neural networks and genetic programming learning machines. They also derived a theoretical explanation of the improved performance of RTQRT-NCL in larger ensembles.

Later, Islam et al. [15] took a constructive approach to building the ensemble, starting from a small group of networks with a minimal architecture. The networks are all partially trained using NCL. The approach can automatically determine weights, network topologies, and ensemble membership. In the following work, Brown et al. [2] formalized NCL, providing a statistical interpretation of its success. Furthermore, for estimators that are linear combinations of other functions, they derive an upper bound on the penalty coefficient, based on properties of the Hessian matrix. Then, Chandra et al. [4,5] proposed a diverse and accurate ensemble learning algorithm, combining evolving neural network and multi-objective algorithm.

In 2007, Chen et al. [7] proposed to incorporate bootstrapping of data, random feature subspaces [14] and evolutionary algorithms with NCL to automatically design accurate and diverse ensembles. The idea promotes the diversity within the ensemble and simultaneously emphasizes the accuracy and cooperation in the ensemble. Dam et al. [10] applied the NCL algorithm to train the neural network ensemble in learning classifier systems, where NCL is shown to improve the generalization of the ensemble.

In [8, 9], Chen and Yao propose the RNCL algorithm and multi-objective regularized negative correlation learning algorithms. The results have shown superior performance with the regularization term when dealing with noisy data. In addition to these development of negatively correlated ensembles, Wang et al. [33–35] have investigated the application of negatively correlated ensembles to imbalanced data, and Minku et al. [26, 27] have investigated the use of negative correlation in incremental learning, and discussed its strong and weak points to incremental and online learning.

Acknowledgements This work has been funded by the European Commission's 7th Framework Program, under grant Agreement INSFO-ICT-270428 (iSense).

References

1. C. M. Bishop. *Neural Networks for Pattern Recognition*. Oxford University Press, USA, 1996.
2. G. Brown, J. Wyatt, R. Harris, and X. Yao. Diversity creation methods: A survey and categorisation. *Journal of Information Fusion*, 6(1):5–20, 2005.

3. G. Brown, J. Wyatt, and P. Tiňo. Managing diversity in regression ensembles. *Journal of Machine Learning Research*, 6:1621–1650, 2005.
4. A. Chandra and X. Yao. Ensemble learning using multi-objective evolutionary algorithms. *Journal of Mathematical Modelling and Algorithms*, 5(4):417–445, 2006.
5. A. Chandra and X. Yao. Ensemble learning using multi-objective evolutionary algorithms. *Journal of Mathematical Modelling and Algorithms*, 5(4):417–445, 2006.
6. H. Chen, P. Tiňo, and X. Yao. Predictive ensemble pruning by expectation propagation. *IEEE Transactions on Knowledge and Data Engineering*, 21(7):999–1013, 2009.
7. H. Chen and X. Yao. Evolutionary random neural ensemble based on negative correlation learning. In *Proceedings of IEEE Congress on Evolutionary Computation (CEC'07)*, pp. 1468–1474, 2007.
8. H. Chen and X. Yao. Regularized negative correlation learning for neural network ensembles. *IEEE Transactions on Neural Networks*, 20(12):1962–1979, 2009.
9. H. Chen and X. Yao. Multiobjective regularized negative correlation learning for neural network ensembles. *IEEE Transactions on Knowledge and Data Engineering*, 22(12): 1738–1751, 2010.
10. H. H. Dam, H. A. Abbass, C. Lokan, and X. Yao. Neural-based learning classifier systems. *IEEE Transactions on Knowledge and Data Engineering*, 20(1):26–39, 2008.
11. P. Darwen and X. Yao. Every niching method has its niche: fitness sharing and implicit sharing compared. In *Proceedings of Parallel Problem Solving from Nature (PPSN) IV*, volume 1141, pp. 398–407, Berlin, Germany, 1996.
12. S. Geman, E. Bienenstock, and R. Doursat. Neural networks and the bias/variance dilemma. *Neural Computation*, 4(1):1–58, 1992.
13. T. Van Gestel, J. A. K. Suykens, G. Lanckriet, A. Lambrechts, B. De Moor, and J. Vandewalle. Bayesian framework for least-squares support vector machine classifiers, gaussian processes, and kernel fisher discriminant analysis. *Neural Computation*, 14(5):1115–1147, 2002.
14. T. K. Ho. The random subspace method for constructing decision forests. *IEEE Transaction on Pattern Analysis and Machine Intelligence*, 20(8):832–844, 1998.
15. M. M. Islam, X. Yao, and K. Murase. A constructive algorithm for training cooperative neural network ensembles. *IEEE Transaction on Neural Networks*, 14(4):820–834, 2003.
16. Y. Jin, T. Okabe, and B. Sendhoff. Neural network regularization and ensembling using multi-objective evolutionary algorithms. In *Proceedings of IEEE Congress on Evolutionary Computation (CEC'04)*, pages 1–8, 2004.
17. A. Krogh and J. A. Hertz. A simple weight decay can improve generalization. In *Advances in Neural Information Processing Systems*, volume 4, pp. 950–957, 1992.
18. A. Krogh and J. A. Hertz. A simple weight decay can improve generalization. *Advances in Neural Information Processing Systems*, pp. 950–950, 1993.
19. A. Krogh and J. Vedelsby. Neural network ensembles, cross validation, and active learning. In *Advances in Neural Information Processing Systems 7*, pp. 231–238, Denver, Colorado, USA, 1995.
20. Y. Liu and X. Yao. Negatively correlated neural networks can produce best ensembles. In *Australian Journal of Intelligent Information Processing Systems 4(3/4)*, pp. 176–185, 1997.
21. Y. Liu and X. Yao. Ensemble learning via negative correlation. *Neural Networks*, 12(10): 1399–1404, 1999.
22. Y. Liu and X. Yao. Simultaneous training of negatively correlated neural networks in an ensemble. *IEEE Transactions on Systems, Man, and Cybernetics, Part B: Cybernetics*, 29(6):716–725, 1999.
23. Y. Liu, X. Yao, and T. Higuchi. Evolutionary ensembles with negative correlation learning. *IEEE Transaction on Evolutionary Computation*, 4(4):380–387, 2000.
24. D. J. C. MacKay. Bayesian interpolation. *Neural Computation*, 4(3):415–447, 1992.
25. R. McKay and H. Abbass. Analyzing anticorrelation in ensemble learning. In *Proceedings of 2001 Conference on Australian Artificial Neural Networks and Expert Systems*, pp. 22–27, 2001.

26. F. L. Minku, H. Inoue, and X. Yao. Negative correlation in incremental learning. *Natural Computing*, 8(2):289–320, 2009.
27. L. L. Minku, A. White, and X. Yao. The impact of diversity on on-line ensemble learning in the presence of concept drift. *IEEE Transactions on Knowledge and Data Engineering*, 22(5): 730–742, 2010.
28. M. F. Møller. A scaled conjugate gradient algorithm for fast supervised learning. *Neural Network*, 6(4):525–533, 1993.
29. R. M. Neal. *Bayesian learning for neural networks*. Springer, New York, 1996.
30. B. D. Ripley. *Pattern Recognition and Neural Networks*. Cambridge University Press, UK, 1996.
31. N. Srinivas and K. Deb. Multiobjective function optimization using nondominated sorting genetic algorithms. *Evolutionary Computation*, 2(3):221–248, 1995.
32. V. N. Vapnik. *The Nature of Statistical Learning Theory*. New York: Springer-Verlag, 1995.
33. S. Wang, H. Chen, and X. Yao. Negative correlation learning for classification ensembles. In *Proceedings of the 2010 International Joint Conference on Neural Networks (IJCNN'10)*, pp. 2893–2900, 2010.
34. S. Wang and X. Yao. Diversity analysis on imbalanced data sets by using ensemble models. In *Proceedings of the 2009 IEEE Symposium on Computational Intelligence and Data Mining (CIDM'09)*, pp. 324–331, 2009.
35. S. Wang and X. Yao. Diversity exploration and negative correlation learning on imbalanced data sets. In *Proceedings of the 2009 International Joint Conference on Neural Networks (IJCNN'09)*, pp. 3259–3266, 2009.
36. X. Yao. Evolving artificial neural networks. *Proceedings of the IEEE*, 87(9):1423–1447, 1999.

Chapter 7
Ensemble Nyström

Sanjiv Kumar, Mehryar Mohri, and Ameet Talwalkar

7.1 Introduction

A common problem in many areas of large-scale machine learning involves manipulation of a large matrix. This matrix may be a kernel matrix arising in Support Vector Machines [9, 15], Kernel Principal Component Analysis [47], or manifold learning [43, 51]. Large matrices also naturally arise in other applications, e.g., clustering, collaborative filtering, matrix completion, and robust PCA. For these large-scale problems, the number of matrix entries can easily be in the order of billions or more, making them hard to process or even store. An attractive solution to this problem involves the Nyström method, in which one samples a small number of columns from the original matrix and generates its low-rank approximation using the sampled columns [53]. The accuracy of the Nyström method depends on the number columns sampled from the original matrix. The larger the number of samples, the higher the accuracy but slower the method.

In the Nyström method, one needs to perform Singular Value Decomposition (SVD) on a $l \times l$ matrix where l is the number of columns sampled from the original matrix. This SVD operation is typically carried out on a single machine. Thus, the maximum value of l used for an application is limited by the capacity

S. Kumar (✉)
76, Ninth Avenue, Google Inc., New York, NY 10011, USA
e-mail: sanjivk@google.com

M. Mohri
Courant Institute, New York, NY, USA
e-mail: mohri@cs.nyu.edu

A. Talwalkar
Division of Computer Science, University of California, Berkeley, CA, USA
e-mail: ameet@eecs.berkeley.edu

C. Zhang and Y. Ma (eds.), *Ensemble Machine Learning: Methods and Applications*,
DOI 10.1007/978-1-4419-9326-7_7, © Springer Science+Business Media, LLC 2012

of the machine. That is why in practice, one restricts l to be less than 20 K or 30 K, even when the size of matrix is in millions. This restricts the accuracy of the Nyström method in very large-scale settings.

This chapter describes a family of algorithms based on mixtures of Nyström approximations called, *Ensemble Nyström algorithms*, which yields more accurate low-rank approximations than the standard Nyström method. The core idea of Ensemble Nyström is to sample many subsets of columns from the original matrix, each containing a relatively small number of columns. Then, Nyström method is performed on each group independently in parallel, and the results are combined yielding high accuracy. These ensemble algorithms naturally fit within distributed computing environments where their computational costs are roughly the same as that of the standard Nyström method. This issue is of great practical significance given the prevalence of distributed computing frameworks to handle large-scale learning problems. Several variants of these algorithms are described, including one based on simple averaging of p Nyström solutions, an exponential weighting method, and a regression based method which consists of estimating the mixture parameters using a few sampled columns.

In Sect. 7.2, we first introduce the notation and basic concepts of low-rank matrix approximation. The standard Nyström method is also described. Then, we present a number of Ensemble Nyström algorithms in Subsect. 7.2.2. In many applications, one needs inverse of a large matrix, e.g., SVM and Gaussian Processes. Deriving approximate inverse using the standard Nyström method is easy but not so for the Ensemble Nyström. We further show in Subsect. 7.2.3 how one can efficiently use Woodbury's approximation with Ensemble Nyström to generate approximate inverses.

Another interesting aspect of the Ensemble Nyström methods is their theoretical properties that give explicit bounds for the reconstruction error for both the Frobenius norm and the spectral norm. In Subsect. 7.3, we give a derivation of these bounds. These arise by developing a different bound for the standard Nyström method as used in practice, i.e., using uniform random sampling of columns without replacement. These novel generalization bounds guarantee a better convergence rate for Ensemble Nyström algorithms in comparison to the standard Nyström method.

Section 7.4 demonstrates the results from Ensemble Nyström algorithms on multiple datasets. A comprehensive comparison against other methods shows clear performance gains over the standard Nyström method. Section 7.4.2 describes a large-scale experiment with 1 M points leading to a matrix of size 1 M $\times$ 1 M. This is a huge dense matrix, containing 1 trillion entries and its explicit storage would require 4 TB space. We show that sampling-based methods can easily handle such matrices and the proposed Ensemble Nyström outperforms other state-of-the-art methods for a fixed computational budget.

To conclude, we provide a summary of the chapter and discuss several open questions in Sect. 7.5. Further, related work is mentioned in Sect. 7.6.

7.2 Algorithms

Let $\mathbf{T} \in \mathbb{R}^{a\times b}$ be an arbitrary matrix. We define $\mathbf{T}^{(j)}$, $j = 1 \ldots b$, as the jth column vector of $\mathbf{T}$, $\mathbf{T}_{(i)}$, $i = 1 \ldots a$, as the ith row vector of $\mathbf{T}$ and $\|\cdot\|$ the l_2 norm of a vector. Furthermore, $\mathbf{T}^{(i:j)}$ refers to the ith through jth columns of $\mathbf{T}$ and $\mathbf{T}_{(i:j)}$ refers to the ith through jth rows of $\mathbf{T}$. If rank($\mathbf{T}$) $= r$, we can write the thin SVD of this matrix as $\mathbf{T} = \mathbf{U}_T \Sigma_T \mathbf{V}_T^\top$ where $\Sigma_T \in \mathbb{R}^{r\times r}$ is diagonal and contains the singular values of $\mathbf{T}$ sorted in decreasing order and $\mathbf{U}_T \in \mathbb{R}^{a\times r}$ and $\mathbf{V}_T \in \mathbb{R}^{b\times r}$ have orthogonal columns that contain the left and right singular vectors of $\mathbf{T}$ corresponding to its singular values. We denote by $\mathbf{T}_k$ the "best" rank-k approximation to $\mathbf{T}$, i.e., $\mathbf{T}_k = \arg\min_{\mathbf{V}\in\mathbb{R}^{a\times b}, \text{rank}(\mathbf{V})=k} \|\mathbf{T}-\mathbf{V}\|_\xi$, where $\xi \in \{2, F\}$ and $\|\cdot\|_2$ denotes the spectral norm and $\|\cdot\|_F$ the Frobenius norm of a matrix. We can describe this matrix in terms of its SVD as $\mathbf{T}_k = \mathbf{U}_{T,k} \Sigma_{T,k} \mathbf{V}_{T,k}^\top$ where $\Sigma_{T,k}$ is a diagonal matrix of the top k singular values of $\mathbf{T}$ and $\mathbf{U}_{T,k}$ and $\mathbf{V}_{T,k}$ are the associated left and right singular vectors.

Now let $\mathbf{K} \in \mathbb{R}^{n\times n}$ be a symmetric positive semidefinite (SPSD) kernel or Gram matrix with rank($\mathbf{K}$) $= r \leq n$, i.e., a symmetric matrix for which there exists an $\mathbf{X} \in \mathbb{R}^{N\times n}$ such that $\mathbf{K} = \mathbf{X}^\top\mathbf{X}$. We will write the SVD of $\mathbf{K}$ as $\mathbf{K} = \mathbf{U}\Sigma\mathbf{U}^\top$, where the columns of $\mathbf{U}$ are orthogonal and $\Sigma = \text{diag}(\sigma_1, \ldots, \sigma_r)$ is diagonal. The pseudo-inverse of $\mathbf{K}$ is defined as $\mathbf{K}^+ = \sum_{t=1}^r \sigma_t^{-1}\mathbf{U}^{(t)}\mathbf{U}^{(t)\top}$, and $\mathbf{K}^+ = \mathbf{K}^{-1}$ when $\mathbf{K}$ is full rank. For $k < r$, $\mathbf{K}_k = \sum_{t=1}^k \sigma_t \mathbf{U}^{(t)}\mathbf{U}^{(t)\top} = \mathbf{U}_k \Sigma_k \mathbf{U}_k^\top$ is the "best" rank-k approximation to $\mathbf{K}$, i.e., $\mathbf{K}_k = \arg\min_{\mathbf{K}'\in\mathbb{R}^{n\times n}, \text{rank}(\mathbf{K}')=k} \|\mathbf{K} - \mathbf{K}'\|_{\xi\in\{2,F\}}$, with $\|\mathbf{K} - \mathbf{K}_k\|_2 = \sigma_{k+1}$ and $\|\mathbf{K} - \mathbf{K}_k\|_F = \sqrt{\sum_{t=k+1}^r \sigma_t^2}$ [23].

We will be focusing on generating an approximation $\widetilde{\mathbf{K}}$ of $\mathbf{K}$ based on a sample of $l \ll n$ of its columns. We assume that l columns are sampled from $\mathbf{K}$ uniformly without replacement. Let $\mathbf{C}$ denote the $n \times l$ matrix formed by these columns and $\mathbf{W}$ the $l \times l$ matrix consisting of the intersection of these l columns with the corresponding l rows of $\mathbf{K}$. Note that $\mathbf{W}$ is SPSD since $\mathbf{K}$ is SPSD. Without loss of generality, the columns and rows of $\mathbf{K}$ can be rearranged based on this sampling so that $\mathbf{K}$ and $\mathbf{C}$ can be written as follows:

$$\mathbf{K} = \begin{bmatrix} \mathbf{W} & \mathbf{K}_{21}^\top \\ \mathbf{K}_{21} & \mathbf{K}_{22} \end{bmatrix} \quad \text{and} \quad \mathbf{C} = \begin{bmatrix} \mathbf{W} \\ \mathbf{K}_{21} \end{bmatrix}. \tag{7.1}$$

7.2.1 *Standard Nyström Method*

The Nyström method uses $\mathbf{W}$ and $\mathbf{C}$ from (7.1) to approximate $\mathbf{K}$. Assuming a uniform sampling of the columns, the Nyström method generates a rank-k approximation $\widetilde{\mathbf{K}}$ of $\mathbf{K}$ for $k < n$ defined by:

$$\widetilde{\mathbf{K}}_k^{\text{nys}} = \mathbf{C}\mathbf{W}_k^+\mathbf{C}^\top \approx \mathbf{K}, \tag{7.2}$$

where $\mathbf{W}_k$ is the best k-rank approximation of $\mathbf{W}$ with respect to the spectral or Frobenius norm and $\mathbf{W}_k^+$ denotes the pseudo-inverse of $\mathbf{W}_k$. The Nyström method thus approximates the top k singular values (Σ_k) and singular vectors ($\mathbf{U}_k$) of $\mathbf{K}$ as:

$$\widetilde{\Sigma}_k^{\text{nys}} = \left(\frac{n}{l}\right)\Sigma_{W,k} \quad \text{and} \quad \widetilde{\mathbf{U}}_k^{\text{nys}} = \sqrt{\frac{l}{n}}\mathbf{C}\mathbf{U}_{W,k}\Sigma_{W,k}^+, \tag{7.3}$$

where $\Sigma_{W,k}$ contains the top k singular values of $\mathbf{W}$, and $\mathbf{U}_{W,k}$ contains the corresponding singular vectors. When $k = l$ (or more generally, whenever $k \geq \text{rank}(\mathbf{C})$), this approximation perfectly reconstructs three blocks of $\mathbf{K}$, and $\mathbf{K}_{22}$ is approximated by the Schur Complement of $\mathbf{W}$ in $\mathbf{K}$:

$$\widetilde{\mathbf{K}}_l^{\text{nys}} = \mathbf{C}\mathbf{W}^+\mathbf{C}^\top = \begin{bmatrix} \mathbf{W} & \mathbf{K}_{21}^\top \\ \mathbf{K}_{21} & \mathbf{K}_{21}\mathbf{W}^+\mathbf{K}_{21} \end{bmatrix}. \tag{7.4}$$

The time complexity of SVD on $\mathbf{W}$ to get top k singular values and vectors is $O(kl^2)$ and matrix multiplication with $\mathbf{C}$ takes $O(kln)$. Hence, the total computational complexity of the Nyström approximation is $O(kln)$ since $n \gg l$.

7.2.2 Ensemble Nyström

In this section, we discuss a meta algorithm called the Ensemble Nyström algorithm. We treat each approximation generated by the Nyström method for a sample of l columns as an *expert* and combine $p \geq 1$ such experts to derive an improved hypothesis, typically more accurate than any of the original experts.

The learning set-up is defined as follows. We assume a fixed kernel function $K: \mathscr{X} \times \mathscr{X} \to \mathbb{R}$ that can be used to generate the entries of a kernel matrix $\mathbf{K}$. The learner receives a set S of lp columns randomly selected from matrix $\mathbf{K}$ uniformly without replacement. S is decomposed into p subsets $S_1, \ldots, S_p$. Each subset S_r, $r \in [1, p]$, contains l columns and is used to define a rank-k Nyström approximation $\widetilde{\mathbf{K}}_r$. Dropping the rank subscript k in favor of the sample index r, $\widetilde{\mathbf{K}}_r$ can be written as $\widetilde{\mathbf{K}}_r = \mathbf{C}_r\mathbf{W}_r^+\mathbf{C}_r^\top$, where $\mathbf{C}_r$ and $\mathbf{W}_r$ denote the matrices formed from the columns of S_r and $\mathbf{W}_r^+$ is the pseudo-inverse of the rank-k approximation of $\mathbf{W}_r$. The learner further receives a sample V of s columns used to determine the weight $\mu_r \in \mathbb{R}$ attributed to each expert $\widetilde{\mathbf{K}}_r$. Thus, the general form of the approximation, $\mathbf{K}^{\text{ens}}$, generated by the Ensemble Nyström algorithm, with $k \leq \text{rank}(\mathbf{K}^{\text{ens}}) \leq pk$, is

$$\widetilde{\mathbf{K}}^{\text{ens}} = \sum_{r=1}^{p} \mu_r \widetilde{\mathbf{K}}_r \tag{7.5}$$

$$= \begin{bmatrix} \mathbf{C}_1 & & \\ & \ddots & \\ & & \mathbf{C}_p \end{bmatrix} \begin{bmatrix} \mu_1\mathbf{W}_1^+ & & \\ & \ddots & \\ & & \mu_p\mathbf{W}_p^+ \end{bmatrix} \begin{bmatrix} \mathbf{C}_1 & & \\ & \ddots & \\ & & \mathbf{C}_p \end{bmatrix}^\top. \tag{7.6}$$

As noted by [36], (7.6) provides an alternative description of the Ensemble Nyström method as a block diagonal approximation of $\mathbf{W}_{\text{ens}}^{+}$, where $\mathbf{W}_{\text{ens}}$ is the $lp \times lp$ SPSD matrix associated with the lp sampled columns.

The mixture weights μ_r can be defined in many ways. The most straightforward choice consists of assigning equal weight to each expert, $\mu_r = 1/p$, $r \in [1, p]$. This choice does not require the additional sample V, but it ignores the relative quality of each Nyström approximation. Nevertheless, this simple *uniform method* already generates a solution superior to any one of the approximations $\widetilde{\mathbf{K}}_r$ used in the combination, as we shall see in the experimental section.

Another method, the *exponential weight method*, consists of measuring the reconstruction error $\hat{\epsilon}_r$ of each expert $\widetilde{\mathbf{K}}_r$ over the validation sample V and defining the mixture weight as $\mu_r = \exp(-\eta\hat{\epsilon}_r)/Z$, where $\eta > 0$ is a parameter of the algorithm and Z a normalization factor ensuring that the vector $\boldsymbol{\mu} = (\mu_1, \ldots, \mu_p)$ belongs to the simplex Δ of $\mathbb{R}^p$: $\Delta = \left\{\boldsymbol{\mu} \in \mathbb{R}^p : \boldsymbol{\mu} \geq 0 \wedge \sum_{r=1}^{p} \mu_r = 1\right\}$. The choice of the mixture weights here is similar to that used in the Weighted Majority algorithm [38]. Let $\mathbf{K}_V$ denote the matrix formed by using the samples from V as its columns and let $\widetilde{\mathbf{K}}_r^V$ denote the submatrix of $\widetilde{\mathbf{K}}_r$ containing the columns corresponding to the columns in V. The reconstruction error $\hat{\epsilon}_r = \left\|\widetilde{\mathbf{K}}_r^V - \mathbf{K}_V\right\|$ can be directly computed from these matrices.

A more general class of methods consists of using the sample V to train the mixture weights μ_r to optimize a regression objective function such as the following:

$$\min_{\boldsymbol{\mu}} \lambda \|\boldsymbol{\mu}\|_2^2 + \left\|\sum_{r=1}^{p} \mu_r \widetilde{\mathbf{K}}_r^V - \mathbf{K}_V\right\|_F^2, \tag{7.7}$$

where $\mathbf{K}_V$ denotes the matrix formed by the columns of the samples V and $\lambda > 0$. This can be viewed as a ridge regression objective function and admits a closed form solution. We will refer to this method as the *ridge regression method*. Note that to ensure that the resulting matrix is SPSD for use in subsequent kernel-based algorithms, the optimization problem must be augmented with standard nonnegativity constraints. This is not necessary, however, for reducing the reconstruction error, as in our experiments. Also, clearly, a variety of other regression algorithms such as Lasso can be used here instead.

The total complexity of the Ensemble Nyström algorithm is $O(pl^3 + plkn + C_\mu)$, where C_μ is the cost of computing the mixture weights, $\boldsymbol{\mu}$, used to combine the p Nyström approximations. In general, the cubic term dominates the complexity since the mixture weights can be computed in constant time for the uniform method, in $O(psn)$ for the exponential weight method, or in $O(p^3 + pls)$ for the ridge regression method. Furthermore, although the Ensemble Nyström algorithm requires p times more space and CPU cycles than the standard Nyström method, these additional requirements are quite reasonable in practice. The space requirement is still manageable for even large-scale applications given that p is typically O(1) and l is usually a very small percentage of n (see Sect. 7.4 for further details). In terms of CPU requirements, we note that this algorithm can be easily parallelized, as all p experts can be computed simultaneously. Thus, with a cluster of p machines, the running time complexity of this algorithm is nearly equal to that of the standard Nyström algorithm with l samples.

7.2.3 Ensemble Woodbury Approximation

In many applications, one needs to invert a matrix $(\mathbf{K} + \lambda\mathbf{I})$, where λ is a positive scalar and $\mathbf{I}$ is the identity matrix. The Woodbury approximation is a useful tool to use alongside low-rank approximations to efficiently (and approximately) invert kernel matrices. We are able to apply the Woodbury approximation since the Nyström method represents $\widetilde{\mathbf{K}}$ as the product of low-rank matrices. This is clear from the definition of the Woodbury approximation:

$$(\mathbf{A} + \mathbf{BCd})^{-1} = \mathbf{A}^{-1} - \mathbf{A}^{-1}\mathbf{B}\left(\mathbf{C}^{-1} + \mathrm{d}\mathbf{A}^{-1}\mathbf{B}\right)^{-1}\mathrm{d}\mathbf{A}^{-1}, \tag{7.8}$$

where $\mathbf{A} = \lambda\mathbf{I}$ and $\widetilde{\mathbf{K}} = \mathbf{BCd}$ in the context of the Nyström method. In contrast, the Ensemble Nyström method represents $\widetilde{\mathbf{K}}$ as the sum of products of low-rank matrices, where each of the p terms corresponds to a base learner. Hence, we cannot directly apply the Woodbury approximation as presented above. There is, however, a natural extension of the Woodbury approximation in this setting, which at the simplest level involves running the approximation p times. Starting with p base learners with their associated weights, i.e., $\widetilde{\mathbf{K}}_r$ and μ_r for $r \in [1, p]$, and defining $\mathbf{T}_0 = \lambda\mathbf{I}$, we perform the following series of calculations:

$$\begin{aligned}
\mathbf{T}_1^{-1} &= \left(\mathbf{T}_0 + \mu_1\widetilde{\mathbf{K}}_1\right)^{-1} \\
\mathbf{T}_2^{-1} &= \left(\mathbf{T}_1 + \mu_2\widetilde{\mathbf{K}}_2\right)^{-1} \\
&\cdots \\
\mathbf{T}_p^{-1} &= \left(\mathbf{T}_{p-1} + \mu_p\widetilde{\mathbf{K}}_p\right)^{-1}.
\end{aligned}$$

To compute $\mathbf{T}_1^{-1}$, notice that we can use Woodbury approximation as stated in (7.8) since we can express $\mu_1\widetilde{\mathbf{K}}_1$ as the product of low-rank matrices and we know that $T_0^{-1} = \frac{1}{\lambda}\mathbf{I}$. More generally, for $1 \le i \le p$, given an expression of T_{i-1}^{-1} as a product of low-rank matrices, we can efficiently compute T_i^{-1} using the Woodbury approximation (we use the low-rank structure to avoid ever computing or storing a full $n \times n$ matrix). Hence, after performing this series of p calculations, we are left with the inverse of $\mathbf{T}_p$, which is exactly the quantity of interest since $\mathbf{T}_p = \lambda\mathbf{I} + \sum_{r=1}^{p}\mu_r\widetilde{\mathbf{K}}_r$. Although this algorithm requires p iterations of the Woodbury approximation, these iterations can be parallelized in a tree-like fashion.

Hence, when working on a cluster, using an Ensemble Nyström approximation along with the Woodbury approximation requires only $\log_2(p)$ more time than using the standard Nyström method.

7.3 Theoretical Analysis

We now present theoretical results that compare the quality of the Nyström approximation to the "best" low-rank approximation, i.e., the approximation constructed from the top singular values and singular vectors of $\mathbf{K}$. This work, related to [18], provides performance bounds for the Nyström method as used in practice, i.e., using uniform sampling without replacement. It holds for both the standard Nyström method as well as the Ensemble Nyström method discussed in Subsect. 7.2.2.

Our theoretical analysis of the Nyström method uses some results previously shown by [18] as well as the following generalization of McDiarmid's concentration bound to sampling without replacement [13].

Theorem 1. *Let $Z_1, \ldots, Z_l$ be a sequence of random variables sampled uniformly without replacement from a fixed set of $l+u$ elements Z, and let $\phi\colon Z^l \to \mathbb{R}$ be a symmetric function such that for all $i \in [1, l]$ and for all $z_1, \ldots, z_l \in Z$ and $z'_1, \ldots, z'_l \in Z$, $\left|\phi(z_1, \ldots, z_l) - \phi(z_1, \ldots, z_{i-1}, z'_i, z_{i+1}, \ldots, z_l)\right| \leq c$. Then, for all $\epsilon > 0$, the following inequality holds:*

$$\Pr\left[\phi - \mathrm{e}[\phi] \geq \epsilon\right] \leq \exp\left[\frac{-2\epsilon^2}{\alpha(l, u)c^2}\right], \tag{7.9}$$

where $\alpha(l, u) = \frac{lu}{l+u-1/2} \frac{1}{1-1/(2\max\{l,u\})}$.

We define the *selection matrix* corresponding to a sample of l columns as the matrix $\mathbf{S} \in \mathbb{R}^{n \times l}$ defined by $\mathbf{S}_{ii} = 1$ if the ith column of $\mathbf{K}$ is among those sampled, $\mathbf{S}_{ij} = 0$ otherwise. Thus, $\mathbf{C} = \mathbf{KS}$ is the matrix formed by the columns sampled. Since $\mathbf{K}$ is SPSD, there exists $\mathbf{X} \in \mathbb{R}^{N \times n}$ such that $\mathbf{K} = \mathbf{X}^\top \mathbf{X}$. We shall denote by $\mathbf{K}_{\max}$ the maximum diagonal entry of $\mathbf{K}$, $\mathbf{K}_{\max} = \max_i \mathbf{K}_{ii}$, and by $d^{\mathbf{K}}_{\max}$ the distance $\max_{ij} \sqrt{\mathbf{K}_{ii} + \mathbf{K}_{jj} - 2\mathbf{K}_{ij}}$.

7.3.1 Standard Nyström Method

The following theorem gives an upper bound on the norm-2 error of the Nyström approximation of the form $\|\mathbf{K} - \widetilde{\mathbf{K}}\|_2 / \|\mathbf{K}\|_2 \leq \|\mathbf{K} - \mathbf{K}_k\|_2 / \|\mathbf{K}\|_2 + O\left(1/\sqrt{l}\right)$ and an upper bound on the Frobenius error of the Nyström approximation of the form $\|\mathbf{K} - \widetilde{\mathbf{K}}\|_F / \|\mathbf{K}\|_F \leq \|\mathbf{K} - \mathbf{K}_k\|_F / \|\mathbf{K}\|_F + O\left(1/l^{\frac{1}{4}}\right)$.

Theorem 2. *Let* $\widetilde{\mathbf{K}}$ *denote the rank-k Nyström approximation of* $\mathbf{K}$ *based on* l *columns sampled uniformly at random without replacement from* $\mathbf{K}$*, and* $\mathbf{K}_k$ *the best rank-k approximation of* $\mathbf{K}$*. Then, with probability at least* $1-\delta$*, the following inequalities hold for any sample of size* l*:*

$$\left\|\mathbf{K}-\widetilde{\mathbf{K}}\right\|_2 \leq \|\mathbf{K}-\mathbf{K}_k\|_2 + \tfrac{2n}{\sqrt{l}}\mathbf{K}_{\max}\left[1+\sqrt{\tfrac{n-l}{n-1/2}\tfrac{1}{\beta(l,n)}\log\tfrac{1}{\delta}}\, d^{\mathbf{K}}_{\max}\Big/\mathbf{K}^{\frac{1}{2}}_{\max}\right]$$

$$\left\|\mathbf{K}-\widetilde{\mathbf{K}}\right\|_F \leq \|\mathbf{K}-\mathbf{K}_k\|_F$$

$$+\left[\tfrac{64k}{l}\right]^{\frac{1}{4}} n\mathbf{K}_{\max}\left[1+\sqrt{\tfrac{n-l}{n-1/2}\tfrac{1}{\beta(l,n)}\log\tfrac{1}{\delta}}\, d^{\mathbf{K}}_{\max}\Big/\mathbf{K}^{\frac{1}{2}}_{\max}\right]^{\frac{1}{2}},$$

where $\beta(l,n)=1-\frac{1}{2\max\{l,n-l\}}$.

Proof. To bound the norm-2 error of the Nyström method in the scenario of sampling without replacement, we start with the following general inequality given by [18, proof of Lemma 4]:

$$\left\|\mathbf{K}-\widetilde{\mathbf{K}}\right\|_2 \leq \|\mathbf{K}-\mathbf{K}_k\|_2 + 2\left\|\mathbf{X}\mathbf{X}^\top-\mathbf{Z}\mathbf{Z}^\top\right\|_2, \tag{7.10}$$

where $\mathbf{Z}=\sqrt{\frac{n}{l}}\mathbf{X}\mathbf{S}$. We then apply the McDiarmid-type inequality of Theorem 1 to $\phi(\mathbf{S})=\left\|\mathbf{X}\mathbf{X}^\top-\mathbf{Z}\mathbf{Z}^\top\right\|_2$. Let $\mathbf{S}'$ be a sampling matrix selecting the same columns as $\mathbf{S}$ except for one, and let $\mathbf{Z}'$ denote $\sqrt{\frac{n}{l}}\mathbf{X}\mathbf{S}'$. Let $\mathbf{z}$ and $\mathbf{z}'$ denote the only differing columns of $\mathbf{Z}$ and $\mathbf{Z}'$, then

$$|\phi(\mathbf{S}')-\phi(\mathbf{S})| \leq \left\|\mathbf{z}'\mathbf{z}'^\top-\mathbf{z}\mathbf{z}^\top\right\|_2 = \left\|(\mathbf{z}'-\mathbf{z})\,\mathbf{z}'^\top+\mathbf{z}\,(\mathbf{z}'-\mathbf{z})^\top\right\|_2 \tag{7.11}$$

$$\leq 2\left\|\mathbf{z}'-\mathbf{z}\right\|_2 \max\left\{\|\mathbf{z}\|_2, \left\|\mathbf{z}'\right\|_2\right\}. \tag{7.12}$$

Columns of $\mathbf{Z}$ are those of $\mathbf{X}$ scaled by $\sqrt{n/l}$. The norm of the difference of two columns of $\mathbf{X}$ can be viewed as the norm of the difference of two feature vectors associated to $\mathbf{K}$ and thus can be bounded by $d_{\mathbf{K}}$. Similarly, the norm of a single column of $\mathbf{X}$ is bounded by $\mathbf{K}^{\frac{1}{2}}_{\max}$. This leads to the following inequality:

$$|\phi(\mathbf{S}')-\phi(\mathbf{S})| \leq \frac{2n}{l} d^{\mathbf{K}}_{\max}\mathbf{K}^{\frac{1}{2}}_{\max}. \tag{7.13}$$

The expectation of ϕ can be bounded as follows:

$$\mathrm{e}[\Phi] = \mathrm{e}\left[\left\|\mathbf{X}\mathbf{X}^\top-\mathbf{Z}\mathbf{Z}^\top\right\|_2\right] \leq \mathrm{e}\left[\left\|\mathbf{X}\mathbf{X}^\top-\mathbf{Z}\mathbf{Z}^\top\right\|_F\right] \leq \frac{n}{\sqrt{l}}\mathbf{K}_{\max}, \tag{7.14}$$

where the last inequality follows Corollary 2 of [34]. The inequalities (7.13) and (7.14) combined with Theorem 1 give a bound on $\left\|\mathbf{X}\mathbf{X}^\top-\mathbf{Z}\mathbf{Z}^\top\right\|_2$ and yield the statement of the theorem.

The following general inequality holds for the Frobenius error of the Nyström method [18]:

$$\left\|\mathbf{K}-\widetilde{\mathbf{K}}\right\|_F^2 \leq \|\mathbf{K}-\mathbf{K}_k\|_F^2 + \sqrt{64k}\ \left\|\mathbf{X}\mathbf{X}^\top - \mathbf{Z}\mathbf{Z}^\top\right\|_F^2 n\mathbf{K}_{ii}^{\max}. \tag{7.15}$$

Bounding the term $\left\|\mathbf{X}\mathbf{X}^\top - \mathbf{Z}\mathbf{Z}^\top\right\|_F^2$ as in the norm-2 case and using the concentration bound of Theorem 1 yields the result of the theorem. □

7.3.2 Ensemble Nyström Method

The following error bounds hold for Ensemble Nyström methods based on a convex combination of Nyström approximations.

Theorem 3. *Let S be a sample of pl columns drawn uniformly at random without replacement from $\mathbf{K}$, decomposed into p subsamples of size l, $S_1, \ldots, S_p$. For $r \in [1, p]$, let $\widetilde{\mathbf{K}}_r$ denote the rank-k Nyström approximation of $\mathbf{K}$ based on the sample S_r, and let $\mathbf{K}_k$ denote the best rank-k approximation of $\mathbf{K}$. Then, with probability at least $1-\delta$, the following inequalities hold for any sample S of size pl and for any $\boldsymbol{\mu}$ in the simplex Δ and $\widetilde{\mathbf{K}}^{\text{ens}} = \sum_{r=1}^p \mu_r \widetilde{\mathbf{K}}_r$:*

$$\left\|\mathbf{K}-\widetilde{\mathbf{K}}^{\text{ens}}\right\|_2 \leq \|\mathbf{K}-\mathbf{K}_k\|_2 + \frac{2n}{\sqrt{l}}\mathbf{K}_{\max}\left[1+\mu_{\max}p^{\frac{1}{2}}\sqrt{\frac{n-pl}{n-1/2}\frac{1}{\beta(pl,n)}\log\frac{1}{\delta}}\,d_{\max}^{\mathbf{K}}\Big/\mathbf{K}_{\max}^{\frac{1}{2}}\right]$$

$$\left\|\mathbf{K}-\widetilde{\mathbf{K}}^{\text{ens}}\right\|_F \leq \|\mathbf{K}-\mathbf{K}_k\|_F + \left[\frac{64k}{l}\right]^{\frac{1}{4}} n\mathbf{K}_{\max}\left[1+\mu_{\max}p^{\frac{1}{2}}\sqrt{\frac{n-pl}{n-1/2}\frac{1}{\beta(pl,n)}\log\frac{1}{\delta}}\,d_{\max}^{\mathbf{K}}\Big/\mathbf{K}_{\max}^{\frac{1}{2}}\right]^{\frac{1}{2}},$$

where $\beta(pl,n) = 1-\frac{1}{2\max\{pl,n-pl\}}$ and $\mu_{\max} = \max_{r=1}^p \mu_r$.

Proof. For $r \in [1, p]$, let $\mathbf{Z}_r = \sqrt{n/l}\,\mathbf{X}\mathbf{S}_r$, where $\mathbf{S}_r$ denotes the selection matrix corresponding to the sample S_r. By definition of $\widetilde{\mathbf{K}}^{\text{ens}}$ and the upper bound on $\left\|\mathbf{K}-\widetilde{\mathbf{K}}_r\right\|_2$ already used in the proof of Theorem 2, the following holds:

$$\left\|\mathbf{K}-\widetilde{\mathbf{K}}^{\text{ens}}\right\|_2 = \left\|\sum_{r=1}^p \mu_r\left(\mathbf{K}-\widetilde{\mathbf{K}}_r\right)\right\|_2 \leq \sum_{r=1}^p \mu_r\left\|\mathbf{K}-\widetilde{\mathbf{K}}_r\right\|_2 \tag{7.16}$$

$$\leq \sum_{r=1}^p \mu_r\left(\left\|\mathbf{K}-\mathbf{K}_k\right\|_2 + 2\left\|\mathbf{X}\mathbf{X}^\top - \mathbf{Z}_r\mathbf{Z}_r^\top\right\|_2\right) \tag{7.17}$$

$$= \left\|\mathbf{K}-\mathbf{K}_k\right\|_2 + 2\sum_{r=1}^p \mu_r\left\|\mathbf{X}\mathbf{X}^\top - \mathbf{Z}_r\mathbf{Z}_r^\top\right\|_2. \tag{7.18}$$

We apply Theorem 1 to $\phi(S) = \sum_{r=1}^{p} \mu_r \left\|\mathbf{X}\mathbf{X}^\top - \mathbf{Z}_r\mathbf{Z}_r^\top\right\|_2$. Let S' be a sample differing from S by only one column. Observe that changing one column of the full sample S changes only one subsample S_r and thus only one term $\mu_r \left\|\mathbf{X}\mathbf{X}^\top - \mathbf{Z}_r\mathbf{Z}_r^\top\right\|_2$. Thus, in view of the bound (7.13) on the change to $\left\|\mathbf{X}\mathbf{X}^\top - \mathbf{Z}_r\mathbf{Z}_r^\top\right\|_2$, the following holds:

$$\left|\phi\left(S'\right) - \phi(S)\right| \le \frac{2n}{l}\mu_{\max} d_{\max}^{\mathbf{K}} \mathbf{K}_{\max}^{\frac{1}{2}}. \tag{7.19}$$

The expectation of Φ can be straightforwardly bounded by:

$$\mathrm{e}[\Phi(S)] = \sum_{r=1}^{p} \mu_r \mathrm{e}\left[\left\|\mathbf{X}\mathbf{X}^\top - \mathbf{Z}_r\mathbf{Z}_r^\top\right\|_2\right] \le \sum_{r=1}^{p} \mu_r \frac{n}{\sqrt{l}}\mathbf{K}_{\max} = \frac{n}{\sqrt{l}}\mathbf{K}_{\max}$$

using the bound (7.14) for a single expert. Plugging in this upper bound and the Lipschitz bound (7.19) in Theorem 1 yields our norm-2 bound for the Ensemble Nyström method.

For the Frobenius error bound, using the convexity of the Frobenius norm square $\left\|\cdot\right\|_F^2$ and the general inequality (7.15), we can write

$$\left\|\mathbf{K} - \widetilde{\mathbf{K}}^{\mathrm{ens}}\right\|_F^2 = \left\|\sum_{r=1}^{p} \mu_r\left(\mathbf{K} - \widetilde{\mathbf{K}}_r\right)\right\|_F^2 \le \sum_{r=1}^{p} \mu_r \left\|\mathbf{K} - \widetilde{\mathbf{K}}_r\right\|_F^2 \tag{7.20}$$

$$\le \sum_{r=1}^{p} \mu_r \left[\left\|\mathbf{K} - \mathbf{K}_k\right\|_F^2 + \sqrt{64k}\left\|\mathbf{X}\mathbf{X}^\top - \mathbf{Z}_r\mathbf{Z}_r^\top\right\|_F n\mathbf{K}_{ii}^{\max}\right]. \tag{7.21}$$

$$= \left\|\mathbf{K} - \mathbf{K}_k\right\|_F^2 + \sqrt{64k}\sum_{r=1}^{p} \mu_r \left\|\mathbf{X}\mathbf{X}^\top - \mathbf{Z}_r\mathbf{Z}_r^\top\right\|_F n\mathbf{K}_{ii}^{\max}. \tag{7.22}$$

The result follows by the application of Theorem 1 to $\psi(S) = \sum_{r=1}^{p} \mu_r \left\|\mathbf{X}\mathbf{X}^\top - \mathbf{Z}_r\mathbf{Z}_r^\top\right\|_F$ in a way similar to the norm-2 case. □

The bounds of Theorem 3 are similar in form to those of Theorem 2. However, the bounds for the Ensemble Nyström are tighter than those for any Nyström expert based on a single sample of size l even for a uniform weighting. In particular, for $\mu_i = 1/p$ for all i, the last term of the ensemble bound for norm-2 is smaller by a factor larger than $\mu_{\max} p^{\frac{1}{2}} = 1/\sqrt{p}$.

7.4 Experiments

In this section, we present experimental results that illustrate the performance of the Ensemble Nyström method. We work with the datasets listed in Table 7.1, and compare the performance of various methods for calculating the mixture weights (μ_r). Throughout our experiments, we measure the accuracy of a low-rank approximation $\widetilde{\mathbf{K}}$ by calculating the relative error in Frobenius and spectral norms, that is, if we let $\xi = \{2, F\}$, then we calculate the following quantity:

$$\% \text{ error} = \frac{\left\|\mathbf{K} - \widetilde{\mathbf{K}}\right\|_\xi}{\left\|\mathbf{K}\right\|_\xi} \times 100. \tag{7.23}$$

7.4.1 *Ensemble Nyström with Various Mixture Weights*

In this set of experiments, we show results for our Ensemble Nyström method using different techniques to choose the mixture weights as previously discussed. We first experimented with the first five datasets shown in Table 7.1. For each dataset, we fixed the reduced rank to $k = 50$, and set the number of sampled columns to $l = 3\% \times n$.[1] Furthermore, for the exponential and the ridge regression variants, we sampled a set of $s = 20$ columns and used an additional 20 columns (s') as a hold-out set for selecting the optimal values of η and λ. The number of approximations, p, was varied from 2 to 30. As a baseline, we also measured the minimum and the mean percent error across the p Nyström approximations used to construct $\widetilde{\mathbf{K}}^{\text{ens}}$. For the Frobenius norm, we also calculated the performance when using the optimal μ, that is, we used least-square regression to find the best possible choice of combination weights for a fixed set of p approximations by setting $s = n$.

The results of these experiments are presented in Fig. 7.1 for the Frobenius norm and in Fig. 7.2 for the spectral norm. These results clearly show that the Ensemble

Table 7.1 Description of the datasets used in our Ensemble Nyström experiments [3, 27, 35, 39, 48]

Dataset	Type of data	# Points (n)	# Features (d)	Kernel
PIE-2.7K	Face images	2,731	2,304	Linear
MNIST	Digit images	4,000	784	Linear
ESS	Proteins	4,728	16	RBF
AB-S	Abalones	4,177	8	RBF
DEXT	Bag of words	2,000	20,000	Linear
SIFT-1M	Image features	1 M	128	RBF

[1] Similar results (not reported here) were observed for other values of k and l as well.

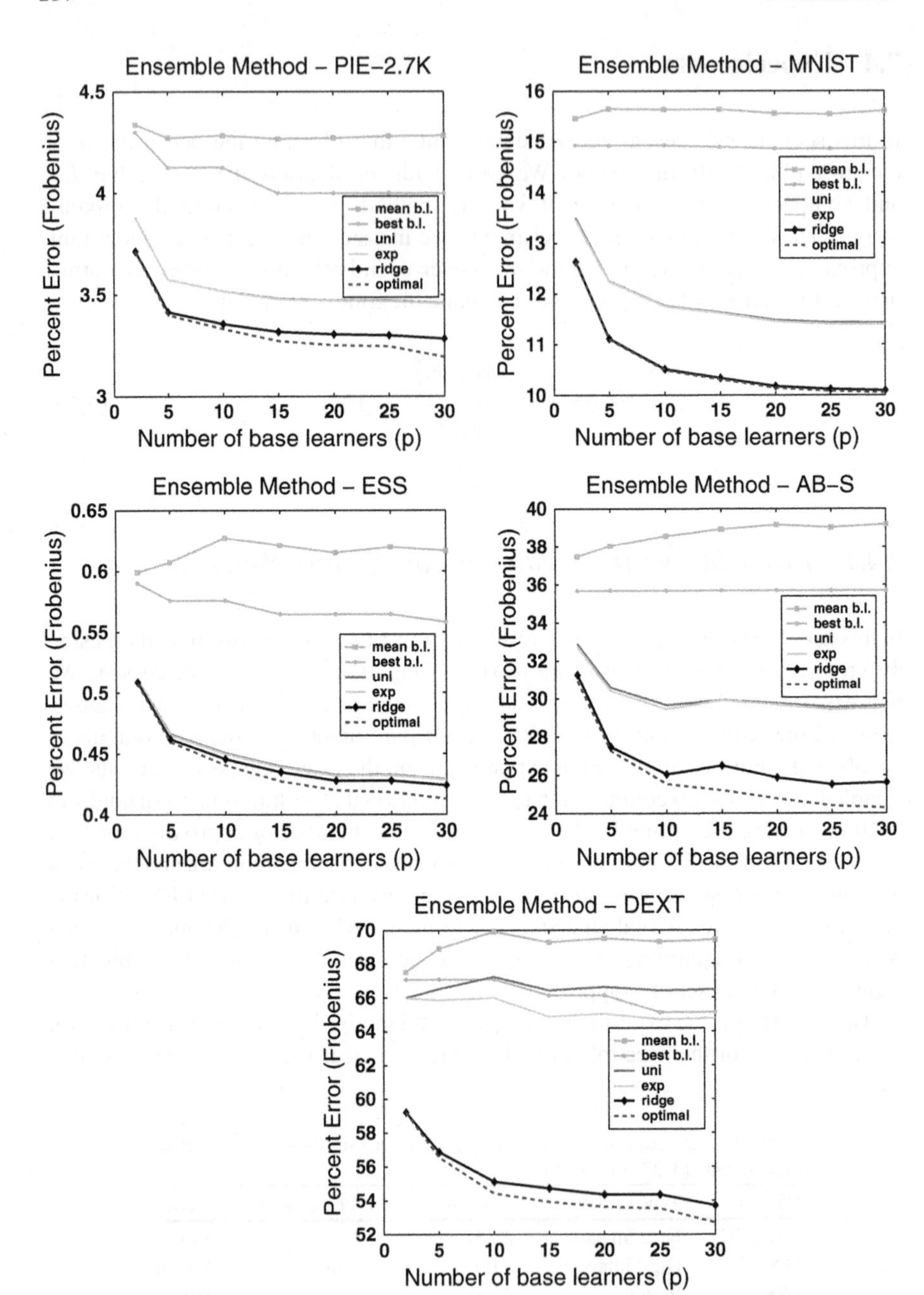

Fig. 7.1 Percent error in Frobenius norm for Ensemble Nyström method using uniform ("uni"), exponential ("exp"), ridge ("ridge"), and optimal ("optimal") mixture weights as well as the best ("best b.l.") and mean ("mean b.l.") of the p base learners used to create the ensemble approximations

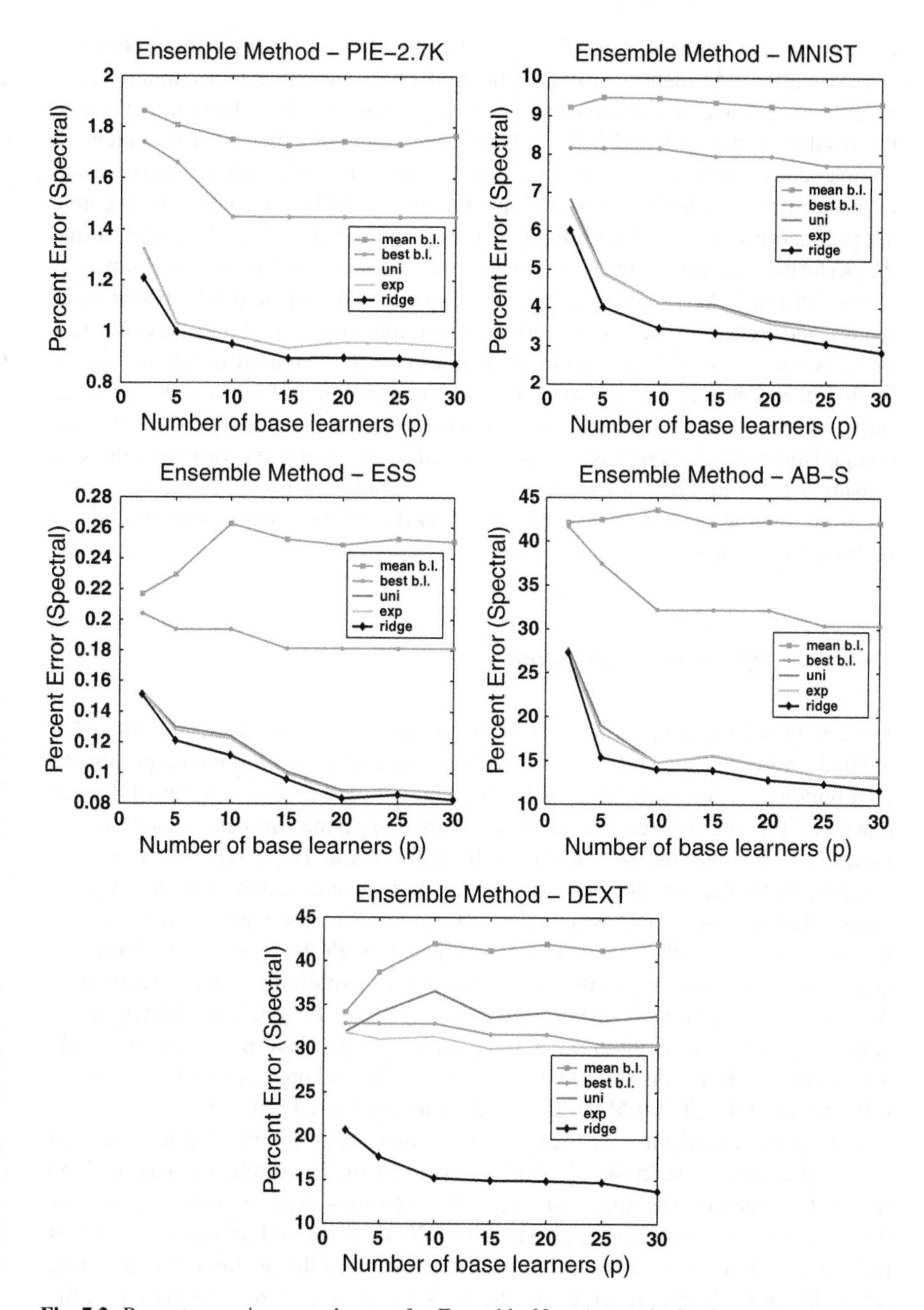

Fig. 7.2 Percent error in spectral norm for Ensemble Nyström method using various mixture weights and the best/mean of the p approximations. Legend entries are the same as in Fig. 7.1

Nyström performance is significantly better than any of the individual Nyström approximations. As mentioned earlier, the rank of the ensemble approximations can be p times greater than the rank of each of the base learners. Hence, to validate the results in Figs. 7.1 and 7.2, we performed a simple experiment in which we compared the performance of the best base learner to the best rank k approximation of the uniform ensemble approximation (obtained via SVD of the uniform ensemble approximation). The results of this experiment, presented in Fig. 7.3, suggest that the performance gain of the ensemble methods is not due to this increased rank.

Furthermore, the ridge regression technique is the best of the proposed techniques and generates nearly the optimal solution in terms of the percent error in Frobenius norm. We also observed that when s is increased to approximately 5–10% of n, linear regression without any regularization performs about as well as ridge regression for both the Frobenius and spectral norm. Figure 7.4 shows this comparison between linear regression and ridge regression for varying values of s using a fixed number of experts ($p = 10$). Finally we note that the Ensemble Nyström method tends to converge very quickly, and the most significant gain in performance occurs as p increases from 2 to 10.

7.4.2 Large-Scale Experiments

We now present an empirical study of the effectiveness of the Ensemble Nyström method on the SIFT-1 M dataset in Table 7.1 containing 1 *million* data points. As is common practice with large-scale datasets, we worked on a cluster of several machines for this dataset. We present results comparing the performance of the Ensemble Nyström method, using both uniform and ridge regression mixture weights, with that of the best and mean performance across the p Nyström approximations used to construct $\widetilde{\mathbf{K}}^{\text{ens}}$. We also make comparisons with the K-means adaptive sampling technique [54, 55]. Although the K-means technique is quite effective at generating informative columns by exploiting the data distribution, the cost of performing K-means becomes expensive for even moderately sized datasets, making it difficult to use in large-scale settings. Nevertheless, in this work, we include the K-means method in our comparison, and present results for various subsamples of the SIFT-1 M dataset, with n ranging from 5 K to 1 M.

For a fair comparison, we performed "fixed-time" experiments. We first searched for an appropriate l such that the percent error for the Ensemble Nyström method with ridge weights was approximately 10%, and measured the time required by the cluster to construct this approximation. We then allotted an equal amount of time (within 1 s) for the other techniques, and measured the quality of the resulting approximations. For these experiments, we set $k = 50$ and $p = 10$, based on the results from the previous section. Furthermore, in order to speed up computation on this large dataset, we decreased the size of the validation and hold-out sets to $s = 2$ and $s' = 2$, respectively.

The results of this experiment, presented in Fig. 7.5, clearly show that the Ensemble Nyström method is the most effective technique given a fixed amount of

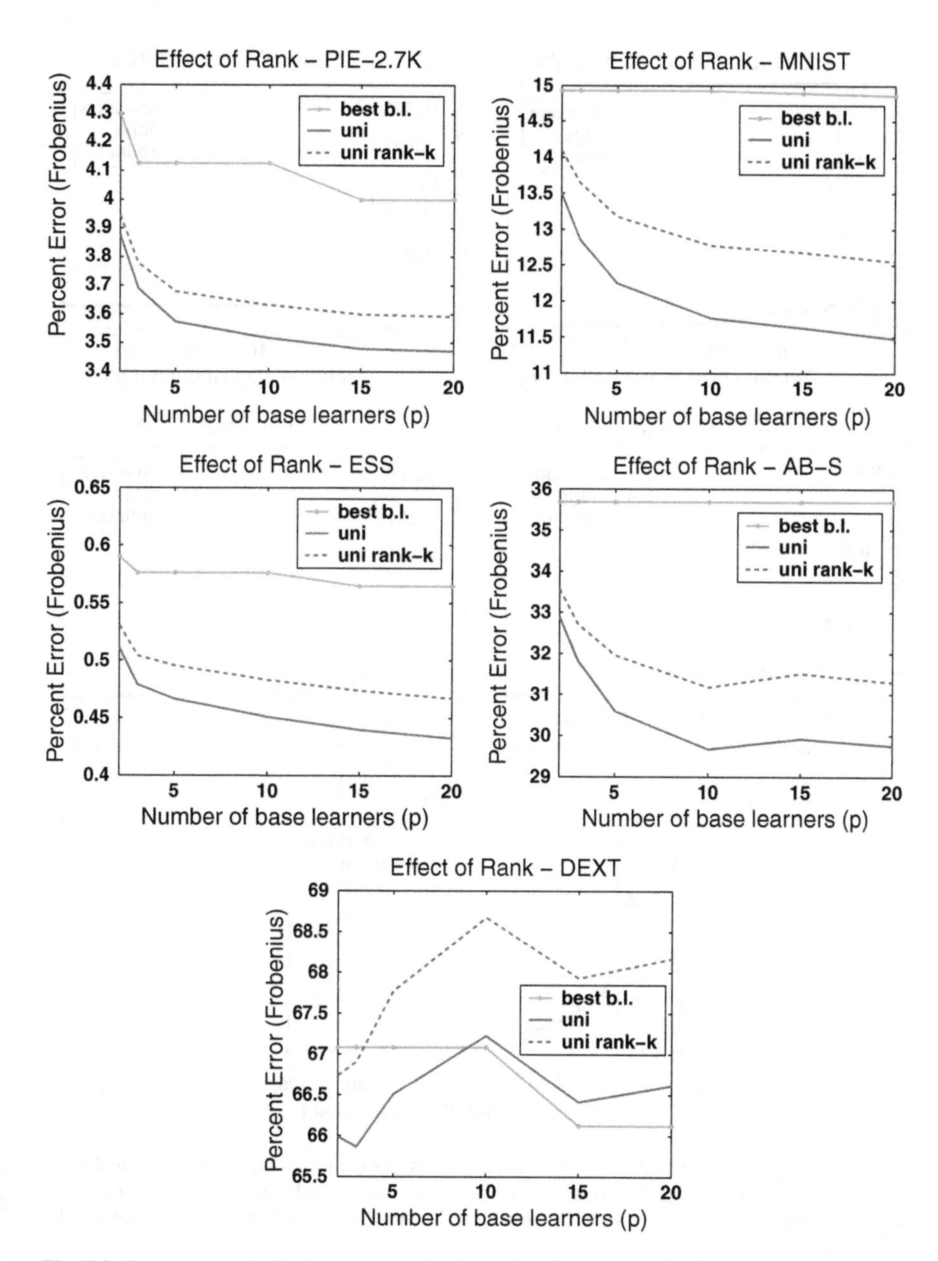

Fig. 7.3 Percent error in Frobenius norm for Ensemble Nyström method using uniform ("uni") mixture weights, the optimal rank-k approximation of the uniform ensemble result ("uni rank-k") as well as the best ("best b.l.") of the p base learners used to create the ensemble approximations

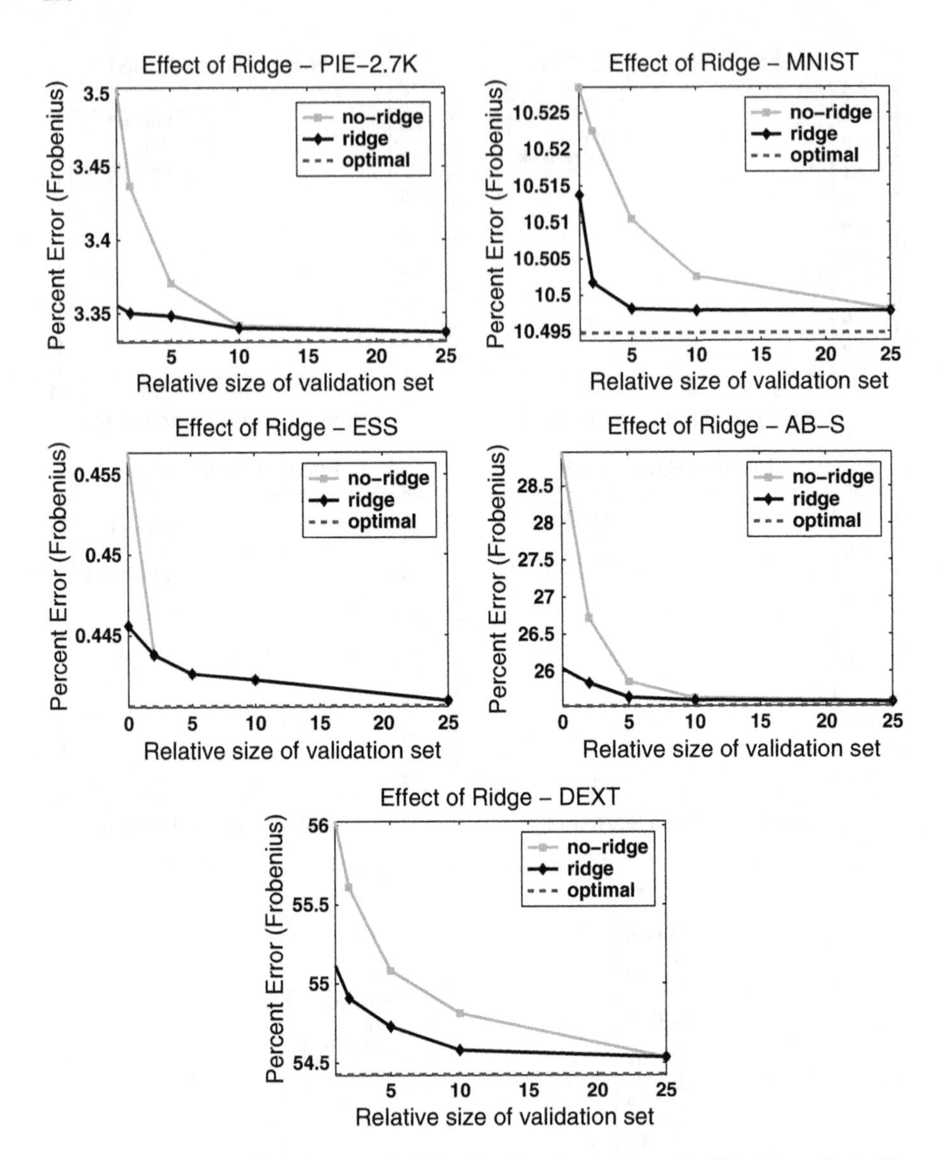

Fig. 7.4 Comparison of percent error in Frobenius norm for the Ensemble Nyström method with $p = 10$ experts with weights derived from linear ("no-ridge") and ridge ("ridge") regression. The dotted line indicates the optimal combination. The relative size of the validation set equals $s/n \times 100$

time. Furthermore, even with the small values of s and s', Ensemble Nyström with ridge-regression weighting outperforms the uniform Ensemble Nyström method. We also observe that due to the high computational cost of K-means for large datasets, the K-means approximation does not perform well in this "fixed-time" experiment. It generates an approximation that is worse than the mean stan-

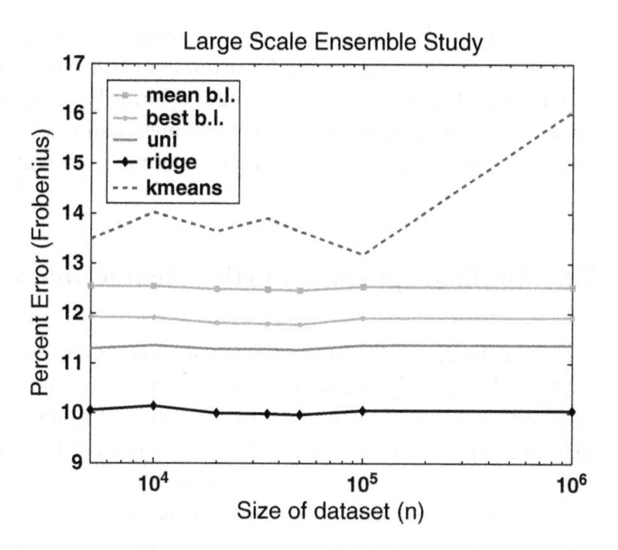

Fig. 7.5 Large-scale performance comparison with SIFT-1 M dataset. For a fixed computational time, the Ensemble Nyström approximation with ridge weights tends to outperform other techniques

dard Nyström approximation and its performance increasingly deteriorates as n approaches 1M. Finally, we note that although the space requirements are 10 times greater for Ensemble Nyström in comparison to standard Nyström (since $p = 10$ in this experiment), the space constraints are nonetheless quite reasonable. For instance, when working with 1 M points, the Ensemble Nyström method with ridge regression weights only required approximately 1% of the columns of $\mathbf{K}$ to achieve an error of 10%.

7.5 Summary and Open Questions

A key element of Nyström approximation is the number of sampled columns used by it. More samples typically result in better accuracy. However, the number of samples that can be processed by a single Nyström approximation is limited due to the computational constraints, restricting its accuracy. In this work, we discussed an ensemble based meta-algorithm for combining multiple Nyström approximations. These ensemble algorithms show consistent and significant performance improvement across a number of different datasets. Moreover, they naturally fit within a distributed computing environment, thus making them quite efficient in large-scale settings. These ensemble algorithms also have better theoretical guarantees than individual Nyström approximation.

One interesting fact revealed by the experiments is that as the number of individual Nyström approximations is increased in the ensemble, the reconstruction error does not go toward zero. The error tends to saturate after a relatively small number of learners and adding more does not benefit the ensemble. Even though this counterintuitive behavior is a good thing in practice since one does not need to use a large number of base learners, it raises intriguing theoretical questions. Why does the error from Ensemble Nyström converge? What is the value to which it is converging? Can this error be brought arbitrarily close to zero? We believe that a better understanding of these questions may lead to even better ways of designing ensemble algorithms for matrix approximation in the future.

7.6 Bibliographical and Historical Remarks

There has been a wide array of work on low-rank matrix approximation within the numerical linear algebra and computer science communities. Most of it has been inspired by the celebrated result of Johnson and Lindenstrauss [31], which showed that random low-dimensional embeddings preserve Euclidean geometry. This result has led to a family of random projection algorithms, which involves projecting the original matrix onto a random low-dimensional subspace [30, 37, 42]. Alternatively, SVD can be used to generate "optimal" low-rank matrix approximations, as mentioned earlier. However, both the random projection and the SVD algorithms involve storage and operating on the entire input matrix. SVD is more computationally expensive than random projection methods, although neither are linear in n in terms of time and space complexity. When dealing with sparse matrices, there exist less computationally intensive techniques such as Jacobi, Arnoldi, Hebbian, and more recent randomized methods [23, 25, 28, 44] for generating low-rank approximations. These iterative methods require computation of matrix-vector products at each step and involve multiple passes through the data. Hence, these algorithms are not suitable for large, dense matrices. Matrix sparsification algorithms [1, 2], as the name suggests, attempt to sparsify dense matrices to speed up future storage and computational burdens, though they too require storage of the input matrix and exhibit superlinear processing time.

Alternatively, sampling-based approaches can be used to generate low-rank approximations. Research in this area dates back to classical theoretical results that show, for any arbitrary matrix, the existence of a subset of k columns for which the error in matrix projection (as defined in [33]) can be bounded relative to the optimal rank-k approximation of the matrix [46]. Deterministic algorithms such as rank-revealing QR [26] can achieve nearly optimal matrix projection errors. More recently, research in the theoretical computer science community has been aimed at deriving bounds on matrix projection error using sampling-based approximations, including additive error bounds using sampling distributions based on leverage scores, i.e., the squared L_2 norms of the columns [17, 22, 45]; relative error bounds using adaptive sampling techniques [16, 29]; and, relative error bounds based on distributions derived from the singular vectors of the input matrix, in work related to the column-subset selection problem [10, 19]. However, as discussed in [33], the task of matrix projection involves projecting the input matrix onto a low-rank subspace, which requires superlinear time and space with respect to n and is not typically feasible for large-scale matrices.

There does however, exist another class of sampling-based approximation algorithms that only store and operate on a subset of the original matrix. For arbitrary rectangular matrices, these algorithms are known as "CUR" approximations (the name "CUR" corresponds to the three low-rank matrices whose product is an approximation to the original matrix). The theoretical performance of CUR approximations has been analyzed using a variety of sampling schemes, although the column-selection processes associated with these analyses often require operating on the entire input matrix [19, 24, 40, 50]. In the context of SPSD matrices, the Nyström method is the most commonly used algorithm to efficiently generate low-rank approximations. The Nyström method was initially introduced as a quadrature method for numerical integration, used to approximate eigenfunction solutions [6, 41]. More recently, it was presented in [53] to speed up kernel algorithms and has been studied theoretically using a variety of sampling schemes [7, 8, 14, 18, 32–34, 49, 52, 54, 55]. It has also been used for a variety of machine learning tasks ranging from manifold learning to image segmentation [21,43,51]. A closely related algorithm, known as the Incomplete Cholesky Decomposition [4, 5, 20], can also be viewed as a specific sampling technique associated with the Nyström method [5]. As noted by [11,52], the Nyström approximation is related to the problem of matrix completion [11, 12], which attempts to complete a low-rank matrix from a random sample of its entries. However, the matrix completion setting assumes that the target matrix is low-rank and only allows for limited access to the data. In contrast, the Nyström method, and sampling-based low-rank approximation algorithms in general, deal with full-rank matrices that are amenable to low-rank approximation. Furthermore, when we have access to the underlying kernel function that generates the kernel matrix of interest, we can generate matrix entries on-the-fly as desired, providing us with more flexibility accessing the original matrix.

References

1. Dimitris Achlioptas and Frank Mcsherry. Fast computation of low-rank matrix approximations. *Journal of the ACM*, 54(2), 2007.
2. Sanjeev Arora, Elad Hazan, and Satyen Kale. A fast random sampling algorithm for sparsifying matrices. In *Approx-Random*, 2006.
3. A. Asuncion and D.J. Newman. UCI machine learning repository. http://www.ics.uci.edu/~mlearn/MLRepository.html, 2007.
4. Francis R. Bach and Michael I. Jordan. Kernel Independent Component Analysis. *Journal of Machine Learning Research*, 3:1–48, 2002.
5. Francis R. Bach and Michael I. Jordan. Predictive low-rank decomposition for kernel methods. In *International Conference on Machine Learning*, 2005.
6. Christopher T. Baker. *The numerical treatment of integral equations*. Clarendon Press, Oxford, 1977.
7. M.-A. Belabbas and P. J. Wolfe. On landmark selection and sampling in high-dimensional data analysis. arXiv:0906.4582v1[stat.ML], 2009.
8. M. A. Belabbas and P. J. Wolfe. Spectral methods in machine learning and new strategies for very large datasets. *Proceedings of the National Academy of Sciences of the United States of America*, 106(2):369–374, January 2009.

9. Bernhard E. Boser, Isabelle Guyon, and Vladimir N. Vapnik. A training algorithm for optimal margin classifiers. In *Conference on Learning Theory*, 1992.
10. Christos Boutsidis, Michael W. Mahoney, and Petros Drineas. An improved approximation algorithm for the column subset selection problem. In *Symposium on Discrete Algorithms*, 2009.
11. Emmanuel J. Candès and Benjamin Recht. Exact matrix completion via convex optimization. *Foundations of Computational Mathematics*, 9(6):717–772, 2009.
12. Emmanuel J. Candès and Terence Tao. The power of convex relaxation: Near-optimal matrix completion. arXiv:0903.1476v1[cs.IT], 2009.
13. Corinna Cortes, Mehryar Mohri, Dmitry Pechyony, and Ashish Rastogi. Stability of transductive regression algorithms. In *International Conference on Machine Learning*, 2008.
14. Corinna Cortes, Mehryar Mohri, and Ameet Talwalkar. On the impact of kernel approximation on learning accuracy. In *Conference on Artificial Intelligence and Statistics*, 2010.
15. Corinna Cortes and Vladimir N. Vapnik. Support-Vector Networks. *Machine Learning*, 20(3):273–297, 1995.
16. Amit Deshpande, Luis Rademacher, Santosh Vempala, and Grant Wang. Matrix approximation and projective clustering via volume sampling. In *Symposium on Discrete Algorithms*, 2006.
17. Petros Drineas, Ravi Kannan, and Michael W. Mahoney. Fast Monte Carlo algorithms for matrices II: Computing a low-rank approximation to a matrix. *SIAM Journal of Computing*, 36(1), 2006.
18. Petros Drineas and Michael W. Mahoney. On the Nyström method for approximating a Gram matrix for improved kernel-based learning. *Journal of Machine Learning Research*, 6:2153–2175, 2005.
19. Petros Drineas, Michael W. Mahoney, and S. Muthukrishnan. Relative-error CUR matrix decompositions. *SIAM Journal on Matrix Analysis and Applications*, 30(2):844–881, 2008.
20. Shai Fine and Katya Scheinberg. Efficient SVM training using low-rank kernel representations. *Journal of Machine Learning Research*, 2:243–264, 2002.
21. Charless Fowlkes, Serge Belongie, Fan Chung, and Jitendra Malik. Spectral grouping using the Nyström method. *Transactions on Pattern Analysis and Machine Intelligence*, 26(2):214–225, 2004.
22. Alan Frieze, Ravi Kannan, and Santosh Vempala. Fast Monte-Carlo algorithms for finding low-rank approximations. In *Foundation of Computer Science*, 1998.
23. Gene Golub and Charles Van Loan. *Matrix Computations*. Johns Hopkins University Press, Baltimore, 2nd edition, 1983.
24. S. A. Goreinov, E. E. Tyrtyshnikov, and N. L. Zamarashkin. A theory of pseudoskeleton approximations. *Linear Algebra and Its Applications*, 261:1–21, 1997.
25. G. Gorrell. Generalized Hebbian algorithm for incremental Singular Value Decomposition in natural language processing. In *European Chapter of the Association for Computational Linguistics*, 2006.
26. Ming Gu and Stanley C. Eisenstat. Efficient algorithms for computing a strong rank-revealing QR factorization. *SIAM Journal of Scientific Computing*, 17(4):848–869, 1996.
27. A. Gustafson, E. Snitkin, S. Parker, C. DeLisi, and S. Kasif. Towards the identification of essential genes using targeted genome sequencing and comparative analysis. *BMC:Genomics*, 7:265, 2006.
28. Nathan Halko, Per Gunnar Martinsson, and Joel A. Tropp. Finding structure with randomness: Stochastic algorithms for constructing approximate matrix decompositions. arXiv:0909.4061v1[math.NA], 2009.
29. Sariel Har-peled. Low-rank matrix approximation in linear time, manuscript, 2006.
30. Piotr Indyk. Stable distributions, pseudorandom generators, embeddings, and data stream computation. *Journal of the ACM*, 53(3):307–323, 2006.
31. W. B. Johnson and J. Lindenstrauss. Extensions of Lipschitz mappings into a Hilbert space. *Contemporary Mathematics*, 26:189–206, 1984.
32. Sanjiv Kumar, Mehryar Mohri, and Ameet Talwalkar. Ensemble Nyström method. In *Neural Information Processing Systems*, 2009.

33. Sanjiv Kumar, Mehryar Mohri, and Ameet Talwalkar. On sampling-based approximate spectral decomposition. In *International Conference on Machine Learning*, 2009.
34. Sanjiv Kumar, Mehryar Mohri, and Ameet Talwalkar. Sampling techniques for the Nyström method. In *Conference on Artificial Intelligence and Statistics*, 2009.
35. Yann LeCun and Corinna Cortes. The MNIST database of handwritten digits. http://yann.lecun.com/exdb/mnist/, 1998.
36. Mu Li, James T. Kwok, and Bao-Liang Lu. Making large-scale Nyström approximation possible. In *International Conference on Machine Learning*, 2010.
37. Edo Liberty. *Accelerated dense random projections*. Ph.D. thesis, computer science department, Yale University, New Haven, CT, 2009.
38. N. Littlestone and M. K. Warmuth. The Weighted Majority algorithm. *Information and Computation*, 108(2):212–261, 1994.
39. David G. Lowe. Distinctive image features from scale-invariant keypoints. *International Journal of Computer Vision*, 60:91–110, 2004.
40. Michael W Mahoney and Petros Drineas. CUR matrix decompositions for improved data analysis. *Proceedings of the National Academy of Sciences*, 106(3):697–702, 2009.
41. E.J. Nyström. Über die praktische auflösung von linearen integralgleichungen mit anwendungen auf randwertaufgaben der potentialtheorie. *Commentationes Physico-Mathematicae*, 4(15):1–52, 1928.
42. Christos H. Papadimitriou, Hisao Tamaki, Prabhakar Raghavan, and Santosh Vempala. Latent Semantic Indexing: a probabilistic analysis. In *Principles of Database Systems*, 1998.
43. John C. Platt. Fast embedding of sparse similarity graphs. In *Neural Information Processing Systems*, 2004.
44. Vladimir Rokhlin, Arthur Szlam, and Mark Tygert. A randomized algorithm for Principal Component Analysis. *SIAM Journal on Matrix Analysis and Applications*, 31(3):1100–1124, 2009.
45. Mark Rudelson and Roman Vershynin. Sampling from large matrices: An approach through geometric functional analysis. *Journal of the ACM*, 54(4):21, 2007.
46. A. F. Ruston. Auerbachs theorem. *Mathematical Proceedings of the Cambridge Philosophical Society*, 56:476–480, 1964.
47. Bernhard Schölkopf, Alexander Smola, and Klaus-Robert Müller. Nonlinear component analysis as a kernel eigenvalue problem. *Neural Computation*, 10(5):1299–1319, 1998.
48. Terence Sim, Simon Baker, and Maan Bsat. The CMU pose, illumination, and expression database. In *Conference on Automatic Face and Gesture Recognition*, 2002.
49. Alex J. Smola and Bernhard Schölkopf. Sparse Greedy Matrix Approximation for machine learning. In *International Conference on Machine Learning*, 2000.
50. G. W. Stewart. Four algorithms for the efficient computation of truncated pivoted QR approximations to a sparse matrix. *Numerische Mathematik*, 83(2):313–323, 1999.
51. Ameet Talwalkar, Sanjiv Kumar, and Henry Rowley. Large-scale manifold learning. In *Conference on Vision and Pattern Recognition*, 2008.
52. Ameet Talwalkar and Afshin Rostamizadeh. Matrix coherence and the Nyström method. In *Conference on Uncertainty in Artificial Intelligence*, 2010.
53. Christopher K. I. Williams and Matthias Seeger. Using the Nyström method to speed up kernel machines. In *Neural Information Processing Systems*, 2000.
54. Kai Zhang and James T. Kwok. Density-weighted Nyström method for computing large kernel eigensystems. *Neural Computation*, 21(1):121–146, 2009.
55. Kai Zhang, Ivor Tsang, and James Kwok. Improved Nyström low-rank approximation and error analysis. In *International Conference on Machine Learning*, 2008.

Chapter 8
Object Detection

Jianxin Wu and James M. Rehg

8.1 Introduction

Over the past twenty years, data-driven methods have become a dominant paradigm for computer vision, with numerous practical successes. In difficult computer vision tasks, such as the detection of object categories (for example, the detection of faces of various gender, age, race, and pose, under various illumination and background conditions), researchers generally learn a classifier that can distinguish an image patch that contains the object of interest from all other image patches. Ensemble learning methods have been very successful in learning classifiers for object detection.

The task of object detection, however, poses new challenges for ensemble learning, which we will discuss in detail in Sect. 8.2. We summarize these challenges into three aspects: scale, speed, and asymmetry.

Various research contributions have been made to overcome these difficulties. In this chapter, we mainly focus on those methods that use the cascade classifier structure together with ensemble learning methods (e.g., AdaBoost). The cascade classifier structure for object detection was first proposed by Viola and Jones [41], who presented the first face detection system that could both run in real-time and achieve high detection accuracy. We will describe this work in Sect. 8.3, with ensemble learning methods being one of the key components in this system.

J. Wu (✉)
School of Computer Engineering, Nanyang Technological University,
N4-2c-82 Nanyang Avenue, Singapore 639798, Singapore
e-mail: jxwu@ntu.edu.sg; wujx2001@gmail.com

J.M. Rehg
School of Interactive Computing, College of Computing,
Georgia Institute of Technology, Atlanta, GA, USA
e-mail: rehg@cc.gatech.edu

C. Zhang and Y. Ma (eds.), *Ensemble Machine Learning: Methods and Applications*,
DOI 10.1007/978-1-4419-9326-7_8, © Springer Science+Business Media, LLC 2012

Various research efforts have been devoted to improve the learning speed of a cascade, which took several weeks in the original version of [41]. In Sect. 8.4, we present several methods that improve the training time by several orders of magnitudes, including a faster implementation of AdaBoost for cascade classifiers [45], an approximate weak classifier training method that is used to form the strong AdaBoost classifier [28], and Forward Feature Selection, an alternative to the AdaBoost learning method [45].

In Sect. 8.5, we present methods that specifically deal with the difficulties associated with the asymmetric learning problem inside a cascade. Two methods are described in detail: the asymmetric AdaBoost method from [40], and the Linear Asymmetric Classifier (LAC) method from [45].

We then move beyond the detection of upright and frontal faces into broader object detection domains in Sect. 8.6. We will, however, still pay attention to methods in the cascade and/or ensemble learning framework. Profile faces and rotated faces are also effectively detected using these techniques [17]. In addition, two methods for pedestrian detection are described: one for detection in still images [54], and the other for detection in low resolution surveillance videos that incorporates motion information [42]. Finally, we show that detection is not only useful for its own sake. It can, for example, be the cornerstone for a visual tracking system (i.e., tracking-by-detection).

Many other methods enable object detection and related tasks by applying novel ensemble learning algorithms, which will be given in Sect. 8.8 as bibliographical notes, after some discussions in Sect. 8.7.

8.2 Brute-Force Object Detection: Challenges

8.2.1 The Brute-Force Search Strategy

When the object of interest has a fixed aspect ratio (e.g., in frontal face detection the height divided by width of all faces are roughly the same), the brute-force search strategy is the most widely used method for object detection.

The first step in the brute-force search approach is to train a classifier that can distinguish between the object of interest and all other image patches. A training dataset is constructed, which consists of positive examples (image patches of the target object) and negative examples (representative image patches from the background and all other objects). Since the object of interest has a fixed aspect ratio, image patches in the training set are normalized to the same size. The next step is then to train a classifier using such a training dataset. In this learning phase, ensemble learning is the most popular choice [41], although other classifiers (e.g., the Support Vector Machine) have also been applied [25].

During the detection phase, we scan all the image patches of a fixed size given by the training patches, by enumerating all possible locations within a testing image.

We usually start from the top-left corner, and apply the learned classifier to determine whether an object of interest exists in this position or not. We then move the scanning window and apply the classifier from left to right, and from top to bottom. That is, the classifier is applied to a regular grid of positions overlaid on the testing image. The step size of the grid can be varied in order to trade off between detection accuracy and running speed.

In order to detect an object bigger or smaller than the training patches, two strategies can be used. One approach is to resize the test images multiple times, so that the true object size equals the detector window size in one of the resized images (e.g., in [45]). It is also viable to resize the detector. For example, in [41] detectors for faces of different sizes are trained.

Postprocessing (or nonmaximum suppression) is the last step in brute-force object detection systems. During the search process, the classifier usually fires at multiple positions and scales around a true object of interest. Thus an object can have more than ten detected positions surrounding it. The postprocessing step merges nearby detections into a single output. In addition, if only a few (e.g., one or two) detections are found around a position, we may want to ignore such a detection. The postprocessing step thus can reduce the number of false detections (false positives), at the cost of potentially missing some target objects.

8.2.2 *Challenges in Learning the Classifier*

Properties that are particular to the learning task in object detection make the classifier learning step very challenging. We summarize such challenges into three aspects: scale, speed, and asymmetry. We will describe relevant machine learning solutions to these challenges in the following sections of this chapter.

8.2.2.1 The Scale Challenge

The number of training examples needed for a learning task grows with the complexity of the problem. Object class detection is a task with high complexity: we need to detect the object under many variations such as pose, illumination, deformation, etc. As a consequence, the training set needs to be quite large, ranging from tens of thousands [25] to several billion [45]. Special considerations are needed for dealing with these large training sets.

8.2.2.2 The Speed Challenge

A huge training set naturally leads to a long training time. For example, the original training method in [41] takes weeks for training a cascade classifier for face detection. It is thus necessary to greatly reduce the training time. However, the testing speed is more important for object detection. Video rate (or even faster) object detection speed has attracted the attention of researchers and substantial progress has been made. Nowadays, real-time face detection is a must-have feature for even entry-level digital cameras.

8.2.2.3 The Asymmetry Challenge

Object detection involves a highly asymmetric learning task. The asymmetry property greatly contributes to the difficulty of classifier learning. Three asymmetries are summarized in [45]:

1. *Uneven class priors.* It is known in machine learning that problems with imbalanced training sets lead to poor accuracy in the minority class (the class with relatively few training examples) [16]. In object detection, the scanning grid generates millions of image patches, but few are the objects of interest. The negative class easily generates a huge set of training examples, but the object class usually has a limited set of training examples. We need to carefully deal with this asymmetry.
2. *Goal asymmetry.* Even a single false detection (false positive) in one testing image may be annoying in object detection and its applications. On one hand, since we are scanning millions of image patches in a medium-sized testing image, we are in essence requiring the classifier to have an extremely low false positive rate (e.g., 10^{-7}). On the other hand, a high detection rate (e.g., 5% false negative/miss rate) is also required. The combination of these requirements and the big difference among these goal numbers make learning in object detection more challenging.
3. *Unequal complexity within the positive and negative classes.* As aforementioned, the positive (object) class is complex because of possible variations. However, the negative class is much more complex because this class contains everything else in the world except the object of interest. It is suggested in [45] that

> It is not hard to distinguish faces from cars. However, it is much harder to distinguish faces from all other objects.

8.3 The Cascade Face Detector

The cascade structured face detector by Viola and Jones [41] is the first system that achieves accurate frontal face detection in real time, with the help of three components: an image feature that can be computed quickly, an efficient classifier structure, and a novel application of the ensemble learning method (discrete AdaBoost in this case).

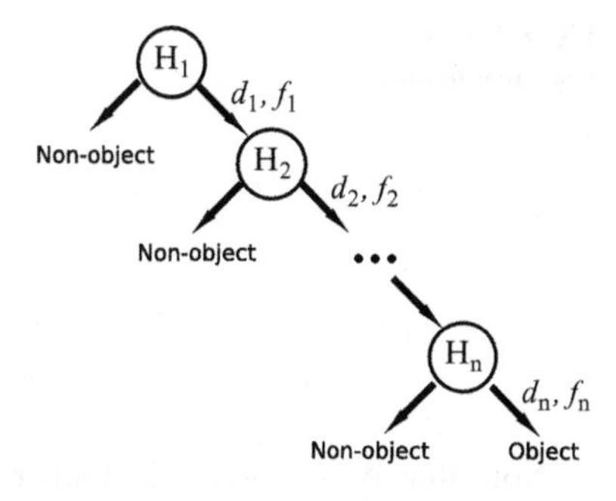

Fig. 8.1 The cascade classifier structure

Algorithm 1 The cascade classifier structure (modified from Algorithm 1 in [45])

1: {Input: a set of positive examples $\mathscr{P}$, a set of initial negative examples $\mathscr{N}$, and a database of bootstrapping negative examples $\mathscr{D}$. }
2: {Input: a learning goal $\mathscr{G}$.}
3: $i \Leftarrow 0$, $H \Leftarrow \emptyset$.
4: **repeat**
5: $i \Leftarrow i + 1$.
6: *NodeLearning* { Learn H_i using $\mathscr{P}$ and $\mathscr{N}$, and add H_i to H. }
7: Remove correctly classified non-face patches from $\mathscr{N}$.
8: Run the current cascade H on $\mathscr{D}$, add any false detection to $\mathscr{N}$ until $\mathscr{N}$ reaches the same size as the initial set.
9: **until** The learning goal $\mathscr{G}$ is satisfied.
10: Output: a cascade

$$(H_1, H_2, \ldots, H_n).$$

8.3.1 The Cascade Classifier Structure

The cascade classifier structure is mainly designed for high testing speed. We have discussed in Sect. 8.2 that a complex classifier is needed for object detection tasks, which also means that testing will be slow. This difficulty is alleviated by the cascade classifier structure [41]. A cascade consists of a sequence of classifiers with binary node classifiers $H_1, H_2, \ldots, H_n$, as illustrated in Fig. 8.1. An image patch is classified as the object of interest if it can pass tests in all the nodes. Since most background patches in a test image are filtered away by the early nodes, only a few image patches will fire all node classifiers (most of which will contain the object of interest). A cascade thus has very fast detection speed, as a consequence of the rarity of target objects.

Let us assume that the errors made by node classifiers are independent of each other. Furthermore, let us assume the cascade has 20 nodes and the node classifiers (refer to Fig. 8.1) have high detection rate $d_i = 99.9\%$ and false positive rate $f_i = 50\%$ for all i. The cascade classifier will be able to detect $\prod_{i=1}^{20} d_i = 98\%$ of the objects, and with a false positive rate $\prod_{i=1}^{20} f_i = 10^{-6}$. The cascade classifier structure is shown in Algorithm 1.

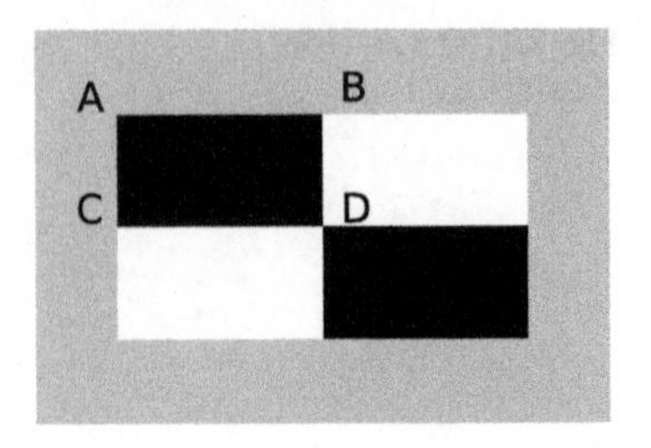

Fig. 8.2 One example Haar-like feature

Note that only a limited number of negative training examples are used for training a node classifier. After T node classifiers have been trained, the current cascade is applied to a huge set of negative image patches to find those "difficult" examples that are wrongly classified by all the existing nodes. These examples are added to the negative training set for training node classifier $T+1$. The database of bootstrapping negative examples $\mathscr{D}$ does not need to be explicitly stored. It can be generated online by applying the current cascade to images that do not contain any object of interest.

8.3.2 The Haar-like Features and the Integral Image

Another factor that makes the Viola–Jones face detector real-time is a set of simple Haar-like visual features. We illustrate one example Haar-like feature in Fig. 8.2. A Haar-like feature corresponds to a mask that has the same size as the training image patches. Elements in the mask can only take three different values: white pixels in the mask corresponding to the value $+1$, the black pixels with the value -1, and the gray pixels with the value 0. One Haar-like feature will have a feature value that equals the dot product between the mask and an image patch. Four types of Haar-like features are proposed in [41].

The Haar-like features can be computed in a constant number of machine instructions, which make evaluation of the node classifiers H_i very fast. As illustrated in Fig. 8.2, the feature value can be quickly computed if we can efficiently compute the sum of pixel values inside each rectangle with the same color (white or black).

Given any image $I(x,y)$, an integral image [41] is defined as an image I' with the same size as I. The values in I' are defined as:

$$I'(x,y) = \sum_{1 \le i \le x, 1 \le j \le y} I(x,y). \tag{8.1}$$

Given I', the sum of pixel values in the rectangle $ABCD$ (Fig. 8.2) is simply $I'(A) + I'(D) - I'(B) - I'(C)$. Thus, a constant number of machine instructions are needed to compute the sum of pixel values in a rectangle. The example feature in Fig. 8.2 requires the sum inside four rectangles, which means that it can be computed in constant time with the help of I'.

The integral image I' for a testing image I also needs to be computed in the testing phase. Fortunately, (8.1) can be computed very efficiently by two recurrence relationships [41]:

$$s(x, y) = s(x, y-1) + I(x, y) \tag{8.2}$$

$$I'(x, y) = I'(x-1, y) + s(x, y), \tag{8.3}$$

in which $s(x, y)$ is the sum of pixel values in the xth row till the yth column. Border values for s and I' can be initialized as $s(x, 0) = 0$ and $I'(0, y) = 0$. In short, only two summations are required to compute the integral image at every pixel position.

8.3.3 AdaBoost Feature Selection and Classification

It is obvious that one single Haar-like feature has weak discrimination capability, far below the requirement for object detection. It is inevitable then to choose (i.e., feature selection) and combine (i.e., classifier learning) a number of these weak features to form a strong classifier that will act as a node classifier H_i.

Feature selection, however, is itself a challenging task. Viola and Jones used training images with size 24×24, which leads to 45,396 Haar-like features [41]. The large number of features makes feature selection difficult in terms of both speed and accuracy considerations. In [41], Viola and Jones proposed a creative idea to use the discrete AdaBoost algorithm [31] to integrate feature selection seamlessly into the classifier learning process. Their method is shown in Algorithm 2.

The key idea in Algorithm 2 is to turn every Haar-like feature into a weak classifier. AdaBoost is then applied to select and combine from the pool of weak classifiers to form a strong classifier.[1]

When equipped with a threshold value, a Haar-like feature can be easily turned into a decision stump (i.e., a decision tree with only one level). Formally, a weak classifier consists of a Haar-like feature (with corresponding mask $\mathbf{m}_j$), a threshold θ_j, and a parity p_j. The resulting weak classifier h_j then classifies an example $\mathbf{x}$ as:

$$h_j(\mathbf{x}) = \begin{cases} 1 & \text{if } p_j \mathbf{m}_j^T \mathbf{x} < p_j \theta_j \\ 0 & \text{otherwise} \end{cases}. \tag{8.4}$$

[1]After training the AdaBoost classifier (i.e., a node classifier in the cascade), one can adjust the threshold θ to meet the learning goal of a node classifier (e.g., a fixed detection rate or a fixed false positive rate.)

Algorithm 2 *NodeLearning* using AdaBoost (modified from Table 1 in [41])

1: {Input: a set of positive examples $\mathscr{P}$, and a set of negative examples $\mathscr{N}$.}
2: Initialize weights $w_{1,i} = \frac{1}{2m}, \frac{1}{2l}$ for $y_i = -1, 1$ respectively, where m and l are the number of negative and positive examples respectively.
3: **for** $t = 1$ to T **do**
4: Normalize the weights w_t so that w_t is a probability distribution.
5: For each feature j, train a weak classifier h_j. The weighted error is $\sum_i w_{t,i} \left| h_j(\mathbf{x}_i) - y_i \right|$.
6: Choose the classifier, h_t, to be the weak classifier with the lowest weighted error ε_t.
7: Update the weights as $w_{t+1,i} = w_{t,i}\beta_t^{1-e_i}$ in which $\beta_t = \frac{\varepsilon_t}{1-\varepsilon_t}$, $e_i = 0$ if $\mathbf{x}_i$ is classified correctly by h_t, and $e_i = 1$ if otherwise.
8: $\alpha_t = \log \frac{1}{\beta_t}$.
9: **end for**
10: Output: a node classifier

$$H(\mathbf{x}) = \begin{cases} 1 & \text{if } \sum_{t=1}^{T} \alpha_t h_t(x) \geq \theta \\ -1 & \text{otherwise} \end{cases},$$

in which θ is an adjustable parameter to control the trade off between detection rate and false positive rate of H.

The parity p_j can take values in $\{+1, -1\}$. This parameter determines which side of the threshold θ_j should be classified as positive. We will describe an efficient method to learn the optimal value for p_j and θ_j in Sect. 8.4.

Although a single Haar-like feature usually has a high error rate (e.g., between 40% and 50% in the face detection task [41]), the discrete AdaBoost algorithm in Algorithm 2 can boost multiple weak classifier into a strong one. Moreover, a cascade of these trained node classifiers can detect faces accurately in real time. The system in [41] detects faces at the speed of 15 frames per second on a slow 700 MHz Pentium III computer. It can detect 89.8% of all faces with only 0.5 false detections per testing image, when evaluated on the benchmark MIT+CMU frontal face detection dataset [30].

8.4 Improving Training Speed of a Cascade

It is reported in [41] that training a complete cascade requires weeks to finish. Thus, it is very important to improve the training speed. In this section, we only consider the *NodeLearning* part.

Algorithm 3 Training a weak classifier (modified from Algorithm 3 in [45])

1: {Input: training examples with labels $\{\mathbf{x}_i, y_i\}_{i=1}^N$ with weights $\{w_{t,i}\}_{i=1}^N$, and a Haar-like feature with corresponding mask $\mathbf{m}$.}
2: Find feature values, $v_1, \ldots, v_N$, where $v_i = \mathbf{x}_i^T \mathbf{m}$.
3: Sort the feature values as $v_{i_1}, \ldots, v_{i_N}$ where $(i_1, \ldots, i_N)$ is a permutation of $(1, \ldots, N)$, and satisfies that $v_{i_1} \leq \cdots \leq v_{i_N}$.
4: $\varepsilon \Leftarrow \sum_{y_i=-1} w_{t,i}$.
5: **for** $k = 1$ to N **do**
6: **if** $y_{i_k} = -1$ **then**
7: $\varepsilon \Leftarrow \varepsilon - w_{t,i_k}, \varepsilon_{t,i} \Leftarrow \varepsilon.$
8: **else**
9: $\varepsilon \Leftarrow \varepsilon + w_{t,i_k}, \varepsilon_{t,i} \Leftarrow \varepsilon.$
10: **end if**
11: **end for**
12: $k = \arg\min_{1 \leq i \leq N} \varepsilon_{t,i}, \tau = \mathbf{x}_{i_k}^T \mathbf{m}.$
13: Output: a weak classifier

$$h(\mathbf{x}) = \operatorname{sgn}\left(\mathbf{x}^T \mathbf{m} - \tau\right).$$

8.4.1 Exact Weak Classifier Learning

Let us denote the number of iterations in Algorithm 2 as T, the number of Haar-like features as M, and the number of training examples as N. The first step to accelerate is the line 5 of Algorithm 2. This line trains a weak classifier from a Haar-like feature, and will be called upon MT times in Algorithm 2. Training weak classifiers needs to be done very efficient because MT is on the order of millions. Algorithm 3 gives a method for accelerating the training process, which is taken from [45].

Suppose there are N training examples, and these examples have feature values $v_1, \ldots, v_N$ for a Haar-like feature. Algorithm 3 first sorts the feature values into $v_{i_1}, \ldots, v_{i_N}$, where $(i_1, \ldots, i_N)$ is a permutation of $(1, \ldots, N)$, and $v_{i_1} \leq \cdots \leq v_{i_N}$. If $v_{i_k} \leq \theta_1, \theta_2 \leq v_{i,k+1}$ is true for some integer k and two different thresholds θ_1 and θ_2, setting the threshold of this Haar-like feature to $\tau = \theta_1$ will have the same accuracy on the training set as that of $\tau = \theta_2$. Thus, we only need to check $N + 1$ possible values for finding the optimal τ. In addition, a sequential update can compute the weighted error rate of different threshold values in $O(N)$ steps. Note that Algorithm 3 only checks the parity $+1$. It is easy to find the optimal weak classifier for both parity values using the idea of Algorithm 3. The complexity of Algorithm 3 is then $O(N \log N)$, dominated by the complexity of sorting feature values to get the permutation. Consequently, the complexity of Algorithm 2 is $O(NMT \log N)$.

However, one does not need to recompute the permutation $(i_1, \ldots, i_N)$ at every iteration inside Algorithm 2 [45]. In AdaBoost learning, weak classifiers need to be re-trained at every iteration because of the updated weights $w_{t,i}$. However, the permutations remain constant throughout Algorithm 2, because they do not depend on $w_{t,i}$. By creating a table to precompute and store the permutations for all

Haar-like features ($O(NM \log N)$), training a weak classifier becomes $O(N)$, and the entire AdaBoost complexity can be reduced to $O(NM(T + \log N))$. In practice this space-for-time strategy leads to two orders of magnitudes speedup.

8.4.2 *Approximate Weak Classifier Learning*

Algorithm 3 seeks to find the optimal threshold that achieves minimum weighted error on the training set. However, the power of AdaBoost resides with the combination of multiple weak classifiers. Theoretically, we only need to guarantee that in each iteration the selected weak classifier h_t has a weighted error rate that is smaller than 0.5. Faster algorithm can be achieved if some approximations are allowed in the weak classifier training step.

One such approximation algorithm was proposed by Pham and Cham [28]. Instead of directly using the raw training examples $\mathbf{x}_i$ and their associated weights $w_{t,i}$, Pham and Cham used statistics of the training set to find the weak classifiers' parameters. Specifically, they assume that feature values for the positive examples follow a normal distribution $N(\mu_+, \sigma_+^2)$. μ_+ is the average weighted feature value for all positive training examples, given a specific Haar-like feature with mask $\mathbf{m}$:

$$\mu_+ = \sum_{y_i=+1} w_{t,i}\mathbf{m}^T\mathbf{x}_i, \tag{8.5}$$

in which $\mathbf{m}$ is the mask corresponding to the Haar-like feature; σ_+ is the standard deviation of the feature value for positive training examples. Similarly, negative examples' feature values are also assumed to follow $N(\mu_-, \sigma_-^2)$. Closed-form and efficient solution exists for finding the optimal separating plane for two one-dimensional normal distributions [11]. We only need to efficiently compute the values $(\mu_+, \sigma_+, \mu_-, \sigma_-)$ when the weights $w_{t,i}$ are updated, in which the integral image once again helps.

An image $\mathbf{x}$ and its integral image $\mathbf{x}'$ are linked together by (8.1), which is obviously a linear transformation. Thus there exists a matrix B such that $\mathbf{x}' = B\mathbf{x}$. The matrix B is constant and invertible, and encodes the linear transformation between $\mathbf{x}$ and $\mathbf{x}'$. The example $\mathbf{x}$ can be expressed as $\mathbf{x} = B^{-1}\mathbf{x}'$. One can in turn use the integral image to compute μ_+ as [28]:

$$\mu_+ = \sum_{y_i=+1} w_{t,i}\mathbf{m}^T B^{-1}\mathbf{x}_i' = \mathbf{m}^T B^{-1}\left(\sum_{y_i=+1} w_{t,i}\mathbf{x}_i'\right). \tag{8.6}$$

Two facts make (8.6) easy to compute. First, the term $\sum_{y_i=+1} w_{t,i}\mathbf{x}_i'$ can be pre-computed and stored whenever $w_{t,i}$ are updated. Second, the vector $\mathbf{m}^T B^{-1}$ is a sparse vector which usually contains less than 10 nonzero entries.

Note that $\sum_{y_i=+1} w_{t,i}\mathbf{x}'_i$ is the weighted average of $\mathbf{x}'$ (integral version of training patches). Let $\Sigma_{\mathbf{x}'}$ be the (weighted) covariance matrix of $\mathbf{x}'$, then we can compute σ_+ as[2]:

$$\sigma_+^2 = \left(\mathbf{m}^T B^{-1}\right) \Sigma_{\mathbf{x}'} \left(\mathbf{m}^T B^{-1}\right)^T . \tag{8.7}$$

The values μ_- and σ_- can be computed using the same trick.

This method has two attractive characteristics. First, the complexity of training a weak classifier for a given feature is independent of N, the number of training examples. When N is big (which is usually the case in object detection), this approximation method is more efficient than Algorithm 3. And its speed advantage increases when N gets bigger. Second, unlike Algorithm 3, which requires additional storage for saving the permutation vectors, this approximation method has a smaller memory footprint. As a direct consequence, more Haar-like features can be used in the training process (and the use of more Haar-like features usually implies higher accuracies).

Empirically, in [28] Pham and Cham reported the training of a cascade classifier using 295,920 Haar-like features (with more feature types than those appearing in [41]) for face detection. The training process finished in 5 h and 30 min. Using Algorithm 3 with 40,000 features, it took 13 h and 20 min to train a cascade, using the same training set and running on the same hardware. With the same number of Haar-like features, both methods achieve very similar detection accuracies, which means that the approximate weak classifier training part does not hinder the final classifier's accuracy. Having the ability to deal with more features, the approximation method reduces the number of false detections at a given detection recall rate in comparison to Algorithm 3.

8.4.3 FFS: Alternative Feature Selection

The AdaBoost-based Algorithm 2 combines the selection of discriminative Haar-like features and the learning of a node classifier into an integrated framework. However, Wu et al. showed that these two components are not necessarily tied together. Alternative feature selection and node classifier learning methods can be applied sequentially, with the benefit of reduced training time and better detection performance.

Forward feature selection (FFS) [44] is a frequently used greedy feature selection method. It can be used effectively to select Haar-like features [45], whose algorithmic details are presented in Algorithm 4.

The first step of Algorithm 4 is to train the weak classifiers for all Haar-like features ($O(NM \log N)$ using Algorithm 3). The same space-for-time strategy is

[2]Special care is required for computing $\Sigma_{\mathbf{x}'}$ efficiently. However, we omit these details. The readers may refer to Sect. 3.2 of [28] for more information.

Algorithm 4 FFS feature selection (modified from Algorithm 2 in [45])

1: {Input: a training set $\{\mathbf{x}_i, y_i\}_{i=1}^N$, a set of Haar-like features $\{h_i\}_{i=1}^M$, where N and M are the number of training examples and Haar-like features, respectively.}
2: **for** $i = 1$ to M **do**
3: Choose appropriate parity and threshold for a Haar-like feature h_i, such that h_i has smallest error on the training set using Algorithm 3.
4: **end for**
5: Make a table V_{ij} such that $V_{ij} = h_i(\mathbf{x}_j), 1 \le i \le M, 1 \le j \le N$.
6: $S \Leftarrow \emptyset, v \Leftarrow \mathbf{0}_{1\times N}$, where $\mathbf{0}_{1\times N}$ is a row vector filled by zeros.
7: **for** $t = 1$ to T **do**
8: **for** $i = 1$ to M **do**
9: $S' \Leftarrow S \cup h_i, v' \Leftarrow v + V_{i:}$, where $V_{i:}$ is the ith row of V.
10: $\{H'(\mathbf{x}) = \text{sgn}\left(\sum_{h\in S'} h(\mathbf{x}) - \theta\right)$ is the classifier associated with S', and we can compute its value using $H'(\mathbf{x}_i) = \text{sgn}(v'_i - \theta)$.}
11: Find the θ that makes H' has the smallest error rate.
12: $\varepsilon_i \Leftarrow$ the error rate of H' with the chosen θ value.
13: **end for**
14: $k \Leftarrow \arg\min_{1\le i\le M} \varepsilon_i$.
15: $S \Leftarrow S \cup h_k, v \Leftarrow v + V_{k:}$.
16: **end for**
17: Output: a node classifier

$$H(\mathbf{x}) = \text{sgn}\left(\sum_{h\in S} h(\mathbf{x}) - \theta\right).$$

used: precompute and save classification results of all such classifiers into a table V. One noticeable difference between Algorithms 2 and 4 is that in FFS these weak classifiers are trained without using the weights. Thus, there is no need to update these weak classifiers.

Both Algorithms 2 and 4 are "wrapper" feature selection methods, in the sense that the effectiveness of a selected subset of features is evaluated by a classifier trained from such a subset. Instead of using weighted voting as that in AdaBoost, FFS uses a simple vote strategy: every weak classifier has the same weight. In every iteration, the FFS algorithm examines every Haar-like feature, temporarily adds it to the selected feature subset, and evaluates the classification accuracy of the updated set of features. The Haar-like feature that leads to the minimum classification error rate is chosen and permanently added to the selected feature subset.

However, although both algorithms are greedy in nature, they solve different optimization objectives. In AdaBoost, a node classifier $H(x)$ is implicitly minimizing the cost function

$$\sum_{i=1}^{N} \exp(-y_i H(\mathbf{x}_i)). \tag{8.8}$$

While in FFS we are explicitly finding the feature that leads to a node classifier with the smallest error rate in the training set.

The classification result of a subset of features can be updated very efficiently using the stored table V. In order to find the optimal threshold value, the trick in

Algorithm 3 is also applicable. Note that at iteration t, the number of votes are always integers within the range $[0 \quad t]$, which means that we only need to consider θ in this range. Two histograms (one for positive examples and one for negative examples) can be built: the cell i contains the number of positive or negative examples that have i votes, respectively. Using both histograms, the error rate at $\theta = 0, \ldots, t$ can be sequentially updated efficiently.

The complexity of Algorithm 4 is $O(NM(T + \log N))$, same as that of Algorithm 2. However, in practice FFS has a smaller constant factor than AdaBoost. It was reported in [45] that the fast AdaBoost implementation usually uses 2.5–3.5 times of the training time of FFS, and the original AdaBoost implementation in [41] needs 50–150 times of that of FFS.

With faster training speed, FFS trains face detectors that have similar results as AdaBoost in terms of both accuracy and testing speed. FFS can be used as an alternative method for AdaBoost in training node classifiers in a cascade object detection system.

8.5 Asymmetric Learning in Cascades

8.5.1 Goal Asymmetry and Asymmetric AdaBoost

As was discussed earlier in Sect. 8.2, at least three major challenges exist in object detection: scale, speed, and asymmetry. Most of these challenges are addressed in the algorithms we have discussed so far. With the cascade classifier structure, testing speed can be improved to video rate or even higher. The bootstrap step in Algorithm 1 can effectively deal with billions of negative training examples. The fast AdaBoost method and the exact weak classifier learning algorithm reduce training time of a complete cascade from weeks to hours, which is further improved by the distribution-based approximation method and the alternative feature selection method (Algorithm 4).

As to the asymmetries, the cascade classifier structure also handles the "uneven class priors" asymmetry. We can choose to use the same number of positive and negative training instances when learning every node classifier. The bootstrap process also implicitly deals with "unequal complexity within the positive and negative classes." Since at every node we only use a small subset of negative examples, the complexity of node negative training set is limited.

We are left with the "goal asymmetry." For the complete cascade, we require high detection rate (e.g., 95%) and extremely low false positive rate (e.g., 10^{-7} [41]. In a node classifier, we try to achieve the asymmetric node learning goal [45]:

> for every node, design a classifier with very high (e.g., 99.9%) detection rate and only moderate (e.g., 50%) false positive rate.

This special requirement demands special learning algorithms. Asymmetric AdaBoost (AsymBoost) was an attempt by Viola and Jones for solving this problem [40]. The idea of AsymBoost is to emphasize false negatives (i.e., classifying faces as nonfaces) more than false positives (i.e., classifying nonfaces as faces). Both false positive and false negative have the same loss in Algorithm 2. This symmetric loss is replaced by an asymmetric loss function (assuming false negatives are k times more important than false positives):

$$A_{\text{Loss}(i)} = \begin{cases} \sqrt{k} & \text{if } y_i = +1, \text{ and } H(\mathbf{x}_i) = -1, \\ \frac{1}{\sqrt{k}} & \text{if } y_i = -1, \text{ and } H(\mathbf{x}_i) = +1, \\ 0 & \text{otherwise.} \end{cases} \tag{8.9}$$

This new loss function can be easily incorporated into Algorithm 2, by pre-weighting training examples using $\exp\left(y_i \log \sqrt{k}\right)$. This strategy, however, is unsuccessful because AdaBoost will quickly absorb this artificial difference in initialization [40]. Instead, Viola and Jones amortize this asymmetric cost into every iteration of AdaBoost. In a node classifier with T AdaBoost iterations, $\exp\left(\frac{1}{T} y_i \log \sqrt{k}\right)$ is multiplied to the example weights $w_{t,i}$ for $t = 1, 2, \ldots, T$, followed by a normalization procedure to make the new weights a distribution.

AsymBoost enforces a higher cost for missing faces ($k > 1$) than false detections in node classifiers. When comparing complete cascades, a cascade trained using AsymBoost usually achieves 1–2% higher detection rate with the same number of false detections on the MIT+CMU benchmark face detection dataset [40].

8.5.2 *Linear Asymmetric Classifier*

Another attempt to address the goal asymmetry is LAC by Wu et al. [45]. LAC trains node classifiers to directly optimize the asymmetric node learning goal: very high (e.g., 99.9%) detection rate and only moderate (e.g., 50%) false positive rate.

8.5.2.1 LAC Formulation

LAC does not perform feature selection. Instead, it assumes that a subset of discriminative features have been selected by other methods (e.g., AdaBoost, FFS, AsymBoost, or any other method). Furthermore, LAC assumes that weak classifiers corresponding to these selected features have also been trained. Given an example $\mathbf{x}$, the classification results of weak classifiers are the input to LAC. In other words, the input to LAC are binary vector ($+1$ or -1) in $\mathbb{R}^d$ if d Haar-like features are selected.

For simplicity in the presentation, we use $\mathbf{x}$ to represent this binary vector for the same training example $\mathbf{x}$ too. These two different meanings should be easily distinguishable from the context. In this section, we use $\mathbf{x}$ and $\mathbf{y}$ to denote positive and negative training examples, respectively. The class labels ($+1$ for $\mathbf{x}$ and -1 for $\mathbf{y}$) are implied in the symbols and thus omitted. An example is denoted as $\mathbf{z}$ if its label is unknown, following the notation of [45].

LAC expresses the asymmetric node learning goal and tries to directly optimize this goal for a linear classifier:

$$\begin{aligned} \max_{\mathbf{a}\neq 0, b} \quad & \Pr_{\mathbf{x}\sim(\bar{\mathbf{x}},\Sigma_{\mathbf{x}})} \left\{\mathbf{a}^T\mathbf{x} \geq b\right\} \\ \text{s.t.} \quad & \Pr_{\mathbf{y}\sim(\bar{\mathbf{y}},\Sigma_{\mathbf{y}})} \left\{\mathbf{a}^T\mathbf{y} \leq b\right\} = \beta. \end{aligned} \tag{8.10}$$

Only the first- and second-order moments are used in the formulation of LAC: $\mathbf{x} \sim (\bar{\mathbf{x}}, \Sigma_{\mathbf{x}})$ denotes that $\mathbf{x}$ is drawn from a distribution with mean $\bar{\mathbf{x}}$ and covariance matrix $\Sigma_{\mathbf{x}}$. The distribution of $\mathbf{x}$, however, is not necessarily Gaussian. Similarly, negative examples are modeled by $\bar{\mathbf{y}}$ and $\Sigma_{\mathbf{y}}$. LAC only considers linear classifiers $H = (\mathbf{a}, b)$:

$$H(\mathbf{z}) = \begin{cases} +1 & \text{if } \mathbf{a}^T\mathbf{z} \geq b \\ -1 & \text{if } \mathbf{a}^T\mathbf{z} < b. \end{cases} \tag{8.11}$$

The constraint in (8.10) fixes the false positive rate to β (and $\beta = 0.5$ when learning a node classifier). The objective in (8.10) is to maximize the detection rate. Thus, (8.10) is a literal translation of the asymmetric node learning goal, under the distribution assumption of training examples.

8.5.2.2 LAC Solution

Let $\mathbf{x_a}$ denote the standardized version of $\mathbf{a}^T\mathbf{x}$ ($\mathbf{x}$ projected onto the direction of $\mathbf{a}$), i.e.,

$$\mathbf{x_a} = \frac{\mathbf{a}^T(\mathbf{x} - \bar{\mathbf{x}})}{\sqrt{\mathbf{a}^T\Sigma_{\mathbf{x}}\mathbf{a}}}. \tag{8.12}$$

$\mathbf{y_a}$ can be defined similarly for negative examples. Equation (8.10) is converted to an unconstrained optimization problem as:

$$\min_{\mathbf{a}\neq 0} \Psi_{\mathbf{x},\mathbf{a}} \left(\frac{\mathbf{a}^T(\bar{\mathbf{y}} - \bar{\mathbf{x}}) + \Psi_{\mathbf{y},\mathbf{a}}^{-1}(\beta)\sqrt{\mathbf{a}^T\Sigma_{\mathbf{y}}\mathbf{a}}}{\sqrt{\mathbf{a}^T\Sigma_{\mathbf{x}}\mathbf{a}}} \right), \tag{8.13}$$

in which $\Psi_{\mathbf{x},\mathbf{a}}$ $(\Psi_{\mathbf{y},\mathbf{a}})$ denotes the cumulative distribution function (c.d.f.) of $\mathbf{x_a}$ $(\mathbf{y_a})$, and $\Psi_{\mathbf{y},\mathbf{a}}^{-1}$ is the inverse function of $\Psi_{\mathbf{y},\mathbf{a}}$. This is, however, a difficult optimization problem because we do not know the properties of $\Psi_{\mathbf{x},\mathbf{a}}$ and $\Psi_{\mathbf{y},\mathbf{a}}^{-1}$.

Two assumptions are made in [45] to simplify (8.13). First, $\mathbf{a}^T\mathbf{x}$ is assumed to follow a scalar normal distribution. Second, the median value of the distribution $\mathbf{y_a}$ is close to its mean (so that we have $\Psi_{\mathbf{y},\mathbf{a}}^{-1}(\beta) \approx 0$ when $\beta = 0.5$). These assumptions align well with the reality in object detection, as shown in Fig. 8.3. $\mathbf{a}^T\mathbf{x}$ fits closely to a normal distribution in the normal probability plot, a visual method to test the normality of a distribution. $\mathbf{a}^T\mathbf{y}$ fits almost perfectly to a normal distribution, which implies that its mean and median are indeed the same.

Under these assumptions, for $\beta = 0.5$ (8.13) can be further approximated by

$$\max_{\mathbf{a}\neq 0} \frac{\mathbf{a}^T(\bar{\mathbf{x}} - \bar{\mathbf{y}})}{\sqrt{\mathbf{a}^T \Sigma_{\mathbf{x}} \mathbf{a}}}, \tag{8.14}$$

which has closed-form solutions:

$$\mathbf{a}^* = \Sigma_{\mathbf{x}}^{-1}(\bar{\mathbf{x}} - \bar{\mathbf{y}}), \tag{8.15}$$

$$b^* = \mathbf{a}^{*T}\bar{\mathbf{y}}. \tag{8.16}$$

When $\Sigma_{\mathbf{x}}$ is positive semi-definite, $\Sigma_{\mathbf{x}} + \lambda I$ can be used to replace $\Sigma_{\mathbf{x}}$, where λ is a small positive number. A summary of applying LAC to train a node classifier in object detection is shown in Algorithm 5.

Algorithm 5 LAC as *NodeLearning* (modified from Algorithm 4 in [45])

1: {Input: a training set composed of positive examples $\{\mathbf{x}_i\}_{i=1}^{n_\mathbf{x}}$ and negative examples $\{\mathbf{y}_i\}_{i=1}^{n_\mathbf{y}}$, a set of Haar-like features, and a feature selection method $\mathscr{F}$.}

2: Select T weak classifiers $\mathbf{h} = (h_1, h_2, \ldots, h_T)$ using $\mathscr{F}$, where $h_i(\mathbf{z}) = \mathrm{sgn}\left(\mathbf{z}^T\mathbf{m}_i - \tau_i\right)$.

3: For each training example, build a feature vector $\mathbf{h}(\mathbf{z}) = (h_1(\mathbf{z}), h_2(\mathbf{z}), \ldots, h_T(\mathbf{z}))$.

4:

$$\bar{\mathbf{x}} = \frac{\sum_{i=1}^{n_\mathbf{x}} \mathbf{h}(\mathbf{x}_i)}{n_\mathbf{x}}, \quad \bar{\mathbf{y}} = \frac{\sum_{i=1}^{n_\mathbf{y}} \mathbf{h}(\mathbf{y}_i)}{n_\mathbf{y}},$$

$$\Sigma_\mathbf{x} = \frac{\sum_{i=1}^{n_\mathbf{x}} (\mathbf{h}(\mathbf{x}_i) - \bar{\mathbf{x}})(\mathbf{h}(\mathbf{x}_i) - \bar{\mathbf{x}})^T}{n_\mathbf{x}},$$

$$\Sigma_\mathbf{y} = \frac{\sum_{i=1}^{n_\mathbf{y}} (\mathbf{h}(\mathbf{y}_i) - \bar{\mathbf{y}})(\mathbf{h}(\mathbf{y}_i) - \bar{\mathbf{y}})^T}{n_\mathbf{y}}.$$

5:

$$\mathbf{a} = \Sigma_\mathbf{x}^{-1}(\bar{\mathbf{x}} - \bar{\mathbf{y}}), \quad b = \mathbf{a}^T\bar{\mathbf{y}}.$$

6: Output: a node classifier

$$H(\mathbf{z}) = \mathrm{sgn}\left(\sum_{t=1}^{T} \mathbf{a}_t \mathbf{h}_t(\mathbf{z}) - b\right) = \mathrm{sgn}\left(\mathbf{a}^T\mathbf{h}(\mathbf{z}) - b\right).$$

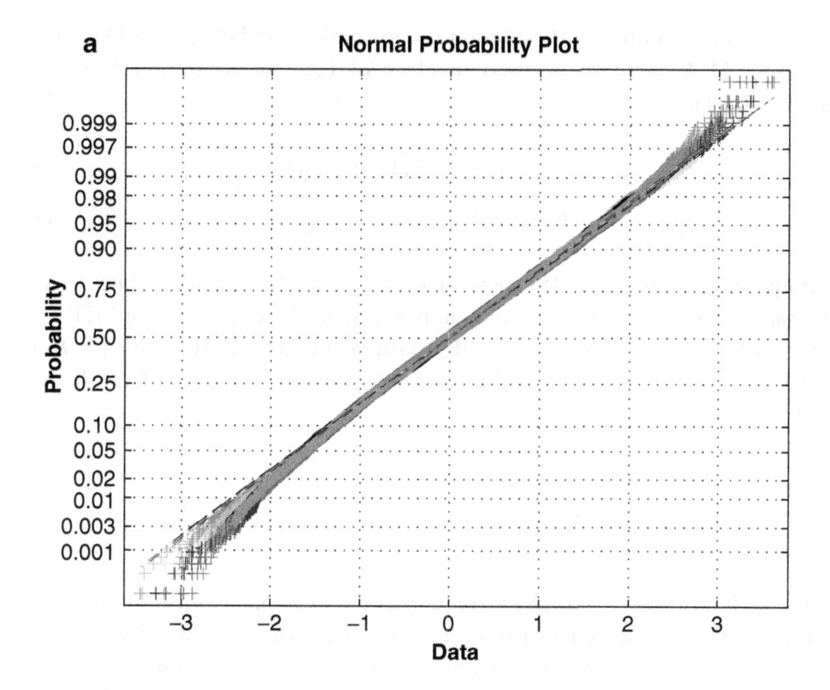

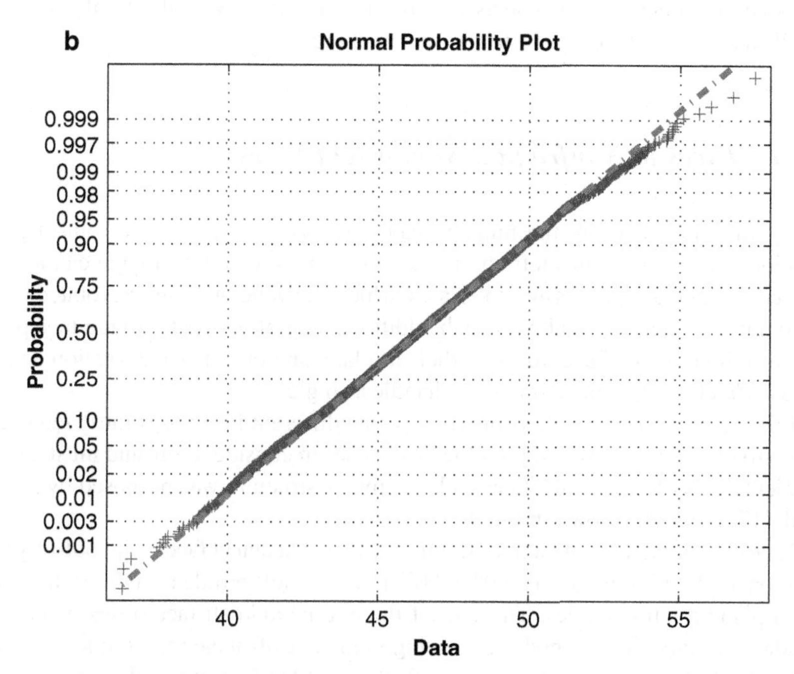

Fig. 8.3 Normality test for $\mathbf{a}^T\mathbf{x}$ and $\mathbf{a}^T\mathbf{y}$. $\mathbf{a}$ is drawn from the uniform distribution $[0\ 1]^T$. Part figure (**a**) shows overlapped results for 10 different $\mathbf{a}$'s. From Wu et al. [45], © 2008 IEEE, with permission. (**a**) $\mathbf{a}^T\mathbf{x}$. (**b**) $\mathbf{a}^T\mathbf{y}$

In fact, the simplified LAC solution is very similar to FDA (Fisher Discriminant Analysis). FDA can also be used in place of LAC in Algorithm 5, using the following equations:

$$\mathbf{a}^* = \left(\Sigma_{\mathbf{x}} + \Sigma_{\mathbf{y}}\right)^{-1} (\bar{\mathbf{x}} - \bar{\mathbf{y}}), \tag{8.17}$$

$$b^* = \mathbf{a}^{*T} \bar{\mathbf{y}}. \tag{8.18}$$

Empirical results from [45] show that both LAC and FDA can effectively deal with the asymmetric node learning goal in the node classifiers. On the MIT+CMU benchmark face detection dataset, both AdaBoost+LAC and AdaBoost+FDA have higher detection rates than that of AsymBoost, when the number of false detections are the same.

8.6 Beyond Frontal Faces

So far we have used frontal face detection as the example application to introduce various methods. Ensemble learning methods and the cascade classifier structure, of course, are useful not only for detecting frontal faces. In this section, we will briefly introduce the detection of objects beyond frontal faces. We will mainly focus on profile and rotated faces, pedestrians, and tracking.

8.6.1 Faces in Nonfrontal, Nonupright Poses

Many variations exist in the human head pose. The head can have out-of-plane rotations, which generates left and right profile faces. One face image can also be rotated using image processing softwares, which can generate in-plane rotated faces. A common strategy to deal with such additional complexity is to divide face poses into different "views" according to their in-plane and out-of-plane rotation angles. A cascade classifier can be trained to handle a single view.

It is, however, not easy to properly separate different face poses into views. The view structure must cover all possible head poses in consideration, and must also be efficient during testing time. One such multiview structure was proposed by Huang et al. [17], which is shown in Fig. 8.4.

Faces are organized into a tree structure, which can detect faces with out-of-plane rotation angles ranging from $-90°$ to $90°$ (i.e., from left profile to right profile), and all in-plane rotation angles. The root of the tree include all face poses, which are divided into three level 2 nodes according to the out-of-plane rotation: left, frontal, and right. Left (right) profile faces are further divided into two nodes according to the rotation angle. Thus, there are 5 nodes in the level 3 in total. Finally, every level

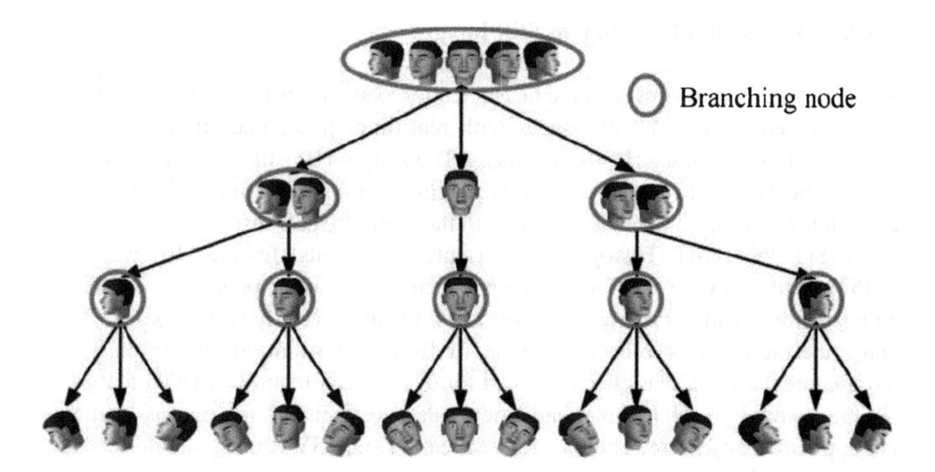

Fig. 8.4 Classifier tree for detecting multiview faces. From Huang et al. [17], © 2007 IEEE, with permission

3 node is divided into three nodes in level 4, corresponding to in-plane rotation angle −45°, 0°, and 45°, respectively.[3] Every level 4 node then detects faces in one specific pose.

One design choice by Huang et al. is to allow multiple nodes at the same level in the classifier tree to be active simultaneously. For example, it is reasonable to activate both the second and the third node in the level 2 for a right profile face with a small out-of-plane rotation angle. This choice increases the possibility that a face is detected. But, it also requires a new kind of node classifier: the node classifier must be a multiple class classifier that allow multiple labels (i.e., next level nodes) to be predicted simultaneously. The challenges that are laid out in Sect. 8.2 are still to be solved by the new node classifier. A Vector Boosting algorithm was proposed by Huang et al. [17] as the new node classifier.

8.6.2 Pedestrians

Pedestrian detection is another area where the cascade classifier structure and ensemble learning have been successful.

[3]In addition to the level 4 nodes shown in Fig. 8.4, Huang et al. rotate their features (called Granule features) by 90°, 180°, and −90° for the level 4 nodes. This strategy effectively covers the entire 360° range in-plane rotation.

8.6.2.1 Pedestrian Detection in Still Images

A system that is very similar to the face detection system was used by Zhu et al. [54] to detect pedestrians in still images with real-time speed. Pedestrians, however, exhibit different characteristics than faces. The simple Haar-like features (and their corresponding simple decision stump weak classifiers) are no longer discriminative enough for pedestrian detection, even with the help of AdaBoost.

Instead, the HOG (Histogram of Oriented Gradients) feature [10] was used in [54]. Within a 64×128 image patch which is typical for pedestrians, different HOG features can be extracted from 5,031 rectangles with various sizes. They constitute the features to be used in an AdaBoost algorithm. Similar to the Haar-like features, a weak classifier is trained for every HOG feature. A HOG feature is 36 dimensional, which automatically means that the decision stump weak learner is not applicable any more. The linear SVM learner (i.e., SVM using the dot-product kernel) is used to train weak classifiers.

HOG features and SVM classifiers do not enjoy the same hyper-speed as Haar-like features and decision stumps, in both the training and the testing phases. Two approaches were used to accelerate pedestrian detection in [54]. First, instead of training all 5,031 weak classifiers, 250 ($\approx$5%) HOG features are randomly sampled and their corresponding linear SVM are trained. Second, the "integral histogram" data structure is used to compute HOG features. The HOG feature used in [54] is a 36 dimensional histogram, and the integral histogram can compute HOG features efficiently, similar to the integral image data structure for the Haar-like features.

Compared to use Haar-like feature directly for pedestrian detection, HOG features can reduce the number of false positives by several orders of magnitudes at the same detection rate. In terms of testing speed, HOG features take about 2–4 times the time of that of Haar-like features, which can still be managed to be real-time with modern computers.

8.6.2.2 Pedestrian Detection in Videos

The pedestrians in [54] are of size 64×128, which is usually larger than pedestrian size in many application domains, e.g., surveillance. Viola et al. proposed to use motion information for pedestrian detection in surveillance videos [42].

For two consecutive video frame I_t and I_{t+1}, five images can be defined in order to capture motion information at time t:

$$\Delta = \text{abs}(I_t - I_{t+1}) \tag{8.19}$$

$$U = \text{abs}(I_t - I_{t+1} \uparrow) \tag{8.20}$$

$$L = \text{abs}(I_t - I_{t+1} \leftarrow) \tag{8.21}$$

$$R = \text{abs}(I_t - I_{t+1} \rightarrow) \tag{8.22}$$

$$D = \text{abs}(I_t - I_{t+1} \downarrow) \tag{8.23}$$

Algorithm 6 Ensemble tracking (modified from Algorithm 1 in [2])

1: {Input: video frames $I_1, \ldots, I_n$, and rectangle r_1 in I_1 that contains the object.}
2: Initialization: Train T weak classifiers and add them to the ensemble.
3: **for** each subsequent frame I_j, $j > 1$ **do**
4: Test all pixels in I_j using the current ensemble strong classifier and create a confidence map L_j.
5: Run mean shift on the confidence map L_j and report new object rectangle r_j.
6: Label pixels inside rectangle r_j as object and all those outside it as background.
7: Keep K best weak classifier.
8: Train new $T - K$ weak classifiers on frame I_j and add them to the ensemble.
9: **end for**
10: Output: Rectangles $r_2, \ldots, r_n$.

in which the operators $\{\uparrow, \leftarrow, \rightarrow, \downarrow\}$ shift an image by one pixel in the corresponding direction. An extended set of Haar-like features can be applied to one of these five images or the difference between Δ and one of the other four images to generate motion features. Motion features, together with the appearance features from I_t, form the new feature set for pedestrian detection. The same integral image trick, the AdaBoost ensemble learning method, and the cascade classifier structure can all be used in this new context.

8.6.3 Tracking

Tracking can also benefit from ensemble learning techniques, sometimes dubbed "tracking-by-detection." In a tracking by detection framework, an object detector (e.g., the pedestrian detector) is continuously applied to every frame of a video. Detection results of consecutive frames are then registered across frames to form a reliable tracking result. In this section, we describe the ensemble tracking approach [2] by S. Avidan, which is shown in Algorithm 6.

AdaBoost is used to train an ensemble classifier that distinguish the target object from the background, and the classifier is continuously updated throughout the tracking process. Initially, positive and negative examples are extracted from the user labels (r_1 as positive and else as negative) from the first frame I_1. The ensemble classifier is then used to classify pixels in the next frame to form a confidence map about the possibility that a pixel belongs to the object or the background. The mean shift mode seeking algorithm [8] is used to find the object from the confidence map. The AdaBoost classifier needs to be updated by adding new weak classifiers using the newly detected object and the background in the new frame.

8.7 Discussions

We have described several object detection methods and applications, centered around the cascade classifier structure and ensemble learning methods (especially AdaBoost) in this chapter. In many applications (for example, face detection, pedestrian detection, and tracking), real-time detection speed, and detection accuracy suitable for practical usage have been achieved.

There are, however, many open questions remain in the object detection task. At least four factors still prevent the methods we discussed above from being applied to detect many other objects: training data, training speed, visual features, and multiclass learning.

- Both face and pedestrian detection require thousands of training image patches from the target object class, which are gathered through the laborious and error-prone human-guided data collection process. A human being need to manually crop the object of interest from the background clutter, and transform the cropped image patch to appropriate size. Similarly, a large set of images that do not contain any object of interest needs to be collected and verified. It is necessary to design new algorithms that only require few positive training images, and do not need negative training images.
- Training a cascade (or ensemble classifier) using a large feature set is time consuming. Although modern algorithms have reduced the training time to a few hours, it is still too long in many applications, e.g., training an object model for image retrieval. In fact, training time is closely related to the training set size. Ultimately we are aiming at an accurate detector that is trained with few examples and within seconds.
- As already seen in this chapter, different feature sets have been used in different tasks [41, 42, 45, 54]. The Haar-like features, although contributing to super fast detection systems, are usually not discriminative enough in the detection of objects beyond frontal faces. It is attractive to obtain a feature set that is capable for detecting many objects, and have an efficient evaluation strategy.
- We have focused on binary classification in this chapter: the object of interest versus the background clutter. However, object detection is a natural multiclass problem because we are usually interested in more than one object. A detector structure that can detect multiple object categories is both desirable and challenging, especially when we are interested in a large number of objects.

8.8 Bibliographical and Historical Remarks

Face detection based on machine learning methods have long been studies, dated back to at least early 1990s. Principal Component Analysis (PCA) was used in early attempts by Turk and Pentland [39], and Moghaddam and Pentland [23]. Sung and Poggio [35] used mixtures of Gaussians to model both faces and nonfaces.

The idea to bootstrap negative examples was also used in [35]. Various other machine learning methods have been used too. Osuna et al. used the Support Vector Machine to face detection in [25]. Yang et al. applied the SNoW learning architecture in [52]. Neural networks were used by Rowley et al. to provide accurate face detection (both frontal and rotated) [30]. Schneiderman and Kanade [32] used Naive Bayes to pool statistics of various image measurements into accurate detectors that detected faces (both frontal and profile) and cars. A survey of early face detection methods can be found in [51].

The cascade detector by Viola and Jones [41] is the first real-time frontal face detector. The cascade classifier structure is a coarse-to-fine search strategy that was previously used by Fleuret and Geman [13], and Amit et al. [1]. Sequential classifier rejections have also been used before, by Baker and Nayar [4], and Elad et al. [12]. The Haar-like features have been used before, by Papageorgiou et al. [27] for object detection. The integral image data structure that accelerates evaluation of Haar-like features was proposed by F. Crow [9]. Both the Haar-like features and the AdaBoost node classifier (which selects features and combines them) were used by Tieu and Viola [36] for image retrieval.

A large number of research efforts focus on improving the cascade framework in various aspects. Training speed of a cascade were greatly enhanced using a fast AdaBoost implementation [45] by Wu et al., using an alternative node learning method [47] by Wu et al., and using an approximate weak classifier training method based on statistical properties by Pham and Cham [28]. A similar idea was used by Avidan and Butman [3].

Features beyond the simple Haar-like features in [41] have been proposed, e.g., in [19,28,42,45]. A type of features called Granule features was proposed by Huang et al. [17]. HOG features were used in [54]. A modified Census Transform feature was used by Froba and Ernst [14]. Local Binary Patterns (LBP) was used by Zhang et al. [53].

Various weak classifiers that are more complex than the decision stump are also helpful. Histogram of feature values was used in [20] by Liu and Shum. Decision trees with more than one internal node was used by Lienhart et al. [19] and Brubaker et al. [7].

Many variants of the boosting algorithm have been used to replace discrete AdaBoost. For example, an empirical study by Lienhart et al. [19] suggested using real AdaBoost and gentle AdaBoost. Boosting variants that deal with asymmetries have also been shown to improve object detection. Viola and Jones used an amortized version of asymmetric AdaBoost [40]. Wu et al. proposed LAC [48]. Cost-sensitive boosting algorithms can also improve the node classifiers [22]. Column generation was used to train boosting classifiers (LACBoost and FisherBoost) by Shen et al. in [33].

Alternative feature selection methods can be used to select features and train node classifiers in a cascade. Forward Feature Selection [47] was used by Wu et al. to form a node classifier. Floating search incorporated into boosting helped eliminate wrong selection, which was shown by Li and Zhang [18]. Sparse linear discriminant analysis was used by Paisitkriangkrai et al. to choose features [26].

Improvements are also made to the cascade structure. Xiao et al. proposed boosting chain to make use of previous trained node classifiers [50]. Cascade with many exits (e.g., one feature is a node) have been shown to improve both detection speed and accuracy by Bourdev and Brandt [5], Pham et al. [29], and Xiao et al. [49].

Choosing optimal operating points in the ROC curve of node classifiers were studied by Sun et al. [34]. A two-point algorithm was proposed by Brubaker et al. [7] to further improve the cascade.

Faces at all poses (including rotation in- and out-of plane) can be detected with improvements to the cascade architectures. Tree structures for detecting multiview faces were proposed by Li and Zhang [18], and Huang et al. [17]. A Vector Boosting algorithm was also proposed in [17].

Cascade and ensemble learning methods achieved a trade off between detection speed and accuracy for pedestrian detection [54]. However, it is worth noting that there exist other machine learning methods that have faster speed [46] or detection accuracy [43]. LAC from [48] was used in pedestrian detection by Mu et al. [24].

Tracking-by-detection is a successful application of object detectors. The Ensemble tracking method by S. Avidan [2] is an example. Tracking-by-detection usually requires online learning. Online boosting was used in [15] by Grabner and Bischof. Boosted particle filter was used by Lu et al. in [21]. A detector confidence particle filter was used by Breitenstein in [6].

Multiclass classification for object detection was studied in [38], using a probabilistic boosting tree. A boosting framework was proposed in [37], which can share features among different object categories.

References

1. Amit, Y., Geman, D., Fan, X.: A coarse-to-fine strategy for multiclass shape detection. IEEE Trans. on Pattern Analysis and Machine Intelligence **26**(12), 1606–1621 (2004)
2. Avidan, S.: Ensemble tracking. IEEE Trans. on Pattern Analysis and Machine Intelligence **29**(2), 261–271 (2007)
3. Avidan, S., Butman, M.: The power of feature clustering: An application to object detection. In: Advances in Neural Information Processing Systems 17, pp. 57–64 (2005)
4. Baker, S., Nayar, S.: Pattern rejection. In: Proc. IEEE Conf. on Computer Vision and Pattern Recognition, pp. 544–549 (1996)
5. Bourdev, L.D., Brandt, J.: Robust object detection via soft cascade. In: Proc. IEEE Conf. on Computer Vision and Pattern Recognition, vol. II, pp. 236–243 (2005)
6. Breitenstein, M.D., Reichlin, F., Leibe, B., Koller-Meier, E., Gool, L.J.V.: Robust tracking-by-detection using a detector confidence particle filter. In: The IEEE Conf. on Computer Vision, pp. 1515–1522 (2009)
7. Brubaker, S.C., Wu, J., Sun, J., Mullin, M.D., Rehg, J.M.: On the design of cascades of boosted ensembles for face detection. International Journal of Computer Vision **77**(1-3), 65–86 (2008)
8. Comaniciu, D., Meer, P.: Mean shift: A robust approach toward feature space analysis. IEEE Trans. on Pattern Analysis and Machine Intelligence **24**(5), 603–619 (2002)
9. Crow, F.C.: Summed-area tables for texture mapping. In: SIGGRAPH, vol. 18, pp. 207–212 (1984)

10. Dalal, N., Triggs, B.: Histograms of oriented gradients for human detection. In: Proc. IEEE Conf. on Computer Vision and Pattern Recognition, vol. 1, pp. 886–893 (2005)
11. Duda, R., Hart, P., Stork, D.: Pattern Classification, second edn. Wiley, New York (2001)
12. Elad, M., Hel-Or, Y., Keshet, R.: Pattern detection using a maximal rejection classifier. Pattern Recognition Letters **23**(12), 1459–1471 (2002)
13. Fleuret, F., Geman, D.: Coarse-to-fine face detection. International Journal of Computer Vision **41**(1-2), 85–107 (2001)
14. Froba, B., Ernst, A.: Face detection with the modified census transform. In: Proc of 16th IEEE Int. Conf. Automatic Face and Gesture Recognition, pp. 91–96 (2004)
15. Grabner, H., Bischof, H.: On-line boosting and vision. In: Proc. IEEE Conf. on Computer Vision and Pattern Recognition, pp. 260–267 (2006)
16. He, H., Garcia, E.A.: Learning from imbalanced data. IEEE Trans. on Pattern Analysis and Machine Intelligence **21**(9), 1263–1284 (2009)
17. Huang, C., Ai, H., Li, Y., Lao, S.: High-performance rotation invariant multiview face detection. IEEE Trans. on Pattern Analysis and Machine Intelligence **29**(4), 671–686 (2007)
18. Li, S.Z., Zhang, Z.: FloatBoost learning and statistical face detection. IEEE Trans. on Pattern Analysis and Machine Intelligence **26**(9), 1112–1123 (2004)
19. Lienhart, R., Kuranov, A., Pisarevsky, V.: Empirical analysis of detection cascades of boosted classifiers for rapid object detection. In: DAGM-Symposium, *Lecture Notes in Computer Science*, vol. 2781, pp. 297–304 (2003)
20. Liu, C., Shum, H.Y.: Kullback-leibler boosting. In: Proc. IEEE Conf. on Computer Vision and Pattern Recognition, vol. I, pp. 587–594 (2003)
21. Lu, W.L., Okuma, K., Little, J.J.: Tracking and recognizing actions of multiple hockey players using the boosted particle filter. Image and Vision Computing **27**(1-2), 189–205 (2009)
22. Masnadi-Shirazi, H., Vasconcelos, N.: Asymmetric boosting. In: Int Conf on Machine Learning (2007)
23. Moghaddam, B., Pentland, A.: Probabilistic visual learning for object representation. IEEE Trans. on Pattern Analysis and Machine Intelligence **19**(7), 696–710 (1997)
24. Mu, Y., Yan, S., Liu, Y., Huang, T., Zhou, B.: Discriminative local binary patterns for human detection in personal album. In: Proc. IEEE Conf. on Computer Vision and Pattern Recognition (2008)
25. Osuna, E., Freund, R., Girosi, F.: Training support vector machines: An application to face detection. In: Proc. IEEE Conf. on Computer Vision and Pattern Recognition, pp. 130–136 (1997)
26. Paisitkriangkrai, S., Shen, C., Zhang, J.: Efficiently training a better visual detector with sparse eigenvectors. In: Proc. IEEE Conf. on Computer Vision and Pattern Recognition, pp. 1129–1136. Miami, Florida (2009)
27. Papageorgiou, C., Oren, M., Poggio, T.: A general framework for object detection. In: The IEEE Conf. on Computer Vision, pp. 555–562 (1998)
28. Pham, M.T., Cham, T.J.: Fast training and selection of haar features using statistics in boosting-based face detection. In: Proc. IEEE Conf. on Computer Vision and Pattern Recognition (2007)
29. Pham, M.T., Hoang, V.D.D., Cham, T.J.: Detection with multi-exit asymmetric boosting. In: Proc. IEEE Conf. on Computer Vision and Pattern Recognition (2008)
30. Rowley, H., Baluja, S., Kanade, T.: Neural network-based face detection. IEEE Trans. on Pattern Analysis and Machine Intelligence **20**(1), 23–38 (1998)
31. Schapire, R., Freund, Y., Bartlett, P., Lee, W.S.: Boosting the margin: A new explanation for the effectiveness of voting methods. Annals of Statististics **26**(5), 1651–1686 (1998)
32. Schneiderman, H.: Feature-centric evaluation for efficient cascaded object detection. In: Proc. IEEE Conf. on Computer Vision and Pattern Recognition, vol. II, pp. 29–36 (2004)
33. Shen, C., Wang, P., Li, H.: LACBoost and FisherBoost: Optimally building cascade classifiers. In: European Conf. Computer Vision, vol. 2, pp. 608–621. Crete Island, Greece (2010)
34. Sun, J., Rehg, J., Bobick, A.: Automatic cascade training with perturbation bias. In: Proc. IEEE Conf. on Computer Vision and Pattern Recognition, vol. II, pp. 276–283 (2004)

35. Sung, K., Poggio, T.: Example-based learning for view-based human face detection. IEEE Trans. on Pattern Analysis and Machine Intelligence **20**(1), 39–51 (1998)
36. Tieu, K., Viola, P.: Boosting image retrieval. International Journal of Computer Vision **56**(1/2), 17–36 (2004)
37. Torralba, A., Murphy, K., Freeman, W.: Sharing features: Efficient boosting procedures for multiclass object detection. In: Proc. IEEE Conf. on Computer Vision and Pattern Recognition, vol. II, pp. 762–769 (2004)
38. Tu, Z.: Probabilistic boosting-tree: Learning discriminative models for classification, recognition, and clustering. In: The IEEE Conf. on Computer Vision, vol. 2, pp. 1589–1596 (2005)
39. Turk, M., Pentland, A.: Eigenfaces for recognition. Journal of Cognitive Neuroscience **3**(1), 71–86 (1991)
40. Viola, P., Jones, M.: Fast and robust classification using asymmetric AdaBoost and a detector cascade. In: Advances in Neural Information Processing Systems 14, pp. 1311–1318 (2002)
41. Viola, P., Jones, M.: Robust real-time face detection. International Journal of Computer Vision **57**(2), 137–154 (2004)
42. Viola, P., Jones, M., Snow, D.: Detecting pedestrians using patterns of motion and appearance. In: The IEEE Conf. on Computer Vision, pp. 734–741 (2003)
43. Wang, X., Han, T.X., Yan, S.: An HOG-LBP human detector with partial occlusion handling. In: The IEEE Conf. on Computer Vision (2009)
44. Webb, A.: Statistical Pattern Recognition. Oxford University Press, New York (1999)
45. Wu, J., Brubaker, S.C., Mullin, M.D., Rehg, J.M.: Fast asymmetric learning for cascade face detection. IEEE Trans. on Pattern Analysis and Machine Intelligence **30**(3), 369–382 (2008)
46. Wu, J., Geyer, C., Rehg, J.M.: Real-time human detection using contour cues. In: Proc. IEEE Int'l Conf. Robotics and Automation (2011)
47. Wu, J., Mullin, M.D., Rehg, J.M.: Learning a rare event detection cascade by direct feature selection. In: Advances in Neural Information Processing Systems (NIPS) 16, pp. 1523–1530 (2004)
48. Wu, J., Mullin, M.D., Rehg, J.M.: Linear asymmetric classifier for cascade detectors. In: Int'l Conf. on Machine Learning, pp. 993–1000 (2005)
49. Xiao, R., Zhu, H., Sun, H., Tang, X.: Dynamic cascades for face detection. In: Proc. IEEE Conf. on Computer Vision and Pattern Recognition (2007)
50. Xiao, R., Zhu, L., Zhang, H.J.: Boosting chain learning for object detection. In: The IEEE Conf. on Computer Vision, pp. 709–715 (2003)
51. Yang, M.H., Kriegman, D., Ahuja, N.: Detecting faces in images: a survey. IEEE Trans. on Pattern Analysis and Machine Intelligence **24**(1), 34–58 (2002)
52. Yang, M.H., Roth, D., Ahuja, N.: A snow-based face detector. In: Advances in Neural Information Processing Systems 12, pp. 862–868 (2000)
53. Zhang, L., Chu, R., Xiang, S., Liao, S., Li, S.Z.: Face detection based on multi-block LBP representation. In: International Conference on Biometrics, pp. 11–18 (2007)
54. Zhu, Q., Yeh, M.C., Cheng, K.T., Avidan, S.: Fast human detection using a cascade of histograms of oriented gradients. In: Proc. IEEE Conf. on Computer Vision and Pattern Recognition, vol. 2 (2006)

Chapter 9
Classifier Boosting for Human Activity Recognition

Raffay Hamid

9.1 Introduction

The ability to visually infer human activities happening in an environment is becoming increasingly important due to the tremendous practical applications it offers [1]. Systems that can automatically recognize human activities can potentially help us in monitoring people's health as they age [7], and to fight crime through improved surveillance [26]. They have tremendous medical applications in terms of helping surgeons perform better by identifying and evaluating crucial parts of the surgical procedures, and providing the medical specialists with useful feedback [2]. Similarly, these systems can help us improve our productivity in office environments by detecting various interesting and important events around us to enhance our involvement in important office tasks [21].

In spite of the plethora of these applications, however, the area of human activity recognition remains mostly an open research field. One of the main challenges in this regard is the fact that there exists no comprehensive feature vocabulary to universally encode all the various human actions and activities. This is unlike some of the other temporal processes where such universal vocabularies of basic information units do exist (e.g., phonemes for speech, words for text, and DNA elements for proteins). The designer of a human activity recognition system therefore needs to have at their disposal a convenient way to formulate events of interest that can be built up from smaller more general components. Since the set of events and their defining features may not be known a priori, a mechanism for combining these smaller units is necessary to produce the final activity detector.

One approach toward such a mechanism is to explore techniques of classifier combination. The main purpose of combining classifiers (from here on referred to as "weak classifiers") is to pool their individual outputs to produce a "strong

R. Hamid (✉)
eBay Research Labs., 2145 Hamilton Ave, San Jose, CA 95125, USA
e-mail: rhamid@ebay.com

C. Zhang and Y. Ma (eds.), *Ensemble Machine Learning: Methods and Applications*, DOI 10.1007/978-1-4419-9326-7_9, © Springer Science+Business Media, LLC 2012

classifier" that is more robust and accurate than each individual weak classifier. This *ensemble learning* view of human activity recognition has recently gained a lot of attention [14, 27, 32, 35, 44, 54] and the results so far have been quite encouraging. While there are many different ways of putting weak classifiers into a stronger ensemble, the particular method we will focus here is called *classifier boosting*.

In this chapter, we present an overview of the various boosting-based learning techniques that have been applied to the problem of human activity recognition. The chapter begins by providing an overview of perceptual characteristics of everyday human activities, and in turn motivating the usefulness of employing ensemble learning approach for the task of automatic activity recognition. This is followed by describing the various features that have previously been used to learn ensemble classifiers for recognizing human activities. Subsequently, the different types of human behaviors that have been approached from an ensemble learning perspective are described. The chapter concludes with some of the challenges that remain open in this field, and the current research directions that are being explored.

9.2 Characterizing Everyday Human Activities

The three classical ways in which scientists have viewed the characterization of human activities [50] are in terms of (1) direct perceptual inputs, (2) using a notion of context sensitive activity descriptors, and (3) considering activities as a set of partially ordered action subsequences. These views facilitate different types of characterizations of human activities, the usefulness of which depends on the dynamics and complexity of the activities being considered. We present here a brief outline of these views of human activities, highlighting their strengths and weaknesses for learning boosting-based models of human behaviors.

9.2.1 Activities from Direct Perceptual Inputs

Consider the activity of a person walking in a room. One way of interpreting this activity may be using the motion properties of the scene detected directly through the raw perceptual cues (see Fig. 9.1a) [39]. In this characterization of walking, there is no notion of time, physical states, or causality, and the activity is coded strictly in terms of low-level sensory stimuli. It is argued that human beings perceive a set of our everyday activities purely on the basis of direct perceptual inputs. The classic demonstration of activity detection by humans using direct perceptual information was done by the "Moving Lights Display" experiment [29] where human subjects were able to distinguish between actions of walking, running, or stair climbing simply from the intensity patterns of the lights attached to the joints of actors.

Not utilizing any semantic information, this characterizations of human activities is generally limited to the class of activities that are quite basic in nature. The main

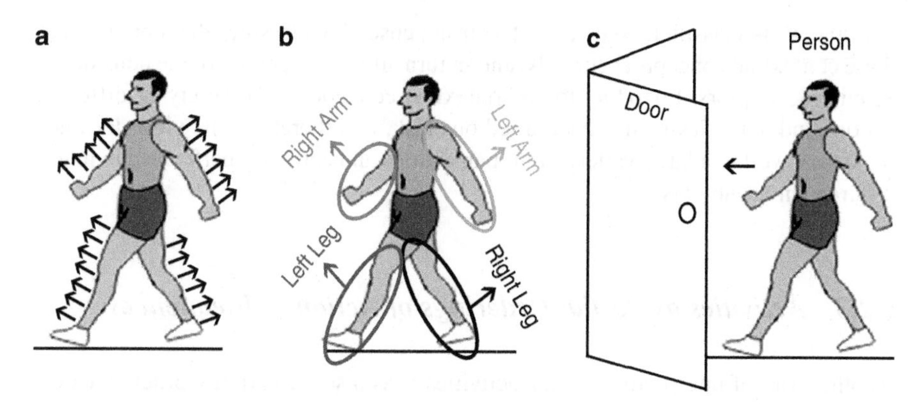

Fig. 9.1 Different Descriptions of an Example Activity of Walking—(**a**) The activity of walking is considered in terms of the very basic perceptual cues. (**b**) Walking activity is considered in terms of mid-level activity descriptors that follow certain temporal and causal constraints such as repetitively placing one foot in front of the other. (**c**) The activity of walking being considered as a function of the person's intent of walking through a door

advantage of this view of human activities is that extracting low-level features from video can be done in an efficient and accurate manner. There is a variety of low-level visual features that can be used to this end. These features can be effectively used in a boosting-based framework to construct an accurate ensemble classifier for detecting human actions. At the same time, however, since this view takes a purely bottom–up approach toward human activities, it usually does not scale well to more involved activities where information about the actor and their context is crucial for recognizing what activities are being performed.

9.2.2 Activities Using Activity Descriptors

Another way to look at our example activity of walking may be in terms of semantically meaningful activity descriptors [45], such as repetitively putting one foot in front of another while keeping the other foot on the ground (see Fig. 9.1b). Such activity descriptors follow basic rules of causality, e.g., the movement of one foot is caused by the other foot having been placed on the ground. Similarly, these activity-descriptors must follow a set of physical constraints, e.g., both feet cannot be apart from each other beyond a certain distance which is a function of the person's physical frame.

This characterization of human activities is context sensitive, i.e., the interpretation of walking using this activity view requires some notion of a person's feet, the difference between left and right, and some notion of the ground [33]. These contextual concepts are usually hard to detect accurately using any one detector. Therefore, boosting-based learning methods can be suitably used to combine multiple weak classifiers to construct a strong ensemble classifier that could detect these contextual concepts accurately, and in turn allow recognition of the considered activity descriptors. Encoding these contextual concepts for a variety of different actions and activities is, however, a tedious task, and therefore this view does not scale very well to a large variety of human actions and activities performed in many different environments.

9.2.3 Activities as Partial Orderings of Action Subsequences

Another way of interpreting human activities is as a set of partially ordered action subsequences that follow certain temporal constraints. For instance, our activity of walking may be interpreted as a set of ordered actions of taking one step after another, such that the current step is not taken until the previous step is complete. This view of human activities is particularly useful when the considered activities take place over a long-duration of time, and involve multiple actors and objects for their completion. Examples of such activities include cooking some dish in a kitchen, or manufacturing a product on a factory floor, etc.

This activity view can be used to encode activities as frequencies of their constituent subsequences. The presence and absence of any one of these subsequences is usually not decisive to infer whether a particular activity is being performed—rather the constituent action subsequences encode the activity signature in a cumulative sense. This view therefore aligns well with the boosting-based learning frameworks, where each constituent subsequence can be considered as a weak classifier of the overall activity.

9.3 Technical Background for Boosting-Based Learning

Classifier boosting is a particular instance of ensemble learning algorithms, where the output of different weak classifiers (weak learners) are combined to produce a more accurate inference. The main intuition behind these methods is to provide a different subset of the training set that is most informative to an individual weak learner, by looking at the training sets provided to the previously trained weak learners. Many boosting algorithms modify the weight for each training sample based on the errors the previous weak learners make, by increasing the weight if there is a classification error for a sample. Likewise, they reduce the weight of a sample if the current ensemble is able to correctly classify a sample. By interpreting the weights as importance, at each boosting iteration the weak learners focus on solving increasingly harder examples.

More formally, boosting [47] provides a simple method to sequentially fit additive models of the form:

$$H(v) = \sum_{m=1}^{M} h_m(v), \tag{9.1}$$

where v is the input feature vector, M is the number of rounds, and

$$H(v) = \log \frac{P(z = 1|v)}{P(z = -1|v)} \tag{9.2}$$

is the log odds of being in class $+1$, where z is the class membership label (± 1). Hence, $P(z = 1|v) = \sigma(H(v))$, where $\sigma(x) = 1/(1 + e^{-x})$ is the sigmoid or the logistic function. The terms h_m are often called weak learners, while $H(v)$ is called a strong learner. Boosting optimizes the following cost function one term of the additive model at a time:

$$J = E\left[e^{-zH(v)}\right], \tag{9.3}$$

where $zH(v)$ is called the "margin," and relates to the generalization or out-of-sample error rate. This cost function can be thought of as a differentiable upper bound on the misclassification rate [41] or as an approximation to the likelihood of the training data under a logistic noise model [17].

There are many different ways to optimize (9.3). One of the more popular of these methods is called "gentleBoost" [16]. In gentleBoost, the optimization of J is done using adaptive Newton steps, which corresponds to minimizing a weighted squared error at each step. Specifically, at each step m, the function H is updated as:

$$H(v) := H(v) + h_m(v), \tag{9.4}$$

where h_m is chosen so as to minimize a second order Taylor approximation of the cost function:

$$\arg\min_{h_m} J(H + h_m) \simeq \arg\min_{h_m} E\left[e^{-zH(v)}(z - h_m)^2\right]. \tag{9.5}$$

Replacing the expectation with an empirical average over the training data, and defining weights $w_i = e^{-z_i H(v_i)}$ for training example i, this reduces to minimizing the weighted squared error:

$$J_{\text{wse}} = \sum_{i=1}^{N} w_i (z_i - h_m(v_i))^2, \tag{9.6}$$

where N is the number of training examples, and for the ith training example, we have $w_i = e^{-z_i} H(v_i)$.

Algorithm 1 Boosting for Binary Classification

Set: $H(v_i), w_i = 0; \forall i = 1 : N$
for $m = 1, 2, 3, \ldots, M$ **do**
Fit: $\forall i = 1 : N, h_m(v_i) := a\delta\left(v_i^f > \theta\right) + b\delta\left(v_i^f \leq \theta\right)$
Update: $\forall i = 1 : N, H(v_i) := H(v_i) + h_m(v_i)$
Update: $\forall i = 1 : N, w_i := w_i e^{-z_i h_m(v_i)}$
end for

Minimizing J_{wse} depends on the specific form of the weak learns h_m. It is common to define the weak learners to be simple functions of the form

$$h_m(v) = a\delta\left(v^f > \theta\right) + b\delta\left(v^f \leq \theta\right), \tag{9.7}$$

where v^f denotes the fth component of the feature vector v, θ is a threshold, δ is the indicator function, and a and b are regression parameters. In this way, the weak learners perform feature selection, since each one picks a single component f. These weak learners are called decision or regression stumps h since they can be viewed as degenerate decision trees with a single node. The best stump can be found just as we would learn a node in a decision tree, i.e., we search over all possible features f to split on, and for each one, we search over all possible thresholds θ induced by sorting the observed values of f; given f and θ, we can estimate the optimal a and b by weighted least squares. Specifically, we have:

$$a = \frac{\sum_i w_i z_i \delta\left(v_i^f > \theta\right)}{\sum_i w_i \delta\left(v_i^f > \theta\right)} \tag{9.8}$$

and

$$b = \frac{\sum_i w_i z_i \delta\left(v_i^f \leq \theta\right)}{\sum_i w_i \delta\left(v_i^f \leq \theta\right)} \tag{9.9}$$

The parameters f, θ, a, and b are picked to have the lowest cost J_{wse}, and the resulting weak learner is added to the classifier ensemble, i.e.:

$$H(v_i) := H(v_i) + h_m(v_i). \tag{9.10}$$

Finally, boosting updates the weight of each training example as:

$$w_i := w_i e^{-z_i h_m(v_i)}. \tag{9.11}$$

The overall algorithm is summarized in Algorithm 1.

9.4 Features Used for Boosting Based Activity Recognition

It is a common practice in ensemble learning methods to extract different types of features from the training videos, and then select them discriminatively to form the final ensemble classifier. These selected features are usually thresholded to form feature stubs, and their outputs are combined together to reach the final classification result. The type of features extracted depends on various factors, including how far is the actor from the capturing camera, how many actors are present in the environment, how much sensor noise does an environment have, are the considered actions short-term or long-term in duration, and do the considered actions have a pronounced motion signature.

There are many different ways to categorize the commonly used features employed in learning ensembles for human activity recognition. In the following, we present one such category of these features that have been shown to be a useful choice to be employed in ensemble learning for activity recognition.

9.4.1 Optical Flow-Based Features

Optical flow of an object is a very informative cue about how the object usually moves over space and time. This measure can be particularly discriminative if the considered actions have observably different motion trajectories. The examples of these action domains include sports actions observed at a distance [14], and human gestures for sign language [9], etc.

In its basic form [24], optical flow methods try to calculate the motion between two image frames which are taken at times t and $t + \delta t$ at every voxel position. For a $2D + t$ dimensional case, a voxel at location (x, y, t) with intensity $I(x, y, t)$ will have moved by ∂x, ∂y, and ∂t between the two image frames, and the following image constraint can be given:

$$I(x, y, t) = I(x + \partial x, y + \partial y, t + \partial t). \tag{9.12}$$

Assuming the movement to be small, the image constraint at $I(x, y, t)$ with Taylor series can be developed to get:

$$I(x + \partial x, y + \partial y, t + \partial t) \approx I(x, y, t) + \frac{\partial I}{\partial x}\partial x + \frac{\partial I}{\partial y}\partial y + \frac{\partial I}{\partial t}\partial t \tag{9.13}$$

which results in

$$\frac{\partial I}{\partial x}V_x + \frac{\partial I}{\partial y}V_y + \frac{\partial I}{\partial t} = 0, \tag{9.14}$$

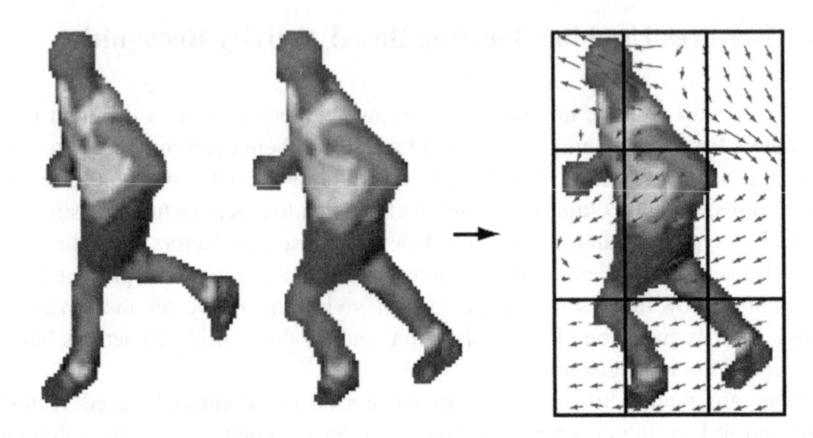

Fig. 9.2 Illustration of flow field for a pair of running action frames. The flow and gradient fields are divided into a 3×3 grid, and for each of the grid-cells is computer to better localize the motion characteristics of the actor

where V_x, and V_y are the x and y components of the velocity or optical flow of $I(x, y, t)$, and $\frac{\partial I}{\partial x}$, $\frac{\partial I}{\partial y}$, and $\frac{\partial I}{\partial t}$ are the derivatives of the image at (x, y, t) in the corresponding directions. The visual representation of optical flow for a pair of images is shown in Fig. 9.2.

As optical flow captures motion information over only a pair of frames, this feature has mostly been used for detecting relatively short duration activities. These include detection of people walking on pavements, players kicking a soccer ball in the field, and dancers performing different dance moves in a performance. The general idea of these applications is to aggregate measures of the optical flow fields in different parts of a frame, and provide them as an input to the boosting algorithm. During training, the boosting algorithm learns the most discriminative features along with their respective thresholds that can disambiguate between the considered actions and everything else.

9.4.2 2-D Shape-Based Features

Shape-based features have traditionally been used for object detection. However, more recently these features have found some use in action recognition problems as well. A popular 2-D shape-based feature that has been very frequently used for action recognition problems is Histogram of oriented gradients or HOG [10]. This feature counts occurrences of gradient orientation in localized portions of an image. This method is similar to that of edge orientation histograms, scale-invariant feature transform descriptors, and shape contexts, but differs in that it is computed on a dense grid of uniformly spaced cells and uses overlapping local contrast normalization for improved accuracy.

The first step to compute HOG feature is to apply the 1-D centered, point discrete derivative mask in one or both of the horizontal and vertical directions of the input image. This is done by filtering the image with the following filter kernels:

$$[-1, 0, 1], \qquad \text{and} \qquad [-1, 0, 1]^T. \tag{9.15}$$

The second step of calculation involves creating the cell histograms. Each pixel within the cell casts a weighted vote for an orientation-based histogram channel based on the values found in the gradient computation. The cells themselves can either be rectangular or radial in shape, and the histogram channels are evenly spread over 0–180° or 0–360°, depending on whether the gradient is "unsigned" or "signed." As for the vote weight, pixel contribution can either be the gradient magnitude itself, or some function of the magnitude.

As with the optical flow features, The general idea of 2-D shape-based features is to aggregate measures of the gradient fields in different parts of a frame, and provide them as an input to the boosting algorithm. As the 2-D shape-based features do not capture the temporal aspect of human activities, their usage for ensemble-based activity classifier is not as popular as, for instance, the optical flow features. However, it has been shown that using the shape-based features along with the optical flow features can improve the overall performance of the ensemble classifier.

9.4.3 3-D Volumetric Features

The 3-D volumetric features extend the 2-D shape-based features in the temporal dimension, and view human actions as three-dimensional shapes induced by the silhouettes in the space–time volume. These methods range from computing a single frame-based representation of the spanned space–time volume to be used in a template matching sense, to computing local space–time features on the space–time volume spanned by an action. There has also been work in matching the space–time volumes spanned by one action class with others. In the following, we briefly explain some of these views on 3-D volumetric features.

9.4.3.1 Temporal Templates

The idea behind temporal templates is to map a spatio temporal patterns of a person's motion to a static spatial pattern, which can thereon be used for discriminating that particular action from others. One such method, proposed in [5] uses the notion of Motion Energy Image (MEI), and Motion History Images (MHI), to encode the motion patterns of various objects into a single static image.

The MEI is the binary cumulative motion content of an action represented in an image. More formally, MEI can be represented as:

$$E_\tau(x, y, t) = \bigcup_{i=0}^{\tau-1} D(x, y, t - i), \tag{9.16}$$

Fig. 9.3 Example space–time volumes spanned by actions of jumping, walking, and running. Figure courtesy [18]

where $D(x, y, t)$ is a binary image sequence indicating regions of motion. For many applications, image differencing is adequate to generate D. Intuitively, the MEI represents *where* in an image was the action performed in a cumulative sense.

The MHI represents *how* the action occurs in an image. In an MHI (H_τ), pixel intensity is a function of the temporal history of motion at that point. More formally, MHI can be represented as:

$$H_\tau(x, y, t) = \begin{cases} \tau & \text{if } D(x, y, t) = 1 \\ \max(0, H_\tau(x, y, t-1) - 1) & \text{otherwise} \end{cases}. \tag{9.17}$$

Temporal template-based representations are fast to compute and robust to sensor noise; however, they are best applicable for simple settings with usually a single object. The reason for this limitation is that these representations focus on the low-level image-signals to encode the activity structure without using any mid-level activity-characterizations that can potentially get at the underlying activity structure in a more explicit way.

9.4.3.2 Local and Global Space–Time Shape Features

This class of features [18] deals with the volumetric space–time shapes induced by human actions, and exploits the solution to the standard Poisson equation [13] to extract various shape properties that are utilized for shape representation and classification. Example space–time shapes generated by different actions in the space–time volume are illustrated in Fig. 9.3.

Consider an action and its space–time shape S surrounded by a simple, closed surface. One way to represent the properties of S is to assign every point on S a value that depends on the relative position of that point within S. More specifically, we can assign each space–time point within S with the mean time required for a particle undergoing a random-walk process starting from the point to hit the boundaries of S. This measure can be computed by solving the Poisson equation [13] of the form:

$$\delta U(x, y, t) = -1 \quad \text{where} \quad (x, y, t) \in S. \tag{9.18}$$

Here, the Laplacian of U is defined as

$$\delta U(x, y, t) = U_{xx} + U_{yy} + U_{tt} \tag{9.19}$$

subject to

$$U(x, y, t) = 0 \quad \text{at the bounding surface} \quad \partial S. \tag{9.20}$$

The solution to the Poisson equation can be used to extract a wide variety of useful shape properties. These include both features that are local in terms of space and time, as well as integral of these to compute more global features that characterize the space–time shapes at a larger scale. A useful local Poisson-based feature is the *space–time saliency* which describes how quickly the space at a certain point in space–time is evolving. Another important local feature is the *space–time orientation* which encodes the direction in which the space–time shape of an action is changing.

9.4.3.3 Space–Time Volumetric Features

Inspired by the 2-D spatial Harr wavelet-based features used in face detection [51], there have been attempts to consider 3-D cubic features to characterize the motion content of a space–time volume [30]. These features span the entire spatio temporal volume of the video at different spatial and temporal scales. They are computed by performing simple arithmetic operations over the content of the volumes they span. These feature values are then provided to the boosting algorithm to select the most discriminative features at the spatial and temporal scale that perform best during training. A graphical illustration for these features is given in Fig. 9.4.

9.4.4 Space–Time Interest Points

This class of features [4, 31, 43] is based on the work done in spatial domain, where points with a significant local variation of image intensities have been extensively investigated [15, 22]. The basic intuition behind these features is that interesting events in video are characterized by strong variations in the data along both the spatial and the temporal dimensions. More generally, points with nonconstant motion correspond to accelerating local image structures that may correspond to

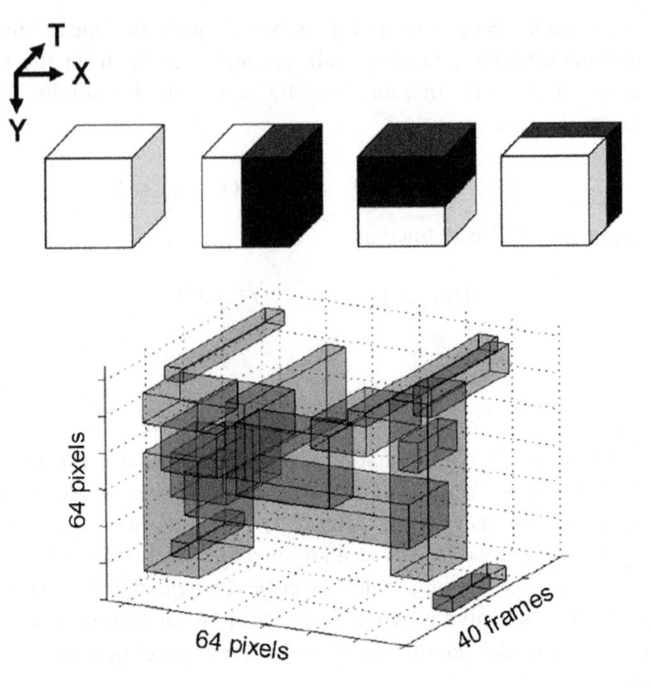

Fig. 9.4 Illustration of space–time volumetric features. The top row illustrates the 3-D volumes used for feature computation. The first feature calculates the volume. The other three features calculate volumetric differences in X, Y, and time. The bottom row shows multiple features learned by using a boosting-based ensemble classifier to recognize a hand-wave action in a space–time volume. Figure courtesy [30]

accelerating objects in the world. Hence, such points can be expected to contain information about the forces acting in the physical environment and changing its structure. Space–time interest point detectors are geared toward automatically detecting such potentially useful points in the spatio temporal video volume. Features based on the statistics of these interest points can be used in an ensemble-based framework to learn discriminative classifiers for different actions. An example of space–time interest points for walking action is illustrated in Fig. 9.5.

9.4.5 Discrete Event-Based Features

Features covered in this chapter so far are mostly quite local both in terms of space as well as time. These types of features work well in a boosting-based framework especially for actions that span over smaller duration of time. Example of such actions include kicking a ball, pointing toward something, and performing a dance step.

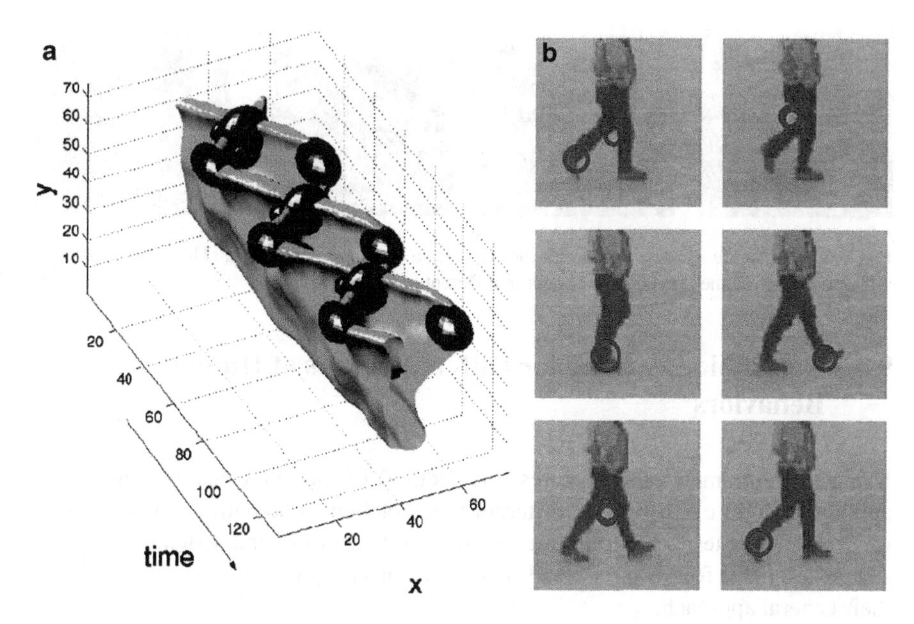

Fig. 9.5 Results of detecting spatio temporal interest points for the motion of the legs of a walking person: (**a**) 3-D plot with a threshold surface of a leg pattern (upside down) and detected interest points; (**b**) interest points overlaid on single frames in the sequence. Figure courtesy [31]

However, such low-level features do not usually scale very well when it comes to recognizing more complex activities, that can span over longer durations of time. The examples of such activities include cooking something in the kitchen, making a delivery at a loading dock, or manufacturing something on a factory floor. Features needed to build classifiers that could recognize such long-duration complex activities generally involve more contextual knowledge and can themselves be considered as discrete short-term events.

The choice of these event-based features depends strongly on the environments in which the activities of interest are being performed. Some of the environments in which the in situ activities have been explored from this perspective include office floors [27], loading dock areas [19], and household kitchens [20]. The examples of the constituent events in one of these environments, say a loading dock, would be a delivery vehicle entering the delivery area, a person opening the back door of the delivery truck, a person pushing a delivery cart, etc. Example frames representing some of these events in a loading dock setting are shown in Fig. 9.6.

An important question in generalizing this parsimonious view of human activity features, that remains far from being solved, is whether there is a minimal set of universal events that could be used to describe the majority, if not all, of human activities [3,38]. The hope is that if we knew such a vocabulary, we could train expert classifiers specifically designed for the different members of this universal set. Until we come up with such a universal set, however, ensemble learning provides a useful alternative to combine different classifiers for a complex activities, that could be combined to achieve relatively high recognition performance.

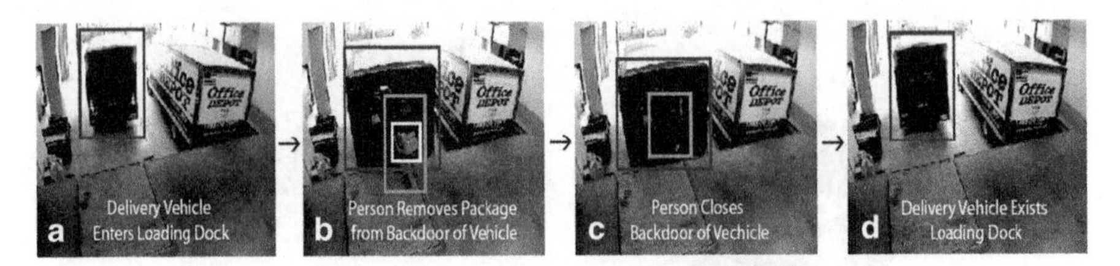

Fig. 9.6 The figure shows a delivery activity in a loading dock area. The objects whose interactions define these events are shown in different colored blocks

9.5 Ensemble Learning for Different Types of Human Behaviors

Using the aforementioned features, the method of ensemble learning has been applied to build classifiers for different types of human activities. These include relatively short-term motions, medium-duration actions, and extended activity sequences. In the following, we will cover some of such previous works and describe their general approach.

9.5.1 Ensemble Learning for Short-Term Motions

Since features based on optical flow measurements are usually short in their temporal duration, such features have been more popular for recognizing short-term motions. Examples of such motions include walking [52], and sports actions [14], etc. The general principle in these methods is to compute some measure of the flow fields in each pair of video frames during training, and use them to learn the most discriminative feature and its corresponding optimal threshold for each round of boosting. Each of these weak classifiers, i.e., the discriminative feature and its corresponding optimal threshold, are then used to classify the training data, and the results of this classification are used to both assign a weight to the weak classifier in direct proportion to how well it did, and to reweigh the training data for the next boosting round in inverse proportion to whether they were correctly classified by the current weak classifier. The intuition here is to give more importance to the training data points that were not classified correctly in the current stage such that the weak classifier selected in the next stage would be aimed more to classify the currently wrongly classified data points. Finally, during testing, a final classification decision is made at each pair of frames by using a linear weighted combination of all the weak classifiers. These frame-based decisions are combined, usually using a voting scheme, to reach the overall video-level motion classification.

Work done in [14] applied this approach of using boosting-based learning to detect short-term actions on the KTH data set [42]. In this data, there were six action classes considered. Each action is performed several times by 25 subjects in four

Fig. 9.7 Sample frames from six action classes used from KTH data-set [42]. Figure courtesy [14]

different conditions, namely outdoors, outdoors with scale variation, outdoors with different clothes, and indoors. Representative frames from this action data set are shown in Fig. 9.7.

The classification results obtained by [14] on the KTH data set are given in Table 9.1. It can be observed that the most confusion is between the last three actions, i.e., running, jogging, and walking. The overall average accuracy that they report is 90.5%. The authors of [14] also report the comparative average classification results that several previously proposed algorithms report on the KTH data set. These results are presented in Table 9.2. Note that the boosting based approach of [14] is comparable with [28], while outperforming other approaches.

9.5.2 Ensemble Learning for Medium-Duration Actions

Similar to short-term motion recognition, the idea of performing frame-pair based classification can be applied for medium-duration actions as well. These include writing on the white board, or cleaning eye-glasses etc. However given the longer

Table 9.1 *Classification Performance*—Results of [14] on KTH data-set [42]: Confusion matrix for per-video classification (overall accuracy of 90.5%). Horizontal rows are ground truth, and vertical columns are predictions

	Boxing	Hand clapping	Hand waving	Jogging	Running	Walking
Boxing	100	0.0	0.0	0.0	0.0	0.0
Hand clapping	2	98	0.0	0.0	0.0	0.0
Hand waving	0.0	0.0	100	0.0	0.0	0.0
Jogging	0.0	0.0	0.0	81	10	9
Running	3	0.0	0.0	24	72	1
Walking	0.0	0.0	0.0	6	1	93

Table 9.2 *Comparative Classification Performance*—Average classification performance of different algorithms reported by [14] on KTH [42] is given. Also provided is the manner in which the data set was divided into testing and training subsets

Methods	Training method	Accuracy
Fatahi et al. [14]	Splits	90.50
Jhuang et al. [28]	Splits	91.70
Nowozin et al. [37]	Splits	87.04
Niebles et al. [36]	Leave one out	81.50
Dollar et al. [12]	Leave one out	81.17
Schuldt et al. [42]	Splits	71.72
Ke et al. [30]	Splits	62.96

duration of actions, it is usually better to incorporate the temporal information of the actions more directly. This can be done either in the form of features used in the feature-pool [40], or the selection mechanism itself [46].

Encoding of temporal information in the features used in the feature pool of a boosting based framework usually involves finding correspondence amongst features over multiple frames [11, 48]. Once these correspondences are found, the matched features are used to learn weak classifiers in the usual boosting framework.

Temporal information regarding the dynamics of actions can also be incorporated in the boosting framework itself. In the classical boosting framework, the decision of a weak classifier is counted based on each frame independently. However, this can be modified to allow a weak classifier to use its previous responses (previous frames in the temporal sense) if it helps decrease the overall error for that classifier. The duration and the manner in which preceding results of a weak classifier contribute to its result in the current frame will impact its current detection and false positive (FP) rates. Learning the optimal combination manner and temporal duration over which the previous decisions impact the current inference of a weak classifier can result in overall improved classification performance.

Work done in [46] applied the approach of incorporating temporal information in the feature selection procedure of boosting to improve classification performance for medium duration actions. In particular, they report results on a data set of 11 actions using a one against-all approach. Data from multiple people was used to

Table 9.3 *Classification Performance of TemporalBoost*—Results of the method proposed in [46] are presented. Column 1 and 2 show the name and number of instances of the considered actions. Column 3 and 4 present the rates of True Positive (TP) and FP. The last column presents the numbers for TP and true negatives for where the algorithm reported the detected action was in the test video

Action	Class size	True Positives (TP) [in %]	False Positives (FP) [in %]	Localize TP/TN
Talking on phone	4	75	0	90/91
Checking voicemail on phone	2	100	0	90/74
Bringing cup to face	5	80.0	20	90/91
Scratching/rubbing face	9	83.3	11.1	92/92
Resting hand on face	12	88.8	16.7	89/94
Taking medication	9	85.7	11.1	87/89
Yawning with hand at mouth	7	100	0	96/96
Yawning with no hand at mouth	6	100	0	93/98
Putting on eyeglasses	9	77.7	33.33	85/86
Putting on earphones	6	66.7	16.7	85/87
Rubbing eyes	8	87.5	12.5	90/1

create this data set. The ensemble classifier for each of the action classes was limited to seven weak-classifiers each. The features used were number of unique segments in the video frames using mean-shift segmentation [8]. The classification and temporal localization performance of the method proposed in [46] on the 11 considered actions is given in Table 9.3.

9.5.3 Ensemble Learning for Extended Activities

The dynamics of long-term activities such as delivering a package in a loading dock [19], or cooking a dish in a household kitchen [20], are usually represented as sequences of shorter duration actions. One way of encoding the structure of such extended activities is in terms of statistics of their action subsequences. In an ensemble-based framework, these subsequences can be used to construct weak classifiers where the selection algorithm would pick the optimal threshold on the counts of the most discriminative feature for each boosting round. This approach toward building ensemble classifiers for long-durational human activities is very similar to how researchers in Natural Language Processing have looked at documents [25]. This similarity between human activities and documents as sequences of discrete information entities can potentially open avenues for further interdisciplinary research.

9.6 Summary

Intelligent systems that can recognize human activities are essential for the progress of many important research areas, e.g., automatic video surveillance, robotics, and healthcare, etc. There are, however, many steep challenges in building such intelligent systems. One of these challenges is the *hitherto* lack of a universal feature set that can discriminatively describe most of the everyday human activities. This underscores the importance of having a mechanism for selecting and combining a subset of visual features from a larger detectable feature set, to form an accurate classifier for the different activities occurring in an environment. Boosting is a class of such feature selection and combination mechanisms that have been shown to be quite effective to learn classifiers for a variety of different types of human activities.

There has been a lot of recent work in exploring the usage of boosting-based ensemble learning technique for human activity recognition [14, 27, 32, 35, 44, 54]. There are, however, some important research questions that still remain to be addressed and further investigated. Some of these directions concern purely with the mechanics of the boosting algorithm itself, while the rest are mostly application centric. Some of these future direction are listed in the following.

9.6.1 Theoretical Research Directions

Better Feature Selection Policy: Recall that in boosting algorithms (Algorithm 1), the weight distribution of the training data depends of how the immediately previous selected weak classifier performed on them. This feature selection mechanism is however quite greedy as it only depends on the last selected weak classifier. There is a substantial potential to improve this selection mechanism such that it is not so greedy in its search policy, and therefore could result in a more global optimum.

More Descriptive Feature Pool: Currently, the feature pool from which weak-classifiers are constructed is created a priori. Therefore, it has a limited representativeness which does not adapt as the boosting algorithm progresses. There is potential to explore if we can *discover* a better feature pool as a function of how the previous feature pools have performed thus far.

Information Sharing Across Multiple Classes: Traditionally, most of the work in boosting based methods has been done on two class classification problems. While there has been some important work done for multiclass classification in a boosting based setting [55], there is ample of research opportunity to exploit the overlap that naturally exists among different classes in order to reduce the sample and computational complexity of the learning problem.

9.6.2 Application-Based Research Directions

Multiagent and Crowd-Based Behaviors: Most of the previous work in using boosting-based methods for activity recognition have focused on single agent activities. However, there are many human behaviors that involve multiple agents who interact with each other to accomplish a task. Similarly there are crowd-based behaviors where the group of people are best considered holistically, and not as individuals. There is plenty of research scope for applying boosting-based methods to learn classifiers that can recognize these types of human activities.

Robust Temporal Features: A crucial part of encoding the dynamics of human activities is the set of features that are used to represent them. Most of the current features used in activity recognition systems only encapsulate the temporal dynamics of human activities over a relatively short duration of time. Therefore, there is significant research opportunity to extend the temporal span of the features used by activity recognition systems. Besides other ways, this can be done by coming up with more robust local feature detection methods and better correspondence algorithms over multiple frames.

9.7 Bibliography and Historical Remarks

The field of automatic understanding of human activities in video has been increasingly active, specially in the last two decades [1, 6, 49]. One of the earliest investigations about the analysis of human motion was done by the contemporary photographers Etienne Marey and Eadweard Muybridge in the 1850s who photographed moving subjects and revealed several interesting and artistic aspects involved in human and animal locomotion. The Moving Light Display experiment of Johansson [29] provided a strong argument for a principled study and analysis of human motion perception, which eventually became a precursor to the current exploration of this problem in Computer Vision. In more recent years, automatic human activity analysis systems have been applied for a variety of different applications, including monitoring people's health as they age [7], to fight crime through improved surveillance [26], to help surgeons perform better by identifying parts of surgical procedures [2], and by detecting important events in office settings to enhance our involvement and participation in our jobs [21].

An important part of the explorative effort of designing systems for automatic human activity analysis has been devising informative and robust visual features or attributes that can be extracted in an efficient manner. These features range from the very low-level ones that attempt to capture the pixel-level characteristics of videos [4, 5, 10, 18, 24, 30, 31, 43], to more high-level ones that incorporate object and scene-level contextual information to represent activities [23, 27, 34].

One of the main challenges in video-based human activity analysis has been a lack of a comprehensive feature vocabulary to universally encode all the various human actions and activities. The designer of a human activity recognition system therefore needs to have at their disposal a convenient way to formulate events of interest that can be built up from smaller more general components. Since the set of events and their defining features may not be known a priori, a mechanism for combining these smaller units is necessary to produce the final activity detector. One approach toward such a mechanism that has recently gained a lot of attention is to use boosting-based schemes to combine multiple weak detectors of human activities to form a more accurate ensemble classifier [14, 27, 32, 44, 54].

In particular, boosting has been successfully applied to learn models for relatively short-term activities such as walking or running [52]. More recently work on the application of boosting-based learning for detecting short-term actions has focused on crowed scenes [30] and group behaviors [53]. Boosting-based learning approach has also found its application in learning short-term actions in sports and performance activities [14] where the movements and capture mechanisms are relatively constrained.

For medium duration actions and activities, e.g., writing on a white board, or cleaning ones eyeglasses, the incorporation of temporal information has been done both at the feature level [40] as well as the selection mechanism itself [46]. An ongoing direction of exploration is the application of boosting-based learning approach for long-term activities, where counts of discrete action subsequences could be used as the feature pool to learn discriminative models of extended activities. This research direction is inspired by the treatment of documents as counts of word subsequences, and the application of boosting-based approach to learn document models [25].

References

1. J.K. Aggarwal and Q. Cai, *Human motion analysis: A review*, Journal of Computer Vision and Image Understanding **73** (1999), no. 3, 428–440.
2. S. Ahmadi, T. Sielhorst, R. Stauder, M. Horn, H. Feussner, and N. Navab, *Recovery of surgical workflow without explicit models*, Conference on Medical Image Computing and Computer Assisted Intervention (MICCAI), 2006, pp. 420–428.
3. U. Akdemir, P. Turaga, and R. Chellappa, *An ontology based approach for activity recognition from video*, International Conference on Multimedia, 2008.
4. H. Bay, A. Ess, T. Tuytelaars, and L. V. Gool, *Surf: Speeded up robust features*, Journal of Computer Vision and Image Understanding **110** (2008), 346–359.
5. A. Bobick and J. Davis, *The representation and recognition of action using temporal templates*, IEEE Transactions on Pattern Analysis and Machine Intelligence (PAMI) **23** (2001), no. 3, 257–267.
6. C. Cedras and M. Shah, *Motion-based recognition: A survey*, Image and Vision Computing **13** (1995), 129–155.
7. M. Choi, B. Jones, J. Shim, K. Hong, and D. Shah, *Enabling clinical decision-making through home monitoring and health information technology*, INFORMS Data Mining and Health Informatics Workshop, 2010.

8. D. Comaniciu and P. Meer, *Mean shift: A robust approach toward feature space analysis*, IEEE Transactions on Pattern Analysis and Machine Intelligence (PAMI).
9. R. Cutler and M. Turk, *View-based interpretation of real-time optical flow for gesture recognition*, IEEE International Conference on Automatic Face and Gesture Recognition, 1998.
10. N. Dalal and B. Triggs, *Histograms of oriented gradients for human detection*, IEEE International Conference on Computer Vision and Pattern Recognition (CVPR), 2005.
11. F. Dellaert, S. Seitz, C. Thorpe, and S. Thrun, *Feature correspondence: A markov chain monte carlo approach*, Advances in Neural Information Processing Systems (NIPS), 2001.
12. P. Dollar, V. Rabaud, G. Cottrell, and S. Belongie, *Behavior recognition via sparse spatio-temporal features*, In Proceedings of Visual Surveillance – PETS., 2005.
13. L. Evans, *Partial differential equations*, second ed., American Mathematical Society, 1998.
14. A. Fathi and G. Mori, *Action recognition by learning mid-level motion features*, IEEE International Conference on Computer Vision and Pattern Recognition (CVPR), 2008.
15. W. Forstner and E. Gulch, *A fast operator for detection and precise location of distinct points, corners and centres of circular features*, Workshop of the International Society for Photogrammetry and Remote Sensing, 1987.
16. J. Friedman, T. Hastie, and R. Tibshirani, *Additive logistic regression: a statistical view of boosting*, Annals of Statistics **28** (1998), 337–407.
17. J. Friedman, T. Hastie, and R. Tibshirani, *Additive logistic regression: a statistical view of boosting*, Annals of statistics **28** (2000), 337374.
18. L. Gorelick, M. Blank, E. Shechtman, M. Irani, and R. Basri, *Actions as space–time shapes*, Transactions on Pattern Analysis and Machine Intelligence **29** (2007), no. 12, 2247–2253.
19. R. Hamid, A. Johnson, S. Batta, A. Bobick, C. Isbell, and G. Coleman, *Detection and explanation of anomalous activities: Representing activities as bags of event n-grams*, IEEE International Conference on Computer Vision and Pattern Recognition (CVPR).
20. R. Hamid, S. Maddi, A. Bobick, and I. Essa, *Structure from statistics: Unsupervised activity analysis using suffix trees*, IEEE International Conference on Computer Vision (ICCV).
21. R. Hamid, S. Maddi, A. Johnson, A. Bobick, I. Essa, and C. Isbell, *Discovery and characterization of activities from event-streams*, Conference on Uncertainty in AI (UAI), 2005.
22. C. Harris and M. Stephens, *A combined corner and edge detector*, Alvey Vision Conference, 1988.
23. S. Hongeng and R. Nevatia, *Multi-agent event recognition*, IEEE International Conference on Computer Vision and Pattern Recognition (CVPR).
24. B. Horn, *Robot vision*, Mcgraw-Hill, New York, 1986.
25. G. Ifrim, G. Bakir, and G. Weikum, *Fast logistic regression for text categorization with variable-length n-grams*, ACM International Conference on Knowledge Discovery and Data Mining (SIGKDD), 2008.
26. Y. Ivanov and A.F. Bobick, *Recognition of multi-agent interaction in video surveillance*, IEEE International Conference on Computer Vision (ICCV), 1999, pp. 169–176.
27. Y. Ivanov and R. Hamid, *Weighted ensemble boosting for robust activity recognition in video*, International Journal on Computer Vision and Graphics **4** (2007).
28. H. Jhuang, T. Serre, L. Wolf, and T. Poggio, *A biologically inspired system for action recognition*, IEEE International Conference on Computer Vision (ICCV), 2007.
29. G. Johansson, *Visual perception of biological motion and a model for its analysis.*, Journal of Perception and Psychophysics **14** (1973), 201–211.
30. Y. Ke, R. Sukthankar, and M. Hebert, *Efficient visual event detection using volumetric features*, IEEE International Conference on Computer Vision (ICCV), 2005, pp. 166–173.
31. I. Laptev and T. Lindeberg, *Space–time interest points*, IEEE International Conference on Computer Vision (ICCV), 2003.
32. C. Liu and P. Yuen, *Human action recognition using boosted eigenactions*, Image and Vision Computing **28** (2010), 825–835.
33. R. Mann, A. Jepson, and J. Siskind, *The computational perception of scene dynamics*, European Conference on Computer Vision (ECCV), 1996.

34. D. Moore, I. Essa, and M. Hayes, *Context management for human activity recognition*, In Proceedings of Audio and Vision-based Person Authentication, 1999.
35. Y. Nejigane, M. Shimosaka, T. Mori, and T. Sato, *Online action recognition with wrapped boosting*, International Conference on Intelligent Robots and Systems (IROS), 2007.
36. J. Niebles, H. Wang, and L. Fei-fei, *Unsupervised learning of human action categories using spatial-temporal words*, British Machine Vision Conference (BMVC), 2006.
37. S. Nowozin, G. Bakar, and K. Tsuda, *Discriminative subsequence mining for action classification*, IEEE International Conference on Computer Vision (ICCV), 2007, pp. 757–766.
38. G. Okeyo, L. Chen, W. Hui, and S. Roy, *Ontology-enabled activity learning and model evolution in smart homes*, International Conference on Ubiquitous Intelligence and Computing (UbiComp), 2010.
39. R. Polana and R. Nelson, *Low level recognition of human motion*, IEEE Workshop on Non-rigid and Articulated Motion, 1994.
40. P. Sand and S. Teller, *Particle video: Long-range motion estimation using point trajectories*, International Journal of Computer Vision (IJCV) **80** (2008).
41. R. E. Schapire, *The boosting approach to machine learning: An overview*, Springer, New York, 2003.
42. C. Schuldt, I. Laptev, and B. Caputo, *Recognizing human actions: A local svm approach*, International Conference on Pattern Recognition (ICPR), 2004, pp. 32–36.
43. P. Scovanner, *A 3-dimensional sift descriptor and its application to action recognition*, International Conference on Multimedia, 2007.
44. M. Shimosaka, Y. Nejigane, T. Mori, and T. Sato, *Fast online action recognition with efficient structured boosting*, IEEE International Conference on Multimedia and Expo (ICME), 2009.
45. J. Siskind, *Grounding the lexical semantics of verbs in visual perception using force dynamics and event logic*, Journal of Artificial Intelligence Research **15** (2001), 31–90.
46. P. Smith, N. D. V. Lobo, and M. Shah, *Temporalboost for event recognition*, IEEE International Conference on Computer Vision (ICCV), 2005.
47. A. Torralba, K. Murphy, and W. Freeman, *Sharing visual features for multiclass and multiview object detection*, IEEE Transactions on Pattern Analysis and Machine Intelligence (PAMI) **29** (2007), no. 5, 854–869.
48. L. Torresani, V. Kolmogorov, and C. Rother, *Feature correspondence via graph matching: Models and global optimization*, European Conference on Computer Vision (ECCV), 2008.
49. P. Turaga, R. Chellappa, V. S. Subrahmanian, and O. Udrea, *Machine recognition of human activities: A survey*, IEEE Transactions on Circuits and Systems for Video Technology (2008).
50. S. Ullman, *The interpretation of visual motion*, MIT Press, Cambridge, MA, 1979.
51. P. Viola and M. Jones, *Robust real-time object recognition*, IEEE International Conference on Computer Vision (ICCV), 2001.
52. P. Viola, M.J. Jones, and D. Snow, *Detecting pedestrians using patterns of motion and appearance*, IEEE International Conference on Computer Vision (ICCV), 2003.
53. Q. Wei, X. Zhang, Y. Kong, W. Hu, and H. Ling, *Group action recognition using space–time interest points*, Proceedings of the 5th International Symposium on Advances in Visual Computing, 2009, pp. 757–766.
54. T. Zhang, J. Liu, S. Liu, Y. Ouyang, and H.Q. Lu, *Boosted exemplar learning for human action recognition*, IEEE Transactions on Circuits and Systems for Video Technology **99** (2011), 538–545.
55. J. Zhu, H. Zou, S. Rosset, and T. Hastie, *Multi-class adaboost*, Statistics and its Inference **173** (2009), 349–360.

Chapter 10
Discriminative Learning for Anatomical Structure Detection and Segmentation

S. Kevin Zhou, Jingdan Zhang, and Yefeng Zheng

10.1 Introduction

Due to the increasing demand for more medical images in clinical practices for better assessment and diagnosis, medical image analysis has gained more importance than ever. In this chapter, we will focus on the subarea of anatomical structure detection and segmentation, which plays an important role in speeding up the diagnostic work flow.

Although remarkable progresses have made in detecting and segmenting anatomical structures, it still confronts a lot of challenges to obtain results that can be used in clinical applications. This is mainly due to significant appearance variation present in the medical images caused by a multitude of factors:

- Sensor noise/artifact. As in any sensor, the medical equipment generates noise/artifact inherent to its own physical sensor and image formation process. The extent of the artifact depends on the image modality. For example, while high-dose Computer Tomography (CT) produces image with less artifacts, low-dose CT is quite noisy. Metal objects (such as implants) can generate a lot of artifacts in CT. Ultrasound imaging has notorious spectral noise and even signal dropout.
- Patient difference. Different patients exhibit different build forms: fat or slim, tall or short, adult or child, etc. As a result, the anatomical structures also exhibit different shapes. All contribute to the creation of different images.
- Machine difference. Machines from different vendors tend to produce different images even for the same patient. This holds even for highly standardized CT machines, although the difference is much more subtle.

S.K. Zhou (✉) • J. Zhang • Y. Zheng
Siemens Corporation, Corporate Research, 755 College Road East, Princeton, NJ 08540, USA
e-mail: shaohua.zhou@siemens.com; jingdan.zhang@siemens.com; yefeng.zheng@siemens.com

C. Zhang and Y. Ma (eds.), *Ensemble Machine Learning: Methods and Applications*,
DOI 10.1007/978-1-4419-9326-7_10, © Springer Science+Business Media, LLC 2012

- Pathology. Pathology can give rise to highly deformed anatomical structures or even missing ones. This makes statistical modeling very difficult.
- Operator experience and preference. For some image modality such as ultrasound, the acquired image quality highly depends on the operator experience of manipulating the ultrasound transducer against the correct anatomy in the correct plane. Again in ultrasound imaging, sonographers have own preferences in adjusting the imaging parameters (such as dynamic range, contrast, etc.). Not to mention that there are a lot of imaging protocols in Magnetic Resonance Imaging (MRI).
- Field of view. Dose radiation is a major concern in CT. In an effort to minimize the dose radiation, only the necessary part of human is imaged. This creates partial scans and narrow field of view, in which the anatomical context is highly weaken or totally gone.
- Soft tissue. Anatomical structures such as internal organs are soft tissues of similar properties. They (such as liver and kidney) might even overlap, forming very weak boundary between them.

In addition, image reading and diagnosis allow almost no room for mistake. While the accuracy requirement is always stringent, the demand for speedy processing does not diminish. A speedy work flow is crucial to any radiology lab. No radiologist or physician has the patience to wait for hours or even minutes to obtain the analysis results.

However, it is still possible to devise algorithms that are used in many real applications. In this chapter, we will present several such algorithms for anatomical structure detection and segmentation. They are based on the principle of learning from a large annotated data set in a discriminative fashion.

10.1.1 Shape Representation

In this chapter, we use an explicit point-based shape representation. The shape C is comprised of two parts: rigid pose P and deformable part S, that is, $C = (P, S)$.

If a similarity transformation is used for a 2D pose P, that is,

$$P = (X, Y, \theta, S_x), \tag{10.1}$$

where (X, Y) for translation, θ for orientation, and S_x for isotropic scale for both x- and y-directions, then the above shape representation reduces to Kendall's interpretation [16].

To better describe the pose, we use for a 2D shape a 5D-parameterization

$$P = (X, Y, \theta, S_x, S_y), \tag{10.2}$$

with (X, Y) for translation, θ for orientation, and (S_x, S_y) for scale in both x- and y-directions, and for a 3D shape a 9D-parameterization

$$P = (X, Y, Z, \psi, \phi, \theta, S_x, S_y, S_z), \tag{10.3}$$

with (X, Y, Z) for translation, (ψ, ϕ, θ) for orientation, and (S_x, S_y, S_z) for anisotropic scale. If no confusion, we also specifically call the orientation and scale parts as pose.

For the deformable part S, we assume that it consists of N points, i.e.,

$$S = (\mathbf{X}_1, \mathbf{X}_2, \ldots, \mathbf{X}_N), \tag{10.4}$$

where $\mathbf{X}_i$ is a point in 2D (i.e., $\mathbf{X} = (X, Y)$) or 3D (i.e., $\mathbf{X}_i = (X, Y, Z)$). In 2D, the cubic spline is used to interpolate the points into a curve. In 3D, the point-based shape representation is equivalent to a 3D mesh or a 3D surface.

Sometime we use the PCA to reduce the dimensionality by keeping a sufficient amount of the total energy. If so,

$$S = (\alpha_1, \alpha_2, \ldots, \alpha_K), \tag{10.5}$$

where α_i is the coefficient for ith principal components.

For anatomical structure detection, we estimate its rigid pose P. For anatomical structure segmentation, we estimate its both rigid pose P and deformable part S. This is usually done sequentially, first estimating P then S, because the detection naturally provides shape initialization. For example, one common way is to transform the mean shape into the pose P as an initial guess.

The chapter is organized as follows. Section 10.2 will present various discriminative learning approaches for efficiently detecting anatomical structures. Section 10.3 will concentrate on discriminative learning approach for accurately segmenting anatomical structures.

10.2 Discriminative Learning for Anatomical Structure Detection

The state-of-the-art object detection method is described in Viola and Jones [26]. They train a binary classifier off-line that differentiates an object of interest from the background and then online exhaustively slide a scanning window over the input image for object instances.

It is challenging to build a real-time detector that incorporates accurate pose estimation using a detector like the one discussed in [26] (i.e., exhaustively scanning the space of all possible combinations of translation and pose). We refer to the

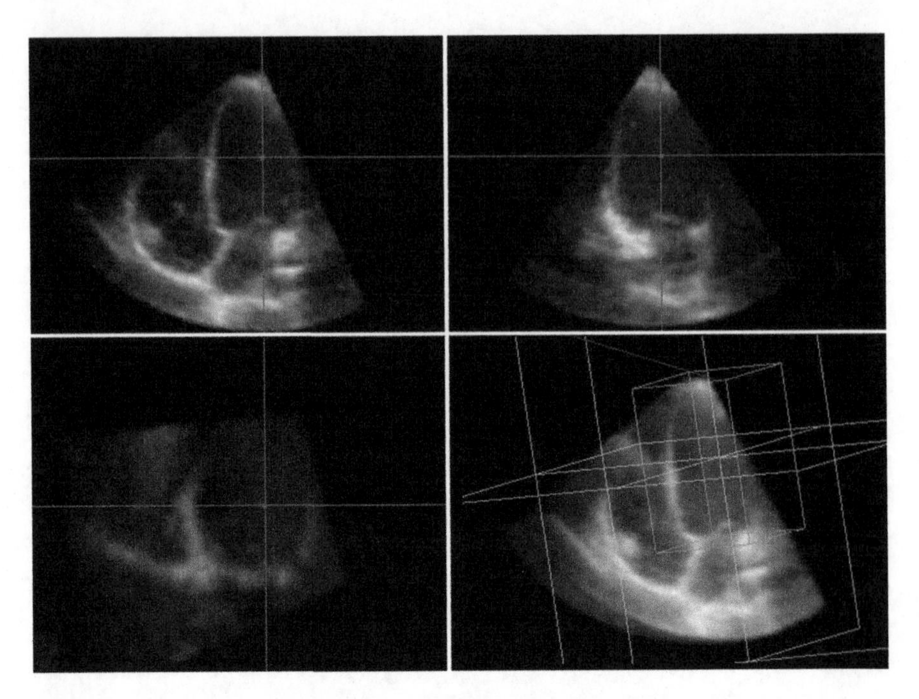

Fig. 10.1 Detecting the LV and its pose in a 3D ultrasound volume is important to automatically navigate multiple canonical planes for clinical practices. Reprinted from [32]. ©2007 IEEE

orientation and scale parameters as *pose* only here and in Sect. 10.2.1. Consider detecting the left ventricle (LV) in a 3D echocardiogram, an ultrasound volume of the human heart (see Fig. 10.1). Discovering the LV configuration is helpful for orienting the 3D volume. For example, from a known LV configuration, one can meet an emerging clinical need to automatically display canonical 2D slices. Because the LV can occur at an arbitrary location and orientation, one needs to search over nine parameters to fully align the LV. When extending the method of [26], the computational cost increases exponentially with the dimensionality of the parameter space. Furthermore, volume rotation and integral volume computations are time consuming because their computation is proportional to the number of voxels. To aggravate the problem, learning one monolithic classifier to handle all possible variations is challenging.

A promising solution that requires only one integral volume/image is to train a collection of binary classifiers to handle different poses. A variety of structures are proposed to combine these classifiers. The most straightforward way is a parallel structure (Fig. 10.2a) that trains a classifier for each discretized pose [28]. In detection, all classifiers are tested for every scanning window. The computation linearly depends on the number of poses. To accelerate the detection speed, the pyramid structure is proposed by Li et al. [17]. For the parallel structure, several classifiers might fire up at the same place when the actual pose is in-between

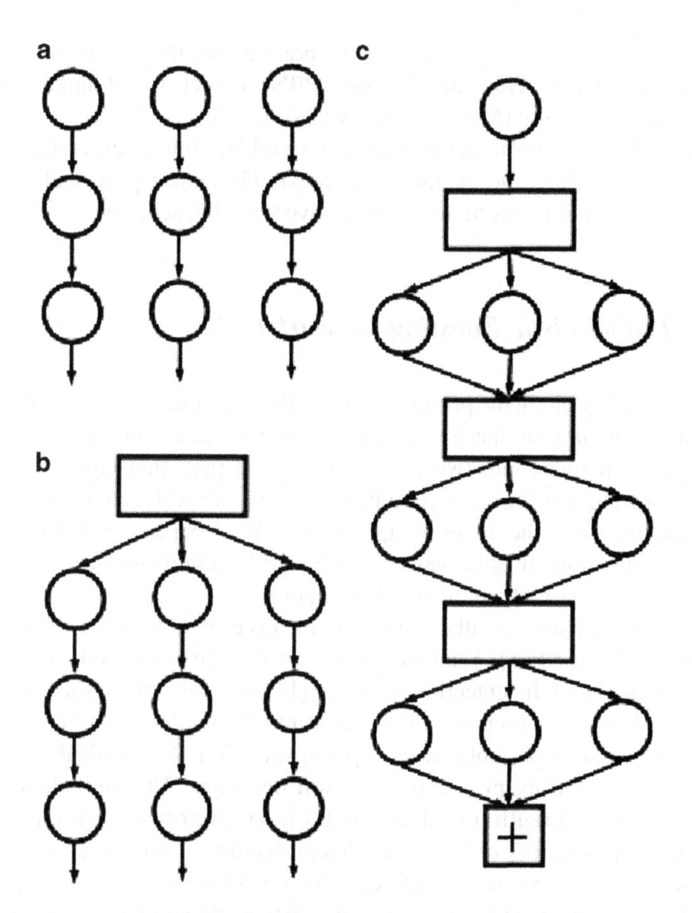

Fig. 10.2 Different structures for multi-pose detection. The circle represents a binary foreground/background classifier. The rectangle represents a multiclass pose classifier. (**a**) Parallel cascade. (**b**) Tree. (**c**) Network. Reprinted from [32]. ©2007 IEEE

the discretized poses. To estimate accurate pose needs additional work due to the difficulty in comparing responses among these classifiers. To discriminate different poses explicitly, a tree structure (Fig. 10.2b) that uses the multiclass classifier as a differentiator is applied. In [15], a decision tree is used to determine the pose of a face, followed by the binary classifier trained for that pose only. Both parallel and tree structures only test a sparse set of orientations and scales to meet the real-time requirement. However, anatomical structures in medical images can possess arbitrary orientations and scales, and the detection task needs to accurately determine them. Under such circumstances, it is challenging to build a rapid detector using the approaches above. In general, the speed is inversely proportional to the number of poses tested. In order to give an accurate estimate of the pose, speed must be sacrificed.

To overcome these bottlenecks, we will present two novel detection methods. The first is Probabilistic Boosting Network (PBN) in Sect. 10.2.1 and the second is Marginal Space Learning (MSL) in Sect. 10.2.2.

In Sect. 10.2.3, we formulate the detection problem into a regression setting in order to better leverage the anatomical context. The object pose will be directly regressed out without resorting to an exhaustive scanning scheme.

10.2.1 Probabilistic Boosting Network (PBN)

In PBN, we explore along the promising line of the tree structure. To avoid searching the whole configuration space and rotating images and volumes, we break up the configuration space into two parts: for the first part, translation, we still use exhaustive search; and for the second part, rotation and scale, we directly estimate the parameters from the image's appearance. We couple exhaustive scanning with pose estimation. In this way, we successfully eliminate the computational dependency proportionally to the number of poses.

For the tree structure, usually only one branch is evaluated based on the decision of the multiclass classifier [15] and hence, the error made by the multiclass classifier has great influence on the detection result. In [14], several branches may be chosen, based on the hard decision provided by the VectorBoost algorithm, for the purpose of reducing the risk of possible errors in pose estimation. We handle the uncertainty of pose estimation using probabilities as soft decisions. The multiclass classifier is trained using the LogitBoost algorithm [10], which provides a sound theory of computing the probability of the pose. This probability is used to choose branches to evaluate. The final probability of being the object is computed by applying the total probability law. In addition, we obtain a better estimation of the pose by using the conditional mean of the pose distribution.

To further increase the efficiency of the tree structure, a graph-structured network (Fig. 10.2c) is proposed to reject background as early as possible. The multiclass classifier is decomposed into several subclassifiers by taking advantage of the additive native of the boosted classifier. The subclassifiers and binary detectors are coupled to form a graph structure that alternates the two task of pose estimation and object detection. Furthermore, we add a binary classifier as a prefilter of the graph to reject the background that can be easily differentiated from the foreground.

10.2.1.1 Pose Estimation

Given a window containing the object of interest, the goal is to estimate the pose parameter(s) of the object based on the image appearance I in the window. We first discuss the algorithm for one-parameter estimation and then use the one-parameter estimation algorithm as a building block for multiple parameters.

The object appearance variation is caused not only by the rigid transformation we want to estimate but it is also influenced by noise, intensity variation, and nonrigid shape deformations (as well as other effects). As a result, a robust algorithm is needed to guarantee the accuracy of the pose estimation. We handle the uncertainty of the pose estimation by learning a probability of the parameter, β, subject to the given input, I, $p(\beta|I)$. This probability can be used to estimate β accurately and to avoid errors in subsequent tasks.

In practice, $p(\beta|I)$ is approximated with discretized distribution $p(\beta_j|I)$, based on a discrete set of parameter values, $\{\beta_1, \beta_2, \ldots, \beta_J\}$. We implemented the image-based multiclass boosting algorithm proposed by Zhou et al. [38]. This algorithm is based on the multiclass version of the influential boosting algorithm proposed by Friedman et al. [10], the so-called LogitBoost algorithm, which fits an additive symmetric logistic model via the maximum-likelihood principle. This fitting proceeds iteratively selecting weak learners and combining them into a strong classifier. The output of the LogitBoost algorithm is a set of J response functions $\{F_j(x);\ j = 1, \ldots, J\}$, where each $F_j(x)$ is a linear combination of a subset of weak learners:

$$F_j^n(x) = \sum_{i=1}^{n} f_{j,i}(x), \tag{10.6}$$

where $f_{j,i}(x)$ is a weak learner, and n is the number of weak learners. LogitBoost provides a natural way to calculate the posterior distribution of class label:

$$p_j^n(x) = \frac{\exp(F_j(x))}{\sum_{k=1}^{J} \exp(F_k(x))}. \tag{10.7}$$

Refer to [38] for more details.

One advantage of the LogitBoost algorithm is that the computed posterior probability asymptotically approximates the ground truth [10]. This means that a trade-off can be made between the approximation accuracy and the computational cost by adjusting the number of weaker learners in the response functions. This property is used to build the network structure to reject background cases more efficiently in the early stages.

We infer the parameter β by using a conditional mean, which is a Minimum Mean Square Error (MMSE) estimator:

$$\hat{\beta}_{\text{MMSE}} = \int_{\beta} \beta p(\beta|I)\, \mathrm{d}\beta \approx \sum_{j} \beta_j p(\beta_j|I). \tag{10.8}$$

This gives a better estimate than a Maximum A Posterior (MAP) estimate from a discrete set of possible values of β because the MMSE estimate can interpolate between values in discrete set.

The multiclass classifier that estimates one parameter can be used as a building block to construct a graph structure for estimating two or more parameters. In this section, we focus on two-parameter estimation, but the same principle can be

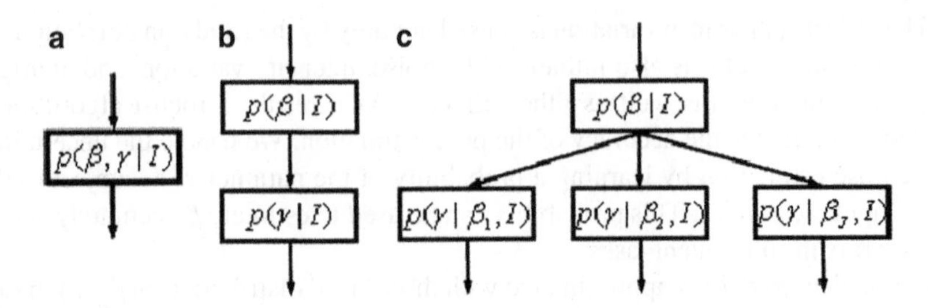

Fig. 10.3 Different structures for estimating probability of two-parameter pose. In the type-*c* structure, the parameter, β, is called the "root" parameter. Reprinted from [32]. ©2007 IEEE

applied to situations with more than two parameters. Suppose that the two unknown parameters are β and γ, the goal is to estimate $p(\beta,\gamma|I)$ from the image I. Figure 10.3 gives a graphical illustration of three possible structures that can be used to model this estimation task.

For the type-*a* structure, we treat the combination of β and γ as a single variable and train $p(\beta,\gamma|I)$ directly. This approach is structurally simple and has good performance when the number of combined states is small. However, when the number of combined states is large, or the appearance variation caused by both parameters are too complex to learn using a single classifier, this approach performs poorly. In this case, a divide-and-conquer strategy is appropriate to estimate parameters sequentially by using multiple multiclass classifiers.

The type-*b* structure assumes β and γ are independent. To train $p(\beta|I)$ (or $p(\gamma|I)$), we treat the variation in γ (or β) as intraclass. The joint distribution is approximated as

$$p(\beta,\gamma|I) \approx p(\beta|I) \times p(\gamma|I). \tag{10.9}$$

The drawback of this approach is the assumption of independence of β and γ is often invalid.

For the type-*c* structure, we apply the exact conditional probability law:

$$p(\beta,\gamma|I) = p(\gamma|\beta,I) \times p(\beta|I). \tag{10.10}$$

This can be represented as a tree structure. A root multiclass classifier is trained to learn $p(\beta|I)$ by treating the variation in γ as intraclass. Each child node corresponds to the conditional probability $p\left(\gamma|\beta_j,I\right)$ for a discrete state β_j. To compute $p(\beta,\gamma|I)$ efficiently, we omit branches whose probability $p(\beta|I)$ is below a specified threshold.

The choice of the root parameter for the type-*c* structure influences the overall performance, because the amount of the image appearance variation caused by the two parameters is not the same. Usually, the parameter that causes larger appearance variation should be the root node. This makes learning $p(\beta|I)$ easier, and leads to a better division of the pose space.

How to choose among these three types is determined by the data properties. An intuitive principle is that, if the number of the poses is small, and the appearance variation can be sufficiently captured by one classifier, we use the type-a structure. If the parameters are justifiably independent, use the type-b structure. Otherwise, use the type-c structure.

10.2.1.2 Probabilistic Boosting Network (PBN)

We now present PBN that integrates evidence from pose estimator and binary detectors. A PBN has three basic features:

1. It is *probabilistic.* A PBN leverages the fundamental total probability law to compute the probability of being object O. Assuming that the pose parameter β is discretized into $\{\beta_1, \beta_2, \ldots, \beta_J\}$, we have

$$p(O|I) = \sum_{j=1}^{J} p(O|I,\beta_j)p(\beta_j|I), \tag{10.11}$$

   where $p(O|I,\beta_j)$ is the binary classifier specific to the parameter β_j. To compute (10.11) efficiently, we ignore branches whose pose probability is smaller than a prespecified threshold p_0:

$$p(O|I) \approx \sum_{j:p(\beta_j|I)\geq p_0} p(O|I,\beta_j)p(\beta_j|I). \tag{10.12}$$

2. It uses *boosting.* As discussed in Subsect. 10.2.1.1, the probability $p(\beta_j|I)$ is implemented using the multiclass LogitBoost algorithm [10]. The classifier $p(O|I,\beta_j)$ is implemented using the cascade of boosted binary classifiers [26], which are able to deal with numerous negative examples and eliminate them as early as possible during testing. To implement the binary classifier in the cascade, one can use AdaBoost [9], binary LogitBoost [10] or other variants. Suppose that the cascade has S_j stages, then $p(O|I,\beta_j)$ is computed as

$$p(O|I,\beta_j) = \prod_{s=1}^{S_j} p_s(O|I,\beta_j), \tag{10.13}$$

   where $p_s(O|I,\beta_j)$ is the binary classifier for the sth cascade. The complexity of the classifier $p_s(O|I,\beta_j)$ increases as the number of stages increases. Without loss of generality, assume that $S_1 = S_2 = \cdots = S_J = S$. If say $S_j < S$, we simply set $p_s(O|I,\beta_j) = 1$ for $s > S_j$.
3. It has a *network* structure. The total probability law:

$$p(O|I) = \sum_{j=1}^{J}\prod_{s=1}^{S} p_s(O|I,\beta_j)p(\beta_j|I), \tag{10.14}$$

can be implemented in a tree-structured network as shown in Fig. 10.2b. Using this structure, negatives are rejected quickly when they flow through the network while the positives traverse through several branches. By combining evidence from these branches, one is able to accurately estimate $p(O|I)$ using (10.14).

In [23] a learning procedure called a Probabilistic Boosting Tree (PBT) was presented. Both PBN and PBT are able to provide object detection probabilities using boosting, and both have a tree structure, but they also differ significantly. In the tree-structured PBN, each node corresponds to a specified parameter, while in PBT there is no specific parameter. PBN also estimates pose parameters explicitly. Finally, PBN provides an efficient graph structure as shown next, which is not the case for PBT.

10.2.1.3 Efficient Graph Structure

The tree-structured PBN is not yet optimal in terms of computation because the overhead of computing the probability $p(\beta|I)$ is necessary for all background windows. These candidate windows are randomly sent to several branches of cascades and rejected by these branches. This creates a dilemma for the tree structure: the purpose of a multiclass classifier is to select proper binary classifiers to reject background, but determining the pose of these background patches wastes computation.

One way to solve this problem is to discard as many background windows as possible via a focus-of-attention mechanism. This can be achieved by pooling together data from positives in different poses to train a pose-invariant classifier as a prefilter. We tune the pose-invariant detector to have a 100% detection rate (although with a large number of false positives) by adjusting the threshold. This detector cannot offer a precise detection, but it is useful for rejecting a large percentage of the background candidates.

Even when the prefilter classifier is used, there are still nonobject windows passing the prefilter, causing unnecessary overhead computations of $p(\beta|I)$. Following the idea of the cascade structure for binary detector $p(O|I, \beta_j)$, which breaks its computation into several stages with increasing complexity, we also decompose the computation of $p(\beta|I)$ into several stages by taking the advantage of the additive model arising from the LogitBoost algorithm. The response functions at the sth stage is

$$F_j^s(x) = F_j^{s-1}(x) + \sum_{i=1}^{s_n} f_{j,i}(x), \tag{10.15}$$

where s_n is the number of weak learners at the sth stage. For the type-b and type-c structures, the computation of the probability $p(\beta, \gamma|I)$ can also be decomposed by distributing the weak learners of the multiclass classifiers to several stages.

To compute the probability $p(O|I)$

1. Let e_j be the number of the binary classifiers already evaluated in the jth branch. Start with $e_j = 0$, response functions $F_j(x) = 0$, and probabilities $p(O|I, \beta_j) = 1$, $j = 1, \ldots, J$.
2. Use the pose-invariant classifier to pre-filter. If I is background, set $p(O|I) = 0$ and exit.
3. Repeat for $s = 1, 2, ..., S$:
   - Add n_s weak learners to $F_j(x)$ and update $p(\beta_j|I)$.
   - For j^{th} branch, $j = 1, \ldots, J$, if $p(\beta_j|I) \geq p_0$ and $p(O|I, \beta_j) > 0$:
     - Compute the probabilities $p_k(O|I, \beta_j)$, $k = e_j + 1, \ldots, s$. If it is background, set $p(O|I, \beta_j) = 0$.
     - Update:
     $$p(O|I, \beta_j) \leftarrow p(O|I, \beta_j) \prod_{k=e_j+1}^{s} p_k(O|I, \beta_j).$$
     - Update $e_j = s$.
   - If $p(O|I, \beta_j) = 0$ for all branches, set $p(O|I) = 0$ and exit.
4. Compute $p(O|I)$ based on the total probability law.

Fig. 10.4 The PBN detection algorithm. Reprinted from [32]. ©2007 IEEE

We organize the whole detector as a graph-structured network as shown in Fig. 10.2c, which alternates the two tasks of pose estimation and background rejection by hierarchically distributing the overhead of computing $p(\beta|I)$. Figure 10.4 shows the detection algorithm of PBN with a single pose parameter. PBN detection with multiple pose parameters can be implemented in a similar way. The network is evaluated top-to-bottom. More weak learners are added to update $p(\beta|I)$ at each new stage, which approximates the true posterior probability more accurately due to its asymptotic convergence property. Based on the newly estimated $p(\beta|I)$, the binary classifiers corresponding to large $p(\beta|I)$ are evaluated at this stage. If a new binary branch unexplored in earlier stages is selected, we trace back to the beginning and re-evaluate the whole branch. If the candidate fails all selected binary classifiers, it is considered as a background window and the computation stops; otherwise, it proceeds to the next stage.

More accurate pose estimation helps to determine whether the candidate window belongs to the foreground, while the binary classifiers help to determine if it is necessary to continue evaluating $p(\beta|I)$. This way, the positives are evaluated at a minimum number of branches and the negatives are quickly rejected by either the prefilter or early detector cascades.

In [32], we applied PBN for real-time detection of the LV from 3D ultrasound volumes and the LA from 2D images. We also compared the PBN to the parallel and tree structure detection approaches. Please refer [32] for more details.

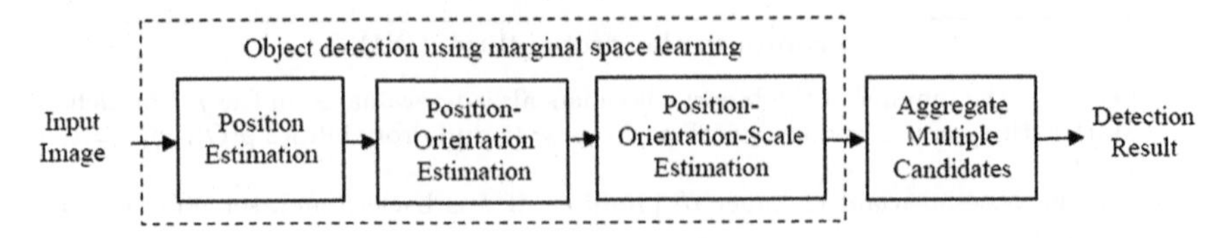

Fig. 10.5 Diagram for object detection using marginal space learning. Reprinted from [34]. ©2008 IEEE

10.2.2 Marginal Space Learning (MSL)

To accurately localize a 3D object, we need to estimate nine pose parameters (three for position, three for orientation, and three for anisotropic scaling). A straightforward extension of a 2D object detection method to 3D is not possible due to the exponential increase of the computation demands by the use of exhaustive search. Here we present a generic learning-based method for efficient 3D object detection, namely MSL [33, 34]. Instead of exhaustively searching the original nine-dimensional pose parameter space, only low-dimensional marginal spaces are searched in MSL to improve the detection speed. To be specific, we split the estimation into three steps: position estimation, position–orientation estimation, and position–orientation–scale estimation, as shown in Fig. 10.5. First, we train a position estimator that can tell us if a position hypothesis is a good estimate of the target object position in an input volume. After exhaustive searching of position marginal space (three-dimensional), we preserve a small number of position candidates (e.g., 100) with the largest detection scores. Second, we do joint position–orientation estimation with a trained classifier that can tell us if a position–orientation hypothesis is good. The orientation marginal space is exhaustively searched for each position candidate preserved after position estimation. Similarly, we only preserve a limited number of position–orientation candidates after this step. Finally, the scale parameters are searched in a similar way. Since after each step we only preserve a small number of candidates, therefore a large portion of search space (which has low posterior probability) is pruned efficiently in the early steps. Complexity analysis shows that MSL can reduce the number of testing hypotheses by six orders of magnitude, compared to the exhaustive full space search. Since the learning and detection are performed in a sequence of marginal spaces, we call our method MSL.

Many modern medical imaging modalities (e.g., CT) can capture a 3D volume with submillimeter resolution. For the localization of an anatomical structure, we do not need such high resolution. Therefore, normally, MSL-based detection is performed on a low resolution (e.g., 3mm) volume. The final boundary delineation is performed on the original high resolution volume.

10.2.2.1 Training of MSL Classifiers

To train a classifier, we need to split a set of hypotheses into two groups, positive and negative, based on their distance to the ground truth. The error in object position and scale estimation is not comparable with that of orientation estimation. Therefore, a normalized distance measure is defined by normalizing the error in each dimension to the corresponding search step size,

$$E = \max_{i=1,\ldots,D} \left| P_i^e - P_i^t \right| / \text{SearchStep}_i, \tag{10.16}$$

where P_i^e is the estimated value for pose parameter i, P_i^t is the corresponding ground truth, and D is the dimension of the pose parameter space. For similarity transformation estimation, the pose parameter space is nine dimensional, $D = 9$. A sample is regarded as a positive one if $E \leq 1.0$ and all the others are negative samples.

Training of Position Estimator: In this step, we want to estimate the position of the object and learning is constrained in a marginal space with three dimensions. Given a hypothesis (X, Y, Z), the classification problem is formulated as whether there is an object centered at (X, Y, Z). Haar wavelet features are fast to compute and have been shown to be effective for many applications [20, 24, 26]. Therefore, we use 3D Haar wavelet features for learning in this step.

The search step for position estimation is one voxel. According to (10.16), a positive sample (X, Y, Z) should satisfy

$$\max \left\{ \left| X - X^t \right|, \left| Y - Y^t \right|, \left| Z - Z^t \right| \right\} \leq 1 \text{ voxel}, \tag{10.17}$$

where (X^t, Y^t, Z^t) is the ground truth of the object center. Given a set of positive and negative training samples, we extract 3D Haar wavelet features and train a classifier using the PBT [23]. After that, we test each voxel in a volume one by one as a hypothesis of the object position using the trained classifier. The classifier assigns each hypothesis a score, and we preserve a small number of candidates (100 as default) with the highest detection score for each volume.

Training of Position–Orientation Estimator: In this step, we want to jointly estimate the position and orientation. The classification problem is formulated as whether there is an object centered at (X, Y, Z) with orientation (ψ, ϕ, θ). After object position estimation, we preserve the top 100 candidates, (X_i, Y_i, Z_i), $i = 1, \ldots, 100$. Since we want to estimate both the position and orientation, we need to augment the dimension of candidates. For each position candidate, we quantize the orientation space uniformly to generate hypotheses. The orientation is represented as three Euler angles in the ZXZ convention, ψ, ϕ, and θ. The distribution range of an Euler angle is calculated from the training data. Each Euler angle is then quantized within the range using a step size of 0.2 radians (11°). For each candidate (X_i, Y_i, Z_i), we augment it with N hypotheses about orientation, $(X_i, Y_i, Z_i, \psi_j, \phi_j, \theta_j)$, $j = 1, \ldots, N$. Some are close to the ground truth (positive) and others are far away (negative). The learning goal is to distinguish the positive and negative samples using trained classifiers. Using the normalized distance measure of (10.16), a hypothesis $(X, Y, Z, \psi, \phi, \theta)$ is regarded as a positive sample if it satisfies both (10.17) and

$$\max\left\{|\psi - \psi^t|, |\phi - \phi^t|, |\theta - \theta^t|\right\} \leq 0.2, \tag{10.18}$$

where $(\psi^t, \phi^t, \theta^t)$ represent the orientation ground truth. All the other hypotheses are regarded as negative samples. To represent the orientation information, we have to rotate either the volume or feature templates. We use the steerable features described below, which are efficient under rotation. Similarly, the PBT is used for training and the trained classifier is used to prune the hypotheses to preserve only a few candidates (50 as default).

Training of Similarity Transformation Estimator: The similarity transformation (adding the scales) estimation step is analogous to position-orientation estimation except learning is performed in the full nine dimensional similarity transformation space. The dimension of each position-orientation candidate is augmented by searching the scale subspace uniformly and exhaustively. The search step is set to two voxels.

10.2.2.2 Steerable Features

Global features, such as 3D Haar wavelet features, are effective to capture the global information (e.g., orientation and scale) of an object. To capture the orientation information of a hypothesis, we should rotate either the volume or the feature templates. However, it is time consuming to rotate a 3D volume and there is no efficient way to rotate the Haar wavelet feature templates. Local features are fast to evaluate but lose the global information of the whole object.

We present steerable features, which can capture the orientation and scale of the object and at the same time be very efficient. In steerable features, we sample a few points from the volume under a sampling pattern. We then extract a few local features for each sampling point (e.g., voxel intensity and gradient) from the original volume. The novelty of our steerable features is that we embed the orientation and scale information into the distribution of sampling points, while each individual feature is locally defined. Instead of aligning the volume to the hypothesized orientation, we steer the sampling pattern. This is where the name "steerable features" comes from.

Figure 10.6 shows how to embed a hypothesis in steerable features using a regular sampling pattern (illustrated for a 2D case for clearance in visualization). Suppose we want to test if hypothesis $(X, Y, Z, \psi, \phi, \theta, S_x, S_y, S_z)$ is a good estimation of the similarity transformation of the object. A local coordinate system is defined to be centered at position (X, Y, Z) (Fig. 10.6a) and the axes are aligned with

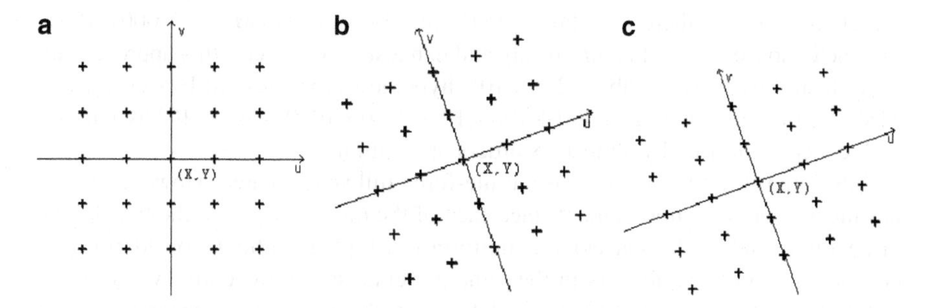

Fig. 10.6 Using a regular sampling pattern to incorporate a hypothesis (X, Y, ψ, S_x, S_y) about a 2D object pose. The sampling points are indicated as "+". (**a**) Move the pattern center to (X, Y). (**b**) Align the pattern to the orientation ψ. (**c**) The final aligned sampling pattern after scaling along each axis, proportional to (S_x, S_y). Reprinted from [34]. ©2008 IEEE

the hypothesized orientation (ψ, ϕ, θ) (Fig. 10.6b). A few points (represented as "+" in Fig. 10.6) are uniformly sampled along each coordinate axis inside a box. The sampling distance along an axis is proportional to the scale of the shape in that direction (S_x, S_y, or S_z) to incorporate the scale information (Fig. 10.6c). The steerable features constitute a general framework, in which different sampling patterns [33] can be defined.

At each sampling point, we extract a few local features. Steerable features are flexible to incorporate different local image features. In our implementation, we extract 24 local features based on the intensity and gradient from the original volume. Please refer to [34] for more details of the local image features. A major reason to select these features is for their efficiency.

10.2.2.3 Object Localization on Unseen Volume

This section provides a summary about the testing procedure on an unseen volume. The input volume is first converted to isotropic low resolution (e.g., 3 mm). All voxels are tested using the trained position estimator and the top 100 candidates, (X_i, Y_i, Z_i), $i = 1, \ldots, 100$, are kept. Each position candidate is augmented with N hypotheses about orientation, $(X_i, Y_i, Z_i, \psi_j, \phi_j, \theta_j)$, $j = 1, \ldots, N$. Next, the trained position–orientation classifier is used to prune these $100 \times N$ hypotheses and the top 50 candidates are retained, $\left(\hat{X}_i, \hat{Y}_i, \hat{Z}_i, \hat{\psi}_i, \hat{\phi}_i, \hat{\theta}_i\right)$, $i = 1, \ldots, 50$. Similarly, we augment each position–orientation candidate with M hypotheses about scaling and use the trained classifier to rank these $50 \times M$ hypotheses. The average of the top K ($K = 100$) candidates is taken as the final aggregated estimate.

In following, we use heart chamber detection in cardiac CT [34] as an example to analyze the efficiency of MSL. At the 3 mm resolution, a typical cardiac CT volume has roughly $64 \times 64 \times 64$ voxels, which corresponds to around 260,000 position hypotheses. The orientation space is discretized under a resolution of 0.2 radians, resulting in about 1000 orientation hypotheses ($N = 1{,}000$). Under two-voxel searching stepsize, there are about 1,000 scale hypotheses ($M = 1{,}000$). If the parameter space is searched uniformly and exhaustively (we call this approach full space learning), there are about 2.6×10^{11} hypotheses to be tested! However, using MSL, we only test about $260{,}000 + 100 \times 1000 + 50 \times 1000 = 4.1 \times 10^5$ hypotheses and reduce the testing by almost six orders of magnitude.

MSL significantly outperforms a brute-force full space search. However, it still has much room for improvement since each of the three subspaces (the translation, orientation, and scale spaces) are uniformly sampled without considering the correlation among parameters in the same marginal space. Recently, we proposed constrained MSL to exploit such correlation, which can further improve the detection speed by an order of magnitude. With the latest improvements, an anatomical structure can be detected in about 0.1 s on a standard personal computer. Please refer to [35] for more details.

Due to the exponential number of hypotheses, full space learning (FSL) does not work on a 3D object detection problem, even after using the coarse-to-fine searching strategy. However, for a 2D object detection problem (estimating five pose parameters), both MSL and FSL methods are applicable. In [36], we performed a thorough comparison experiment on LV detection in MRI images. Experiments show MSL significantly outperforms FSL on both speed and accuracy. For more details, please refer to [36].

10.2.3 Shape Regression Machine (SRM)

Figure 10.7a demonstrates the basic idea of the regression-based medical anatomy detection using the 2D B-mode echocardiogram, which is a 2D image slice of the heart acquired by an ultrasonic imaging device. In particular, we focus on the canonical view of apical four chamber (A4C) acquired using the transthoracic transducer. An A4C echocardiogram contains all four heart chambers, namely LV, right ventricle (RV), left atrium (LA), and right atrium (RA). For illustrative purpose, we address only the translation parameter in P as in Fig. 10.7a. In other words, we are only interested in finding the center position $P_0 = (X_0, Y_0)$ of the LV in an A4C echocardiogram, assuming that the orientation of the LV is upright and the scale/size of the LV is fixed. It is straightforward to extend the 2D case to the 5D-parameterization.

Suppose that, during running time, we randomly sample an image patch $I(P)$ centered at position $P = (X, Y)$. If there exists a function $\mathtt{F}_1$ that does the following: given an image patch $I(P)$ as input, it outputs the difference vector $\mathrm{d}P$ between the current position P and the target position P_0, *i.e.*, $\mathrm{d}P = P_0 - P$, then we achieve the detection using *just one scan*. Mathematically, through the function $\mathtt{F}_1$ that defines a mapping $\mathtt{F}_1 : I \rightarrow \mathrm{d}P$, the ground truth position P_0 is calculated as follows:

$$\mathrm{d}P = \mathtt{F}_1[I(P)], \quad P_0 = P + \mathrm{d}P = P + \mathtt{F}_1[I(P)]. \tag{10.19}$$

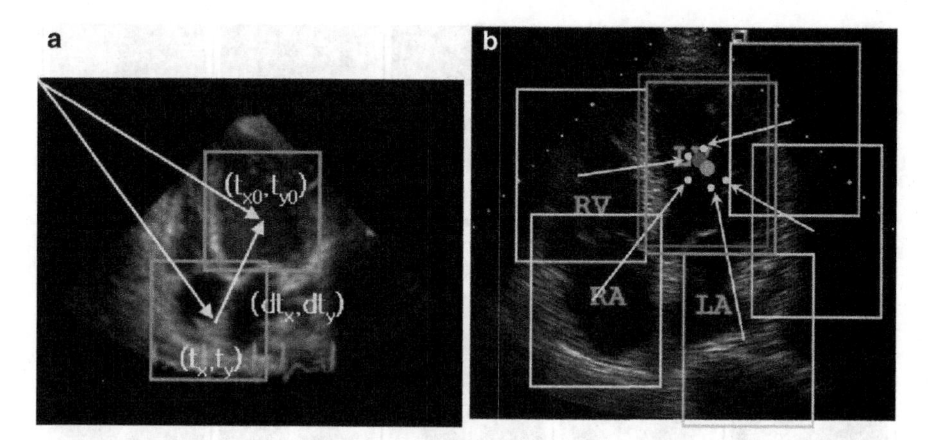

Fig. 10.7 **(a)** A graphical illustration of regression-based detection using a 2D translation parameterization: the learned regressor predicts the difference vector. **(b)** A robust fusion algorithm for regression-based object detection. The white denote scanning boxed and predicted difference vector, the light gray is the fused box, and the dark gray is the ground truth box. Reprinted from [37]. ©2010 Elsevier

Learning the function $F_1[I(P)]$ is referred to as *regression* in machine learning.

From human anatomical atlas, we know that in the A4C echocardiogram there is only one target LV available, whose relation with other anatomies such as LA, RV, and RA is geometrically regularized (that is why they are called left/right ventricle/atrium). Also there exists a strong correlation among their appearances, defining the so-called anatomical image context. By knowing where the LA, RV, or RA is, we can predict the LV position quite accurately. In principle, by knowing the current position (i.e., knowing P) and then looking up the map/atlas that tells the difference from the target (i.e., telling $\mathrm{d}P$ through the function), one can reach the target without exhaustive search.

Medical atlases are widely used in the literature [1, 2, 19, 22, 25]. However, the methods in [1, 2, 19, 22, 25] use the atlases as an *explicit* source of prior knowledge about the location, size, and shape of the anatomic structures and deform it to match the image content for registration, segmentation, tracking, etc. In this paper, we take an *implicit* approach, that is, embedding the atlases in a learning framework. After learning, the atlas knowledge is fully absorbed for the specific task and the atlases are no longed kept.

We collect training data from an annotated database. As in Fig. 10.8 (again using 2D translation for illustration), we form the input–output pairs as training data. By randomly varying the location within a prior range, we crop out different local image patches while recording their corresponding difference vectors. Similarly, we can extract the training data for a 5D parameterization. Note that the number of training pairs is not limited by the number of training images as we can arbitrarily vary the location within the prior range. In theory, we wish to form the training pairs as

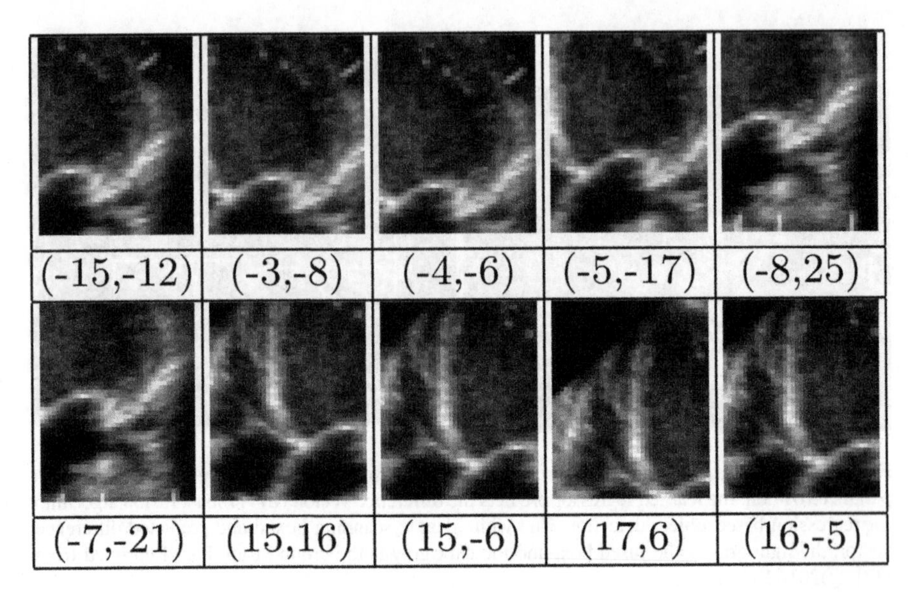

Fig. 10.8 Training image examples (generated based on the image in Fig. 10.7a): image I and its associated difference vector $\mathrm{d}P = (\mathrm{d}X, \mathrm{d}Y)$. Reprinted from [37]. ©2010 Elsevier

many as possible in order to learn a robust regressor; in practice, the actual number of training pairs is limited by the computer storage and memory space.

We now confront a multiple regression setting with a multidimensional output, i.e., the input variable is an image patch, depicted by a multidimensional vector, and the output variable is a multidimensional displacement vector. This regression setting has not been well addressed in the machine learning literature [13]. We leverage the boosting principle to fulfill the learning challenge, which results in the IBRR algorithm. Please refer to [37] for implementation details.

10.2.3.1 Detection Algorithm

In theory, only one scan is needed to find the target; in practice, we conduct a sparse set of random scans and then estimate the parameter using fusion for a robust solution. Suppose that M random samples with parameters $\{P^{\langle 1\rangle}, P^{\langle 2\rangle}, \ldots, P^{\langle M\rangle}\}$ are scanned. For each $P^{\langle m\rangle}$, we invoke the regressor to predict the difference parameter $\mathrm{d}P^{\langle m\rangle}$ and, subsequently, the target parameter $P_0^{\langle m\rangle}$ as follows:

$$P_0^{\langle m\rangle} = P^{\langle m\rangle} + \mathrm{d}P^{\langle m\rangle} = P^{\langle m\rangle} + \mathrm{F}_1\left[I\left(P^{\langle m\rangle}\right)\right], \quad m = 1, 2, \ldots, M. \quad (10.20)$$

A simple fusion strategy is to take the sample mean, assuming that the predicted difference parameters are i.i.d.

$$\hat{P}_0 = \frac{\sum_{m=1}^{M} P_0^{\langle m \rangle}}{M}. \tag{10.21}$$

While taking the sample mean is quite effective, we empirically find that some of the predicted difference parameters are far away from the ground truth, which could compromise the final estimation. A better strategy is to associate the predicted difference parameter with a confidence score that calibrates the goodness of the prediction. Unfortunately, our IBRR algorithm is a black-box approach and currently does not provide such a score.

To address the above, we learn a binary classifier (or detector) $\mathtt{D}$ that separates the object from the background and use its posterior probability of being positive, denoted by p_d, as a confidence score. After finding the mth prediction $P_0^{\langle m \rangle}$, we apply the detector $\mathtt{D}$ to the image patch $I\left(P_0^{\langle m \rangle}\right)$. If the detector $\mathtt{D}$ fails, we discard the mth sample; otherwise, we keep the sample and its confidence score $q_d^{\langle m \rangle}$. This way, we have a weighted set of valid scans $\left\{\left(P_0^{\langle j \rangle}, q_d^{\langle j \rangle}\right); j = 1, 2, \ldots, J\right\}$ (note that $J \leq M$ as samples might be dropped), from which we calculate the weighted mean as the final estimate $\hat{P}_0$,

$$\hat{P}_0 = \frac{\sum_{j=1}^{J} q_d^{\langle j \rangle} P_0^{\langle j \rangle}}{\sum_{j=1}^{J} q_d^{\langle j \rangle}}. \tag{10.22}$$

In practice, we stop scanning when $J \geq J_{\text{valid}}$ in order to further save computation. If there is no sample $P_0^{\langle m \rangle}$ passing $\mathtt{D}$, then we use the unweighted mean of $P_0^{\langle m \rangle}$ as the final estimate as in (10.21). Figure 10.7b illustrates the scanning and fusion processes and Fig. 10.9 summarizes the proposed regression-based detection algorithm.

Combining the regressor and binary detector yields an effective tool for medical anatomy detection. When compared with the method using only the regressor, it needs only a smaller number of scans to reach a better precision. Figure 10.10 demonstrates this improvement using the 2-D translational case. Three images are shown along with their 100 predicted positions (the dots). The majority of the prediction is close to the ground truth (the neighboring point) although outliers exist. Figure 10.10 also shows the predicted points passing the detector: All the outliers are eliminated, thereby significantly reducing the uncertainty of the estimate as evidenced by the smaller region bounded by the 95% confidence curve.

10.2.3.2 Classification-Based Versus Regression-Based Object Detection

A successful object detection approach based on machine learning must harness the learning complexity in its offline learning and the computational complexity in its online inference from a test image.

[Learning the regressor F_1]

- Input: images $\{I_1, I_2, \ldots, I_C\}$ and their according rigid box parameters $\{P_1, P_2, \ldots, P_C\}$.
- Training data generation:
  Loop over $c = 1, 2, \ldots, C$ and $m = 1, 2, \ldots, M$
  - Create a random sample of $P_c^{<m>}$.
  - Crop the image patch $I_c(P^{<m>})$.
  - Record the pair $(I_c(P^{<m>}), dP_c^{<m>} = P_c - P_c^{<m>})$ into the training data set.
- Learning:
  Using the training data set $\{(I_c(P^{<m>}), dP_c^{<m>});\ c = 1, 2, \ldots, C,\ m = 1, 2, \ldots, M\}$, invoke the IBRR algorithm [37] to learn the regressor F_1.

[Learning the binary detector D]

- We follow one of [11, 27, 32].

[Testing]

- Input: test image I. Output: the estimated rigid parameter $\hat{P}_0$.
- Set $J = 0$.
- Loop over $m = 1, 2, \ldots, M$
  - Create a random sample of $P^{<m>}$.
  - Crop the image patch $I(P^{<m>})$.
  - Invoke the regressor F_1 to estimate the difference vector $dP^{<m>} = F_1[I(P^{<m>})]$.
  - Check if the estimated parameter $P_0^{<m>} = P^{<m>} + dP^{<m>}$ is valid by passing the image patch $I(P_0^{<m>})$ to the detector D. If valid, set $J = J + 1$.
  - (optional) If the valid sample iterator $J > J_{valid}$, then break out the iteration; otherwise, continue the iteration.
- Take weighted average using Eq. (22). If there is no valid sample, then take the unweighted average using Eq. (21).

Fig. 10.9 The proposed regression-based object detection algorithm. Reprinted from [37]. ©2010 Elsevier

- *Learning complexity.* In the classification-based approach, the main challenge lies in handling the number of negatives—anything other than positive is negative, apart from the large image appearance variations in positives and negatives. In theory, one image contributes one positive (assuming the single presence of the anatomy) but innumerable negatives. The dominance of negatives poses a significant challenge for learning an effective classifier of good separability between positives and negatives. In the regression-based approach, the challenge is aggravated because we have to associate a real-valued output or vector for each sample, rather than a binary variable in the classification-based approach.
- *Computational complexity in inference.* This is related to the running-time detection speed. In the classification-based approach, brute force exhaustive search is time consuming as its computation is exponential in the dimensionality of the parameter space. In the regression-based approach, the exponential nature of the computation in inference no longer exists. Also, the learned model complexity affects the inference complexity: the more sophisticated the model is,

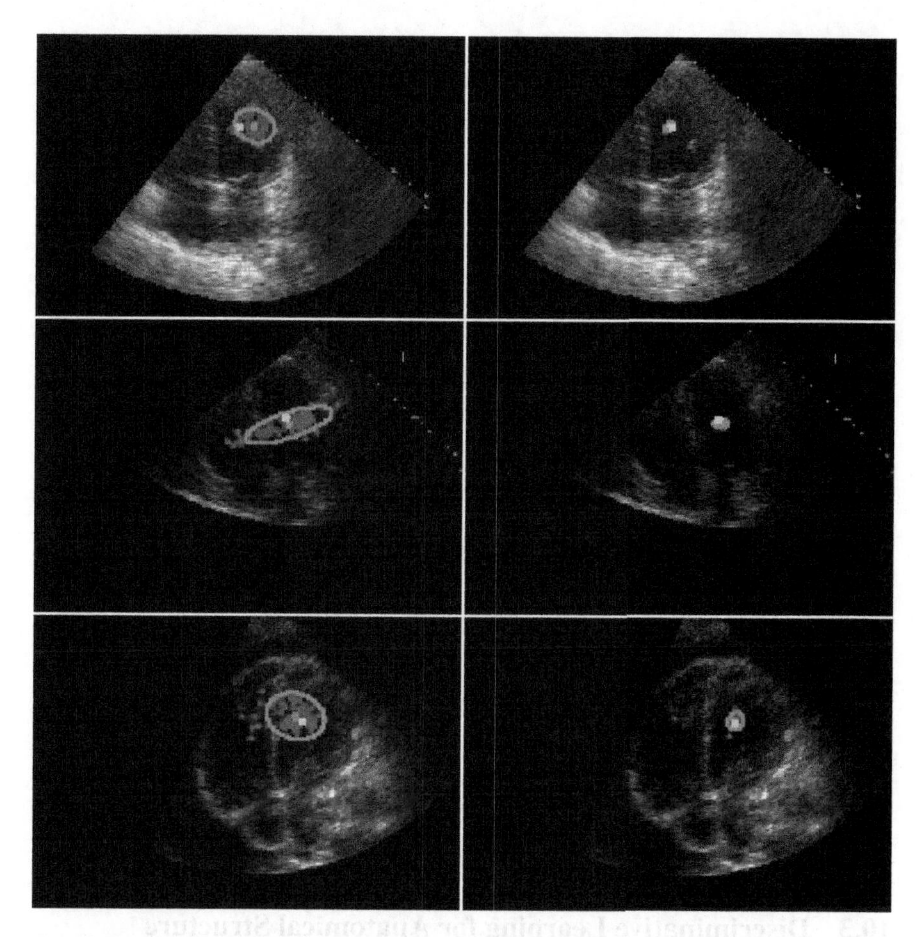

Fig. 10.10 The left row shows the 100 predicted target outputs (the dots) and the right row shows only the predicted target outputs (the dots) passing the detector. The curve is the 95% confidence curve whose center is the final estimate of the target position and the neighboring point indicates the ground truth position. The regions bounded the 95% confidence curves on the images in the top row are significantly smaller than those in the bottom row. Reprinted from [37]. ©2010 Elsevier

the slower is the inference. Although the learned regressor is more complex than the binary detector, its overall computational complexity in inference is much less than that of the binary detector because of the avoidance of exhaustive search.

Clearly, there is a *trade-off* between the learning complexity and computational complexity in inference. The classification-based approach learns a less complex model and runs slower; the regression-based approach learns a more complex model and runs faster. However, for the regression-based detection to work, the image has to possess the anatomical context (or some kind of geometric context). Table 10.1 presents a summary of comparison between the classification-based and regression-based object detection approaches.

Table 10.1 Comparison of the classification- and regression-based object detection approaches. Reprinted from [37]. ©2010 Elsevier

Detection approach	Regression-based	Classification-based
Representative work	This work	Viola and Jones [26]
Where applicable	Medical anatomy detection	Generic object detection
Use of context	Use anatomical context	Use no context information
Number of target objects	Known	Unknown
Learning method	Regression	Binary classification
Learning complexity	High	Low
Inference method	Sparse scanning and sample averaging	Exhaustive scanning and ad hoc grouping
Detection speed	Extremely fast	Fast

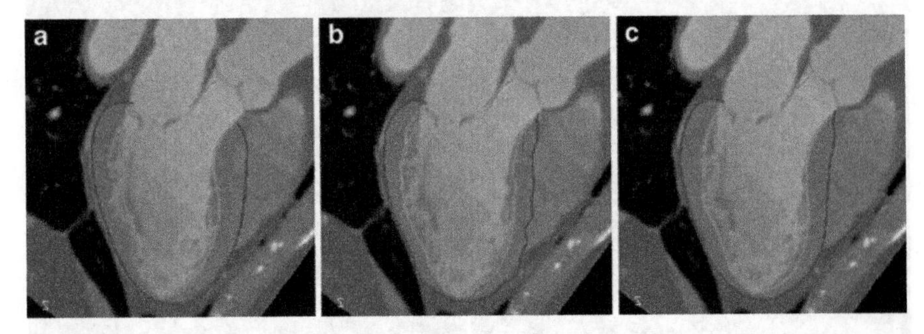

Fig. 10.11 Nonrigid deformation estimation for the left ventricle in cardiac CT with the inner contour for endocardium and the outer contour for epicardium. (**a**) Mean shape. (**b**) After boundary adjustment. (**c**) Final delineation by projecting the adjusted shape onto a shape subspace (50 dimensions). Reprinted from [34]. ©2008 IEEE

10.3 Discriminative Learning for Anatomical Structure Segmentation

In this section, we present three approaches for segmenting anatomical structures. In the first approach of discriminative active shape model (ASM), we replace the edge response that guides the ASM fitting with learned boundary detectors. In the second method, we continue to present the SRM used for shape segmentation this time. Finally, we present a discriminative formulation to learn a shape fitting function.

10.3.1 Discriminative Active Shape Model

After automatic object localization, we align the mean shape with the estimated pose. We then deform the mean shape to fit the object boundary. ASMs are widely used to deform an initial estimate of a nonrigid shape under the guidance of the image evidence and the shape prior. The nonlearning-based generic boundary detector in the original ASM [4] does not work in our application due to the complex background and weak edges. Learning-based methods have been demonstrated to have better performance on 2D images [7, 12, 18] since they can exploit more image evidences to achieve robust boundary detection. In the previous work [7, 18], a detector was trained to detect boundary with a specific orientation (e.g., horizontal boundary). In order to detect boundary with different orientations, we need to perform detection on a set of rotated images.

Here we extend learning-based methods to 3D and completely avoid time-consuming volume rotation using our efficient steerable features. Here, boundary detection is formulated as a classification problem: whether there is a boundary passing point (X, Y, Z) with orientation (ψ, ϕ, θ). This problem is similar to the classification problem we solved for position–orientation estimation. Therefore, the same approach is used to train a boundary detector using the probabilistic boosting-tree (PBT) [23] and steerable features.

Our nonrigid deformation estimation approach is within the ASM framework. The major difference is that we use a learning-based 3D boundary detector, which is more robust under complex background. The trained boundary detector is used to move each mesh point along the mesh surface normal to the optimal position where the estimated boundary probability is maximized. Since more accurate delineation of the shape boundary is desired, this stage is performed on the original high resolution volume. Figure 10.11b shows the adjusted shape of LV in a cardiac CT volume, which follows the boundary well but is not smooth and unnatural shape may be generated. Shape constraint is enforced by projecting the adjusted shape onto a shape subspace to get the final result [4], as shown in Fig. 10.11c.

10.3.2 Shape Regression Machine (SRM)

After the first stage that finds the bounding box (parameterized by $\hat{P}_0$) to contain the object, we have the object rigidly aligned. This solves an initialization problem. For example, one common way is to transform the mean shape into the bounding box as an initial guess. In the second stage, we are interested in inferring the deformable part S.

10.3.2.1 Basic Idea

We formulate the deformable shape inference again as a regression problem. In other words, we seek a function F_2 that tells the shape S based on the image patch $I\left(\hat{P}_0\right)$ that is assumed to contain the object of interest.

$$S = \mathrm{F}_2\left[I(\hat{P}_0)\right]. \tag{10.23}$$

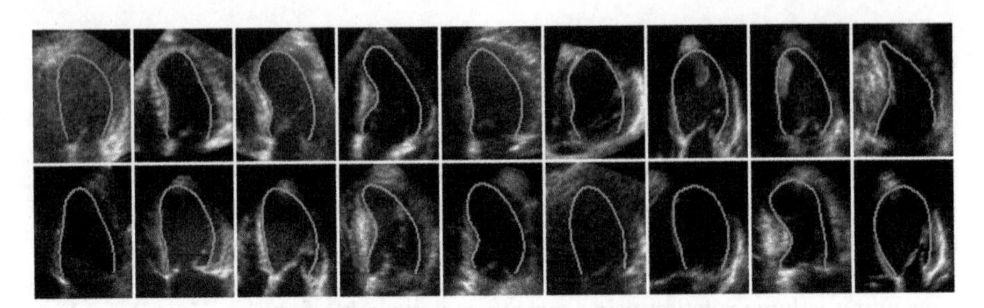

Fig. 10.12 Training image examples: image I and its associated deformable shape S. The normalized image size is 104 by 80. Reprinted from [37]. ©2010 Elsevier

In [5], Covell presented one of the earliest linear models to predict point positions from intensity. A so-called coupled manifold model based on a joint Gaussian distribution was proposed to combine the control point information and the image intensity vector within a neighborhood of the control point. Recently, Cristinacce and Cootes [6] used boosted regression to predict the displacement from the true feature location based on the local neighborhood appearance in the ASM framework. The AAM [3] also uses a linear regression prediction form to some extent. In a recent work of Saragih and Goecke [21], a nonlinear regressor was used to replace the linear parameter updating module in the AAM fitting, which enabled better fitting accuracy.

10.3.2.2 Learning the Regression Function $\mathtt{F}_2$

Because we deal with one particular anatomical structure (say LV) imaged by one particular medical device (say ultrasound), there exists regularity in its appearance and shape although the variations can be quite significant. Figure 10.12 displays several images with corresponding shapes that are rigidly aligned to the mean shape. As mentioned earlier, a linear modeling of the appearance and shape is insufficient. Here we attempt nonlinear modeling.

Given an annotated database, we extract corresponding pairs of (already rigidly aligned) shape and appearance as in Fig. 10.12. We also slightly perturb the rigid parameter to accommodate imperfect localization derived from the first SRM stage. We now again confront a multiple regression setting with a multidimensional output; this time the output cardinality is much higher.

10.3.2.3 Inference Algorithm

To improve robustness, we slightly perturb the estimated bounding box $\hat{P}_0$ to generate L image patches $\{I^{\langle 1\rangle}, I^{\langle 2\rangle}, \dots, I^{\langle L\rangle}\}$ and apply the regressor to obtain shape estimates $\{S^{\langle 1\rangle}, S^{\langle 2\rangle}, \dots, S^{\langle L\rangle}\}$, where

$$S^{\langle l\rangle} = \mathtt{F}_2\left[I^{\langle l\rangle}\right];\ l = 1, 2, \dots, L. \tag{10.24}$$

We also build a nonparameteric kernel density $q_s(S)$ based on the prior shape examples and use it as a confidence score. The density $q_s(S)$ is estimated as

$$q_s(S) = \frac{\sum_{c=1}^{C} h_\sigma(S; S_c)}{C}, \tag{10.25}$$

where $\{S_n;\ c = 1, 2, \ldots, C\}$ is the set of training shapes and $h_\sigma(S; S_c)$ is the radial basis function (RBF) kernel,

$$h_\sigma(S; S_c) = \mathrm{rbf}_\sigma(S; S_c) = \exp\left(-\frac{\|S - S_c\|^2}{2\sigma^2}\right), \tag{10.26}$$

whose parameter σ^2 is set empirically as the sample variance:

$$\sigma^2 = \frac{1}{C}\sum_{c=1}^{C} \|\bar{S} - S_c\|^2\ \bar{S} = \frac{1}{C}\sum_{c=1}^{C} S_c. \tag{10.27}$$

Finally, we output the weighted mean as the final estimate $\hat{S}$ for the shape parameter (we empirically choose $L = 10$):

$$\hat{S} = \frac{\sum_{l=1}^{L} q_s\left(S^{\langle l \rangle}\right)\ S^{\langle l \rangle}}{\sum_{l=1}^{L} q_s\left(S^{\langle l \rangle}\right)}. \tag{10.28}$$

Figure 10.13 summarizes the proposed regression-based shape inference algorithm.

In [37], we used the SRM algorithm to automatically detect and segmentation the LV endocardium in A4C echocardiogram. It takes the SRM approach with the IBRR regression implementation about 120 ms to finish both detection and segmentation. The SRM detection speed is much faster than the brute-force search while yielding almost the same detection precision. The SRM segmentation accuracy outperforms the AAM method and other conventional regression methods such as nonparameteric kernel regression (NPR), linear methods, and their nonlinear kernel variants such as kernel ridge regression (KRR), and support vector regression (SVR).

10.3.3 Discriminative Shape Fitting

Deformable shape segmentation can be considered as searching through a model space for the model that best represents a target shape. In order to measure the fitness of a hypothesis model, fitting functions are built for characterizing the relationship between the model and image appearance. A desired fitting function should differentiate correct models from their background in the model space.

[Learning the regressor F_2]

- Input: images $\{I_1, I_2, \ldots, I_C\}$ and their according rigid box parameters $\{P_1, P_2, \ldots, P_C\}$ and deformable shape parameters $\{S_1, S_2, \ldots, S_C\}$.
- Training data generation:
  Loop over $c = 1, 2, \ldots, C$ and $m = 1, 2, \ldots, M$
  - Create a random small perturbation $dP_c^{<m>}$.
  - Crop the image patch $I_{c,m} = I_c(P_c + dP_c^{<m>})$.
  - Compute the shape parameter $S_{c,m}$ according to the image patch $I_{c,m}$.
  - Record the pair $(I_{c,m}, S_{c,m})$ into the training data set.
- Learning:
  Using the training data set $\{(I_{c,m}, S_{c,m});\ c = 1, 2, \ldots, C,\ m = 1, 2, \ldots, M\}$, invoke the IBRR algorithm in [37] to learn the regressor F_2.

[Learning the nonparametric density p_s]

- Input the training shapes $\{S_c;\ c = 1, 2, \ldots, C\}$.
- Learn the variance σ^2 as in (27).
- Form the nonparametric density as p_s in (26).

[Testing]

- Input: test image I and the estimated rigid box parameter $\hat{P}_0$. Output: the estimated rigid parameter $\hat{S}$.
- Loop over $l = 1, 2, \ldots, L$ (say $L = 10$)
  - Create a random sample of $dP^{<l>}$.
  - Crop the image patch $I^{<l>} = I(\hat{P}_0 + dP^{<l>})$.
  - Invoke the regressor to estimate the shape parameter $S^{<l>} = F_2[I^{<l>}]$.
  - Compute the nonparametric density $q_s(S^{<l>})$ using (26).
- Take weighted average using Eq. (28). (21).

Fig. 10.13 The proposed regression-based deformable shape inference algorithm. Reprinted from [37]. ©2010 Elsevier

In this section, we present a comparative study on how to apply three discriminative learning approaches—classification, regression, and ranking—to learn fitting functions from training images with expert annotations [31]. By using discriminative learning in the model space, the fitting function can be learned in a steerable manner. We discuss how to extend the classification approach from object detection to deformable object segmentation. We also propose a regression-based and a ranking-based approach for learning the fitting functions. The fitting function is trained to produce the highest score around the ground truth solution, and it also possesses a suitable shape to guide optimization algorithms to this solution. To address the high-dimensional learning challenges presented in the learning framework, we apply a multilevel approach to learn all discriminative models.

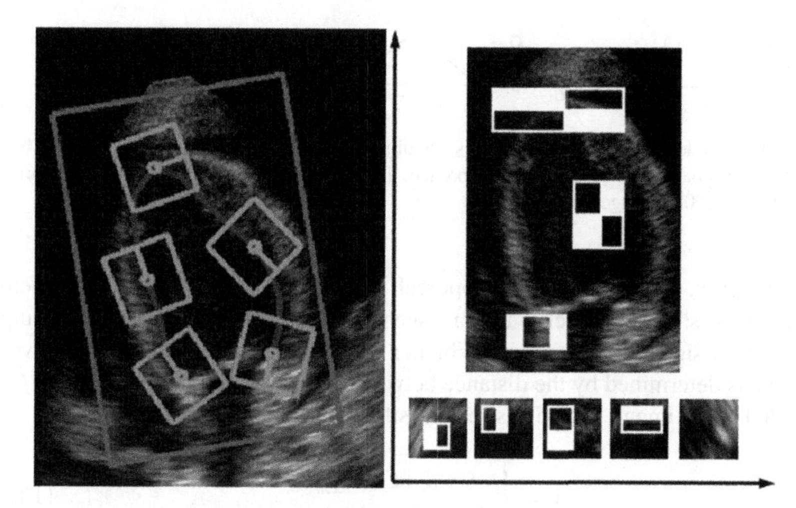

Fig. 10.14 The feature image x associated with a hypothesis model C. The contour represented by the model C is plotted as blue line. The image patch enclosed by the big box contains global fitness information. The image patches enclosed by the small boxes contain the local fitness information. The image x is composed by image patches with normalized orientation as shown on the right. Examples of Haar-like features are also shown in x. Reprinted from [30]. ©2008 IEEE

10.3.3.1 Discriminative Learning Approaches

A shape in an image I can be parameterized by a set of continuous model parameters, C, which contains both rigid and nonrigid components. Given an image, I, and a hypothesis model, C, a feature image, $x(I, C)$, can be extracted to describe the image appearance associated with C. For conciseness, we use x instead of $x(I, C)$ when there is no confusion in the given context.

There are a variety of ways of building shape models and computing feature images [29]. The discriminative learning approaches presented are not bound to a specific shape model. A simple way is to represent a shape by a set of control points and used Point Distribution Model (PDM) [4]. A PDM is built by aligning the training shapes using the generalized Procrustes analysis [4] and applying PCA to the aligned shapes. The model C is defined as $(X, Y, \theta, S_x, \alpha_1, \alpha_2, \ldots)$, including pose parameters (2D translation, rotation, and scale) and shape parameters corresponding to a reduced set of eigenvectors associated with the largest eigenvalues. For a shape C within an image I, the feature image $x(I, C)$ can be extracted efficiently by combining global and local image information, as shown in Fig. 10.14. Then the Haar-like image features can be computed rapidly as features for discriminative fitting functions.

A supervised learning approach attempts to train a fitting function $f(x(I, C))$ based on a set of training images $\{I\}$ and their corresponding ground-truth shape models $\{\underline{C}\}$. The desired output of f is specific to the discriminative approach.

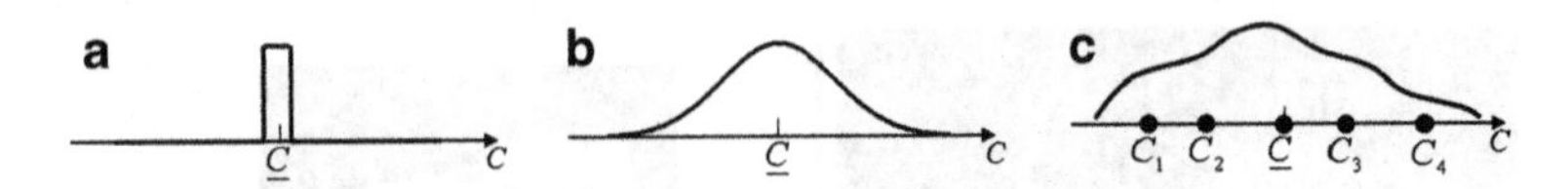

Fig. 10.15 The learned $f(I, C)$ when C is one dimensional: (**a**) a classification approach, (**b**) the regression approach, and (**c**) a ranking approach. The ground truth of the model is $\underline{C}$. Reprinted from [31]. ©2008 Springer

Classification. The classification approach learns a classifier f to indicate whether a hypothesis shape C matches the one seen in image I or not. The desired output y of f is a signed binary value. Whether a feature image $x(I, C)$ is positive or negative is determined by the distance between C and the ground truth model $\underline{C}$ as provided in the image I. The desired classification output is

$$y = \begin{cases} 1 & \text{if } \|C - \underline{C}\| \leq \epsilon \\ -1, & \text{otherwise} \end{cases}, \tag{10.29}$$

where ϵ is a threshold that determines the aperture of f. The learned $f(x(I, C))$ is a boxcar function around the ground truth. Figure 10.15a shows an ideal learned function f when C is one dimensional. Because the learned f only provides binary indication, an exhaustive search is necessary to estimate the solution, which is computationally prohibitive when the dimensionality of the model C is high.

Regression. The regression approach [30] learns a regression function f with real-valued output, which indicates the fitness of a hypothesis model C to an image I. In order to facilitate searching, the desired output y of f is constrained to be a normal distribution:

$$y = \mathcal{N}(C; \underline{C}, \Sigma), \tag{10.30}$$

where Σ is a covariance matrix determining the aperture of f. The learned f has a smooth and unimodal shape, e.g., a 1D example as shown in Fig. 10.15b. The regression function f learned in this way can be effectively optimized by general-purpose local optimization techniques, such as gradient descent or simplex, due to f's single maximum and smoothness. However, when compared with a classification approach, the desired output is more complicated and, hence, more information needs to be learned at the training stage as it requires the regressor to produce a desired real value for each point in the model space. Boosting principle is employed to learn the regressor by selecting relative features to form an additive committee of weak learners. Each weak leaner, based on a Haar-like feature that can be computed rapidly, provides a rough fitness measurement of the object to the image's appearance. The learned regressor computes a robust measurement of fitness by integrating the measurements of selected weak learners. Refer [30] for implementation details.

Ranking. Discriminative learning via ranking was originally proposed to retrieve information based on user preference [8]. We propose a ranking approach for learning a partial ordering of points in the model space [31]. The ordering learned by the ranking function provides the essential information to guide the optimization algorithm at the testing stage. Unlike a regression approach, which forces the regressor to produce an exact value at each point in the model space, ranking only tries to learn relative relations between point pairs in the model space. Let (C_0, C_1) be a pair of points in the model space, and (x_0, x_1) be its associated feature image pair. The ordering of x_0 and x_1 is determined by their shape distance to the ground truth: The one closer to the ground truth has a higher rank. We learn a ranking function f to satisfy the following constraint:

$$\begin{cases} f(x_0) > f(x_1) & \text{if } \|C_0 - \underline{C}\| < \|C_1 - \underline{C}\| \\ f(x_0) < f(x_1) & \text{if } \|C_0 - \underline{C}\| > \|C_1 - \underline{C}\| \\ f(x_0) = f(x_1), & \text{otherwise} \end{cases} \tag{10.31}$$

Figure 10.15c illustrates the basic idea of the ranking approach. There are five points in the 1D model space, and $\underline{C}$ is the ground truth. At the training stage, a ranking function f is learned to satisfy the ordering constrains: $f(x(I,\underline{C})) > f(x(I,C_2))$, $f(x(I,C_2)) > f(x(I,C_1))$, $f(x(I,\underline{C})) > f(x(I,C_3))$, and $f(x(I,C_3)) > f(x(I,C_4))$. Similar to the regression approach, the learned ranking function f is unimodal, which is desirable for local optimization techniques. However, the model of ranking is simpler, in the sense that the amount of information to be learned for ranking is less than the amount for regression. The regression approach learns a full ordering of points in the model space, while the ranking approach only learns a partial, pairwise ordering.

Similar to the boosting-based regression, we employ the boosting principle to learn the ranking function by selecting relative features to form an additive committee of weak learners. Refer [31] for implementation details.

10.3.3.2 Learning in a High-Dimensional Space

The first step toward learning a discriminative function is to sample training examples in the model space. As a result of the curse of dimensionality, the number of training examples should be an exponential function of the model's dimensionality to ensure training quality. This poses a huge challenge to applying discriminative learning to deformable segmentation applications, where the dimensionality of the model space is usually high. Another challenge is the increasing difficulty of discriminating the correct solution from its background when the background points get closer to the solution. In this situation the image appearance of the background points becomes more and more similar to that of the correct solution. As a result of these two challenges, learning a single function across the whole model space to accurately distinguish the optimal solution from its background is ineffective.

We use a multilevel approach [30] to learn a series of discriminative functions $f_k, k = 1, \ldots, K$, each of which focusing on a region that gradually narrows down to the ground truth. Let Ω_k be the focus region of f_k in the model space, which is defined within an ellipsoid centered at the ground truth:

$$\Omega_k = \left\{ C = (c_1, c_2, \ldots, c_Q) \middle| \sum_{q=1}^{Q} \left(c_q - \underline{c}_q \right)^2 / r_{k,q}^2 \leq 1 \right\}, \tag{10.32}$$

where Q is the dimensionality of the model space and $R_k = (r_{k,1}, \ldots, r_{k,Q})$ defines the range of the focus region. The focus regions are designed to have a nested structure gradually shrinking to the ground truth:

$$\Omega_1 \supset \Omega_2 \supset \cdots \supset \Omega_K \supset \underline{\Omega} \ni \underline{C}, \tag{10.33}$$

where Ω_1 defines the initial region of the model parameters. It should be big enough to include all the possible solutions in the model space. The final region $\underline{\Omega}$ defines the desired segmentation accuracy.

In segmentation applications, the initial focus region Ω_1 is highly elongated due to the variation in parameter range. It is desirable to first decrease the range of the parameters with a large initial range. The evolution of the range is designed as:

$$r_{k+1,q} = \begin{cases} r_k^{\max}/\delta & \text{if } r_{k,q} > r_k^{\max}/\delta \\ r_{k,q} & \text{otherwise} \end{cases}, \tag{10.34}$$

where $r_k^{\max}$ is the largest value in R_k and δ is a constant controlling the shrinking rate of focus regions (we empirically set $\delta = 2.9$ for all experiments). Geometrically, the region gradually shrinks from a high-dimensional ellipsoid to a sphere, and then shrinks uniformly thereafter. The top figure in Fig. 10.16 shows the evolution of the focus regions in a 2D example.

At the testing stage, we apply optimization algorithms sequentially to the learned functions to refine the segmentation results. At the kth stage, we want the solution fallen within the region Ω_k to be pushed into the region Ω_{k+1}. In order to achieve this, the learned function f_k should be able to differentiate the instances in the region Ω_{k+1} from those in the region $\Omega_k - \Omega_{k+1}$ and provide effective guidance to the optimization algorithms especially in the region $\Omega_k - \Omega_{k+1}$. Data sampling strategies should be accordingly designed.

Classification. For the classification approach, the learned function f_k should be able to differentiate the instances in the region Ω_{k+1} from those in the region $\Omega_k - \Omega_{k+1}$. In order to achieve this, the positive examples are sampled from the region Ω_{k+1} and the negatives from the region $\Omega_k - \Omega_{k+1}$. Figure 10.16 shows an example of two-dimensional sampling.

Regression. For the regression approach, a gradient sampling is proposed in [30]. The learned regressors provide guidance to optimization algorithms based on the local gradient. Because the regressor f_k has a large gradient in the region $\Omega_k - \Omega_{k+1}$, more training examples are drawn from the region $\Omega_k - \Omega_{k+1}$ to ensure the training quality.

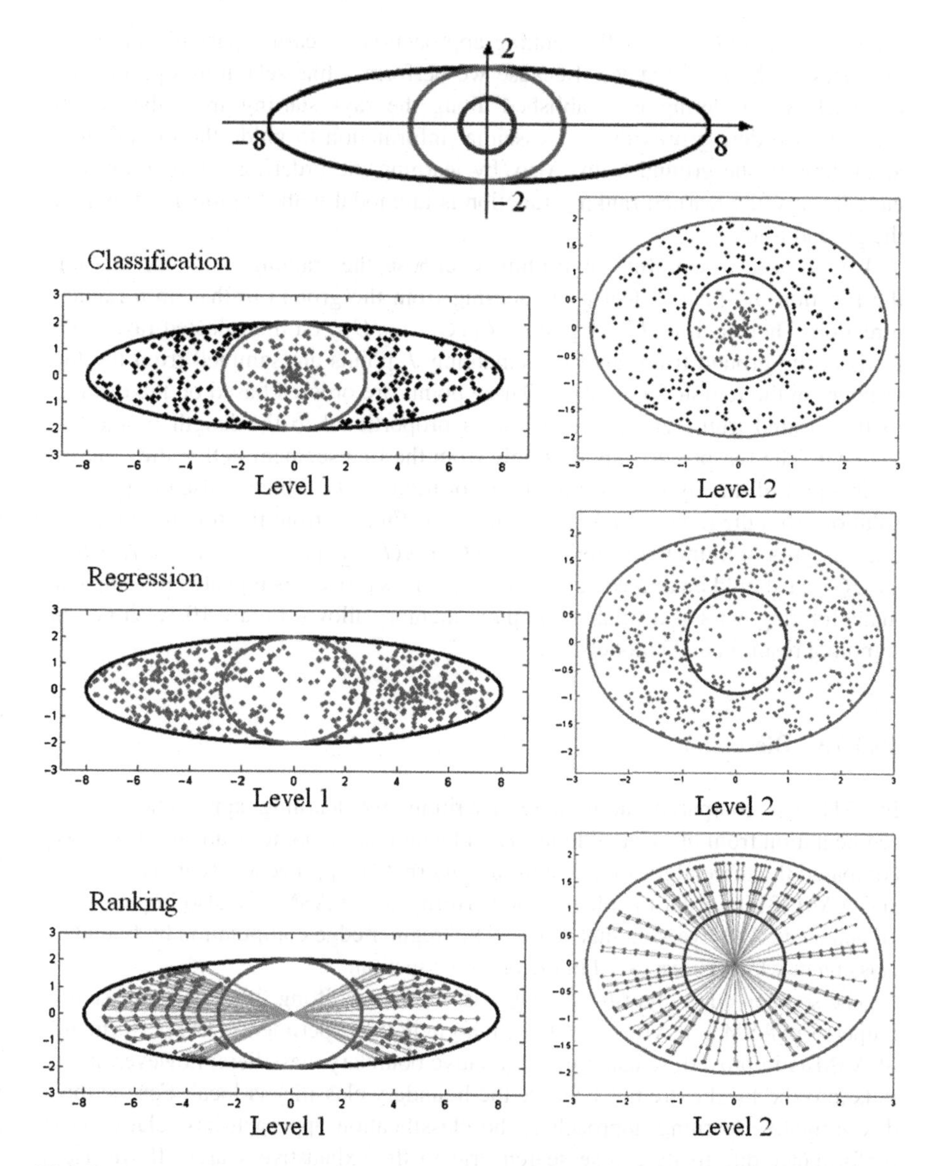

Fig. 10.16 An example of sampling training data in a 2D model space. The first row shows the three nested focus regions defined by R_1 (the large region), R_2 (the medium region), and R_3 (the small region). The second row shows the sampling results of classification, where positive examples are drawn as dots in the small region and negative examples as dots outside the small region. The third row shows the sampling results of regression. The fourth row shows the sampling results of ranking. Reprinted from [31]. ©2008 Springer

Ranking. The objective of the ranking approach is to learn a partial ordering of instances in the model space. Because we perform a line-searching optimization approach, the ordering is established along the rays starting from the ground truth. This ordering provides the essential information to guide the optimization algorithms to the ground truth. Also, by learning the ordering information from enough rays, the learned ranking function is unimodal with its global optimum at the ground truth.

We propose a sampling algorithm to choose the training pairs for learning the function f_k. First, select a ray starting from the ground truth with a random direction. Then sample $J + 1$ points $\{C_0, C_1, \ldots, C_J\}$ on the selected ray, where C_0 is at the ground truth, and the remaining J points are sampled from the line segment in the region $\Omega_k - \Omega_{k+1}$. These points are ordered based on the distance to the ground truth. The parameter J is proportional to the length of the line segment. The reason for sampling only from the line segment is that the ordering on this part of the ray is most important for training f_k, which is used to push the solution from the region $\Omega_k - \Omega_{k+1}$ to Ω_{k+1}. Finally, from the training image I, draw J pairs of training examples $\{(x(I, C_j), x(I, C_{j-1})), j = 1, \ldots, J)\}$, where $x(I, C_{j-1})$ should be ranked above $x(I, C_j)$. This process is repeated to sample as many training pairs as available computer memory allows. Figure 10.16 shows the sampling result in a 2D model space.

10.3.3.3 Discussion

In [31], we compared these three discriminative learning approaches on LV segmentation from ultrasound images and facial feature point localization. We also compared the three algorithms with other alternative approaches, such as ASM [4] and AAM [3]. In order to enhance the performance of ASM, we also implemented an enhanced ASM version that replaces the regular edge computation by boundary classifiers, which is discussed in the previous section.

The segmentation algorithms using discriminative fitting functions consistently outperform ASM and AAM by a large margin in the experiments. The performance of ASM is improved by using discriminative boundary classifiers; however, it still falls into the local extremes because the boundary classifier is local. For the three discriminative learning approaches, the classification approach has relative poor performance due to its coarse search grid in the exhaustive search. If we use a fine search grid, the segmentation accuracy is expected to improve. The ranking approach converges to the correct solution faster than the regression approach, as indicated in the benchmarks of the first-level and the second-level refinement. The main reason might be that ranking only attempts to learn a partial ordering information in the model space, and hence its learning complexity is lower than regression. The learned ranking functions are more effective in guiding the search algorithm to the correct solution.

Like all discriminative learning problems, the discriminative learning approaches suffer from the problem of overfitting, especially when the variation of training data does not cover the full variability. Furthermore, the number of sampled data points is hardly sufficient when the dimension of the model space is high. Because of these problems, the fitting function does not have the desired shape on some test data, and the local optimization algorithm fails to converge to the ground truth.

References

1. Adreasenm, N., Rajarethinam, R., Cizadlo, T., Arndt, S., Swayze II, V., Flashman, L., O'Leary, D., Enrhardt, J., Yuh, W.: Automatic atlas-based volume estimation of human brain regions from MR images. Journal of Computer Assisted Tomography **20**(1), 98–106 (1996)
2. Cootes, T., Beeston, C., Edwards, G., Taylor, C.: A unified framework for atlas matching using active appearance models. In: Proc. Information Processing in Medical Imaging (1999)
3. Cootes, T.F., Edwards, G.J., Taylor, C.J.: Active appearance models. IEEE Trans. Pattern Anal. Machine Intell. **23**(6), 681–685 (2001)
4. Cootes, T.F., Taylor, C.J., Cooper, D.H., Graham, J.: Active shape models—their training and application. Computer Vision and Image Understanding **61**(1), 38–59 (1995)
5. Covell, M.: Eigen-points: Control-point location using principal component analysis. In: International Conference on Automatic Face and Gesture Recognition, pp. 122–127. Killington, USA (1996)
6. Cristinacce, D., Cootes, T.: Boosted regression active shape models. In: Proc. British Machine Vision Conference, vol. 2, pp. 880–889 (2007)
7. Dollár, P., Tu, Z., Belongie, S.: Supervised learning of edges and object boundaries. In: Proc. IEEE Conf. Computer Vision and Pattern Recognition, pp. 1964–1971 (2006)
8. Freund, Y., Iyer, R., Schapire, R., Singer, Y.: An efficient boosting algorithm for combining preferences. J. Machine Learning Research **4(6)**, 933–970 (2004)
9. Freund, Y., Schapire, R.E.: A decision-theoretic generalization of on-line learning and an application to boosting. J. Computer and System Sciences **55**(1), 119–139 (1997)
10. Friedman, J., Hastie, T., Tibbshirani, R.: Additive logistic regression: A statistical view of boosting. The Annals of Statistics **28**(2), 337–407 (2000)
11. Georgescu, B., Zhou, X.S., Comaniciu, D., Gupta, A.: Database-guided segmentation of anatomical structures with complex appearance. In: Proc. IEEE Conf. Computer Vision and Pattern Recognition (2005)
12. van Ginneken, B., Frangi, A.F., Staal, J.J., ter Haar Romeny, B.M., Viergever, M.A.: Active shape model segmentation with optimal features. IEEE Trans. Medical Imaging **21**(8), 924–933 (2002)
13. Hastie, T., Tibshirani, R., Friedman, J.: The Elements of Statistical Learning. Springer (2001)
14. Huang, C., Ai, H., Li, Y., Lao, S.: Vector boosting for rotation invariant multi-view face detection. In: Proc. ICCV (2005)
15. Jones, M., Viola, P.: Fast multi-view face detection. MERL-TR2003-96 (July 2003)
16. Kendall, D., Barden, D., Carne, T., Le, H.: Shape and Shape Theory. Wiley (1999)
17. Li, S., Zhang, Z.: FloatBoost learning and statistical face detection. PAMI **26**, 1112–1123 (2004)
18. Martin, D., Fowlkes, C., Malik, J.: Learning to detect natural image boundaries using local brightness, color and texture cues. IEEE Trans. Pattern Anal. Machine Intell. **26**(5), 530–549 (2004)
19. Mazziotta, J., Toga, A., Evans, A., Lancaster, J., Fox, P.: A probabilistic atlas of the human brain: Theory and rational for its development. Neuroimage **2**, 89–101 (1995)
20. Oren, M., Papageorgiou, C., Sinha, P., Osuna, E., Poggio, T.: Pedestrian detection using wavelet templates. In: Proc. IEEE Conf. Computer Vision and Pattern Recognition, pp. 193–199 (1997)

21. Saragih, J., Goecke, R.: A nonlinear discriminative approach to AAM fitting. In: Proc. Int'l Conf. Computer Vision. Rio de Janerio, Brazil (2007)
22. Thompson, P., Toga, A.: A framework for computational anatomy. Comput Visual Sci **5**, 13–34 (2002)
23. Tu, Z.: Probabilistic boosting-tree: Learning discriminative methods for classification, recognition, and clustering. In: Proc. Int'l Conf. Computer Vision, pp. 1589–1596 (2005)
24. Tu, Z., Zhou, X.S., Barbu, A., Bogoni, L., Comaniciu, D.: Probabilistic 3D polyp detection in CT images: The role of sample alignment. In: Proc. IEEE Conf. Computer Vision and Pattern Recognition, pp. 1544–1551 (2006)
25. Vemuri, B., Ye, J., Chen, Y., Leonard, C.: Image registration via level-set motion: Applications to atlas-based segmentation. Medical Image Analysis **7**, 1–20 (2003)
26. Viola, P., Jones, M.: Rapid object detection using a boosted cascade of simple features. In: Proc. IEEE Conf. Computer Vision and Pattern Recognition, pp. 511–518 (2001)
27. Viola, P., Jones, M.: Robust real-time face detection. Int. J. Computer Vision **57**(2), 137–154 (2004)
28. Wu, B., AI, H., Huang, C., Lao, S.: Fast rotation invariant multi-view face detection based on real AdaBoost. In: Proc. Auto. Face and Gesture Recognition (2004)
29. Xu, C., Pham, D.L., Prince, J.L.: Medical image segmentation using deformable models. Handbook of Medical Imaging – Volume 2:Medical Image Processing and Analysis pp. 129–174 (2000)
30. Zhang, J., Zhou, S., Comaniciu, D., McMillan, L.: Conditional density learning via regression with application to deformable shape segmentation. In: Proc. CVPR (2008)
31. Zhang, J., Zhou, S., Comaniciu, D., McMillan, L.: Discriminative learning for deformable shape segmentation: A comparative study. In: Proc. ECCV (2008)
32. Zhang, J., Zhou, S., McMillan, L., Comaniciu, D.: Joint real-time object detection and pose estimation using probabilistic boosting network. In: Proc. CVPR (2007)
33. Zheng, Y., Barbu, A., Georgescu, B., Scheuering, M., Comaniciu, D.: Fast automatic heart chamber segmentation from 3D CT data using marginal space learning and steerable features. In: Proc. Int'l Conf. Computer Vision (2007)
34. Zheng, Y., Barbu, A., Georgescu, B., Scheuering, M., Comaniciu, D.: Four-chamber heart modeling and automatic segmentation for 3D cardiac CT volumes using marginal space learning and steerable features. IEEE Trans. Medical Imaging **27**(11), 1668–1681 (2008)
35. Zheng, Y., Georgescu, B., Ling, H., Zhou, S.K., Scheuering, M., Comaniciu, D.: Constrained marginal space learning for efficient 3D anatomical structure detection in medical images. In: Proc. IEEE Conf. Computer Vision and Pattern Recognition (2009)
36. Zheng, Y., Lu, X., Georgescu, B., Littmann, A., Mueller, E., Comaniciu, D.: Robust object detection using marginal space learning and ranking-based multi-detector aggregation: Application to automatic left ventricle detection in 2D MRI images. In: Proc. IEEE Conf. Computer Vision and Pattern Recognition (2009)
37. Zhou, S.K.: Shape regression machine and efficient segmentation of left ventricle endocardium from 2D B-mode echocardiogram. Medical Image Analysis **14(4)**, 563–581 (2010)
38. Zhou, S.K., Park, J.H., Georgescu, B., Simopoulos, C., Otsuki, J., Comaniciu, D.: Image-based multiclass boosting and echocardiographic view classification. In: Proc. CVPR (2006)

Chapter 11
Random Forest for Bioinformatics

Yanjun Qi

11.1 Introduction

Modern biology has experienced an increased use of machine learning techniques for large scale and complex biological data analysis. In the area of Bioinformatics, the Random Forest (RF) [6] technique, which includes an ensemble of decision trees and incorporates feature selection and interactions naturally in the learning process, is a popular choice. It is nonparametric, interpretable, efficient, and has high prediction accuracy for many types of data. Recent work in computational biology has seen an increased use of RF, owing to its unique advantages in dealing with small sample size, high-dimensional feature space, and complex data structures.

The aim of this chapter is twofold. First, to provide a review of notable extensions of RF in bioinformatics, whereby promising direction such as RF-based feature selection is discussed. Second, to briefly introduce the applications of RF and its extensions. RF has been applied in a broad spectrum of biological tasks, including, for example, to classify different types of samples using gene expression of microarrays data, to identify disease associated genes from genome wide association studies, to recognize the important elements in protein sequences, or to identify protein–protein interactions (PPIs).

11.2 Random Forest and Extensions in Bioinformatics

Random forest provides a unique combination of prediction accuracy and model interpretability among popular machine learning methods. The random sampling and ensemble strategies utilized in RF enable it to achieve accurate predictions as

Y. Qi (✉)
Machine Learning Department, NEC Labs America, Princeton, NJ, USA

4 Independence Way, Suite 200, Princeton, NJ, USA
e-mail: yanjun@nec-labs.com; qiyanjun07@gmail.com

C. Zhang and Y. Ma (eds.), *Ensemble Machine Learning: Methods and Applications*,
DOI 10.1007/978-1-4419-9326-7_11, © Springer Science+Business Media, LLC 2012

well as better generalizations. This generalization property comes from the bagging scheme which improves the generalization by decreasing variance, while similar methods like boosting achieve this by decreasing bias [47].

Three features of RF receive the main focus [6]:

- It provides accurate predictions on many types of applications;
- It can measure the importance of each feature with model training;
- Pairwise proximity between samples can be measured by the trained model.

Extending random forest is currently a very active research area in the computational biology community, where most previous efforts focused on extending the features above. Several notable techniques among them are briefly introduced in the sections that follow.

11.2.1 Classification Purpose

Random forest retains many benefits of decision trees while achieving better results through the usage of bagging on samples, random subsets of variables, and a majority voting scheme [6]. It handles missing values, a variety of variables (continuous, binary, categorical), and is well suited to high-dimensional data modeling. Unlike classical decision trees, there is no need to prune trees in RF since the ensemble and bootstrapping schemes help RF overcome overfitting issues. Motivated by the excellent performance of RF, developing RF variants is an active research topic in computational biology [47].

One category of extension tried to revise how to construct trees in RF. For instance, Zhang et al. [48] proposed a deterministic procedure to form a forest of classification trees to maintain scientific interpretability in the structure of the trees. The procedure screens trees by selecting a prespecified number, say 20, of top splits of the root node and another prespecified number, say 3, of the top splits of the two daughter nodes of the root node. This protocol of top nodes gives rise to a total of 180 possible ($20 \times 3 \times 3$) trees (Fig. 11.1), among which, those with perfect or near perfect classification precision are of particular interests. Finally, a fixed number of available trees are selected to form a deterministic forest. Their experiments claimed that the deterministic forest performs similar to RFs, but with better reproducibility and interpretability.

Researchers also tried to extend RF by considering special properties in biological data sets, e.g., too many noisy features in DNA microarray data. Amaratunga et al. [2] designed so-called "enriched random forest" for when the number of features is huge and the percentage of truly informative features is small. To reduce the contribution of trees whose nodes are populated by noninformative features, enriched RF used a simple adjustment to choose the eligible subsets at each node by weighted random sampling instead of simple random sampling. When the feature space is huge and the ratio of noisy features is large, the performance of the base classifiers degrades. This is because, almost all eligible features at each node,

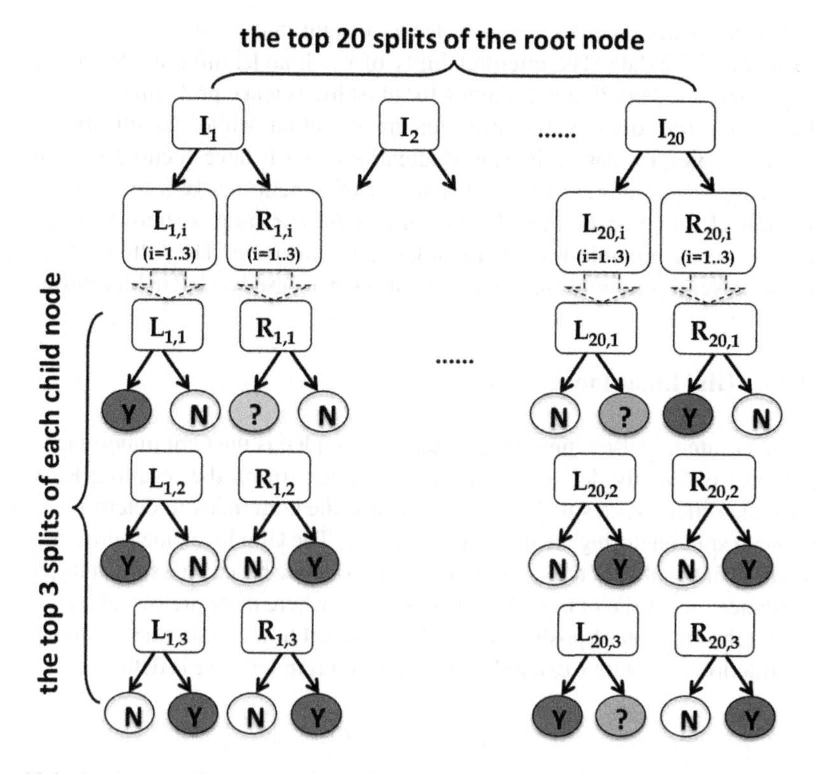

Fig. 11.1 A schematic illustration of the "deterministic forest" method for binary classification (proposed in [48]). $I_1, \ldots, I_{20}$ are the top 20 splits of the root node. Each of these top splits leads to a left (L) and a right (R) child node. The child nodes have their own splits ($L_{j,i}$ and $R_{j,i}$, where $j \in \{1, \ldots, 20\}$ and $i \in \{1, 2, 3\}$). Three top splits are drawn underneath each of them. Based on the combinations of the root splits ($I_{1,\ldots,20}$) and the child splits ($L_{j,i}$ and $R_{j,i}$), the method made multiple trees with different terminal nodes (circles). The terminal nodes are color coded based on the counts of two classes. The more positive examples a terminal node has, the more black it is. Nodes with "?" contain examples from both classes

are predominated by noninformative ones. This issue can be remedied by using weighted, instead of simple, random sampling. By utilizing weights tilted in favor of informative features, the odds of trees containing more informative features being included in the forest increases. Consequently, the resultant enriched RF might contain a higher number of better base classifiers, resulting in a better prediction model.

11.2.2 Measuring Feature Importance

The high-dimensional nature of many tasks in bioinformatics has created urgent needs [37] for feature selection techniques. The goal of feature selection in this field are manifold, where the two most important are: (a) to avoid overfitting and improve model performance, and (b) to gain a deeper insight into the underlying processes that generated the data. The interpretability of machine learning models is treated as important as the prediction accuracy for most life science problems.

Random forest directly performs feature selection while classification rules are built. In bioinformatics, increased attentions of RF have focused on using it for variable selection, e.g., to select a subset of genetic markers relevant for the prediction of a certain disease. Feature importance is used to rank features and there exist many possible ways [11] to define the measure. The following section discusses several commonly used feature importance based on RF in bioinformatics.

11.2.2.1 Gini Importance

The first commonly used importance measure from RF is the Gini importance. Gini importance is directly derived from the Gini index [6] on the resulting RF trees. The RF classifier uses a splitting function called the Gini index to determine which attribute to split on during the tree learning phase. The Gini index measures the level of impurity/inequality of the samples assigned to a node based on a split at its parent. For instance, under the binary classification case, where there are two classes, let p represent the fraction of positive examples assigned to a certain node k and $1 - p$ as the fraction of negative examples. Then, the Gini index at m is defined as:

$$G_k = 2p(1 - p). \tag{11.1}$$

The purer a node is, the smaller the Gini value is. Every time a split of a node is made using a certain feature attribute, the Gini value for the two descendant nodes is less than the parent node. A feature's Gini importance value in a single tree is then defined as the sum of the Gini index reduction (from parent to children) over all nodes in which the specific feature is used to split. The overall importance in the forest is defined as the sum or the average of its importance value among all trees in the forest.

Learning on biological data is often characterized by a large number of features and few available examples. As a simple estimate of the feature importance for the prediction task, RF Gini feature importance is a popular choice used in biological data mining tasks [37]. However, recent reports [41] pointed out that Gini measures are biased in favor of variables taking more categories if predictors are categorical.

11.2.2.2 Permutation Based Variable Importance

RF permutation importance [11] is another important feature ranking measure when using RF for feature selections. Before introducing this concept, the term of "out-of-bag (OOB) samples" need to be explained. RF does not use all training samples when constructing an individual tree. This leaves a set of OOB samples, which could be used to derive the validated classification accuracy from the tree. RF permutation importance is measured by randomly permuting the feature variables and computing the increase in OOB estimate of the accuracy loss. Specifically, to measure a feature k's importance in RF trees, the values of this feature is randomly shuffled in the OOB samples. If we use V_k to describe the difference of the classification accuracy between the intact OOB samples and the OOB samples with the particular feature permutated, RF "permutation importance" [6] for feature k is then defined as the average of V_k over all trees in the forest.

RF permutation importance covers the impact of each variable individually while considering multivariate interactions with other features at the same time. It uses an intuitive permutation strategy, and is utilized more frequently than Gini importance in the general "random forest" literature. However, it is time consuming to compute and its magnitude does not have a bounded value range which can be negative. Similar to Gini importance, RF permutation importance was also shown to unreliable when potential variables vary in their scale of measurement or their number of categories [41].

11.2.2.3 Revised RF Feature Importance

The shortcomings mentioned in above two subsections led to several recent variants of RF feature importance from bioinformatics community. Chen et al. [9] proposed the so-called "depth importance" measure to reflect the quality of the node split which is similar to the Gini importance. The major difference is that the depth importance takes into account the position of the node in the trees. It is claimed to be effective in identifying risk genes responsible for complex diseases.

In another notable work, Strobl et al. [41] proposed a revised RF model based on conditional inference trees [21] (pruned trees using stopping criteria based on multiple test procedures). The revised RF provides unbiased variable selection in each individual classification tree. Using subsampling without replacement, the resultant variable importance was claimed to provide reliable variable selection even when the potential variables vary in their scales or vary in the number of categories.

Later, Strobl et al. [40] pointed out another issue of RF variable importance which shows a bias toward correlated predictor variables. The issue of correlated feature variables happens commonly in high-dimensional bioinformatics tasks, e.g., genomics. This paper [40] developed a conditional permutation scheme which used the partition automatically provided by the fitted model as a conditioning grid. The resulting measure was claimed to reflect the true impact of each predictor (variable) better than the original, marginal approach. Simulation results proved that even though the conditional permutation cannot entirely eliminate the preference of correlated predictor variables, it provides a more fair way of comparison that can help to identify the truly relevant feature variables.

Most RF importance measures reflect the average contributions among all trees in a forest. Recently measures based on extreme statistic in a forest are proposed as well. A good example is the "maximal conditional chi-square importance" from [44]. For a specific feature it is defined as the maximal chi-square statistic among all nodes' splits in a forest. This score was shown to improve the performance of RF when using top-ranked features to refit RF. It was claimed to be more powerful in identifying feature interactions based on simulation studies [44].

More recently, Altmann et al. [1] introduced a heuristic scheme for normalizing feature importance measures that can correct the feature importance bias. The method normalizes the biased RF measure based on a permutation test and returns significance *P*-values for each feature. The repeated permutations are applied on the response vector to preserve the relations between features. The *P*-value of the observed importance provides a corrected measure that addresses the importance bias issue. An improved RF model was then retrained to use top-ranked significant variables with respect to the proposed new importance and was shown to improve the prediction accuracy.

11.2.3 Random Forest Proximity

RF could provide the measure of pairwise proximity between examples using the trained forest. More specifically, for a given forest f and two samples x_i and x_j, the RF similarity is calculated by the following procedure. First, we propagate the value of each sample down all trees within f. Next, the terminal node position for each sample in each of the trees is recorded. Let $z^{(i)} = (z_1^{(i)}, \ldots, z_K^{(i)})$ be these tree node positions for x_i and similarly define $z^{(j)}$ for sample x_j. Then the similarity between x_i and x_j is set to:

$$S(x_i, x_j) = \frac{\sum_{k=1}^{K} I\left(z_k^{(i)} == z_k^{(j)}\right)}{K}, \tag{11.2}$$

where $I(\cdot)$ is the indicator function. As proposed by [6], the sample proximity from RF could be utilized to remove outlier data samples. The noise issue commonly exists in bioinformatics data sets. This strategy has been proved successful in predicting drug response for cell-line gene expression data by removing outlier cell lines in [36].

RF proximity in bioinformatics can also be used for certain classification tasks where the train set provides no negative examples and exhibits a highly skewed distribution between positive and negative classes. For these prediction tasks, relative ranking among predictions normally matter and the cost associated with various classes are different. In order to overcome the issue of problematic training sets and achieve good relative ranking, Qi et al. [35] converted the classification into a ranking task and handled it with a two-step approach using RF proximity. First, it computes a similarity measure between a pair of samples. Then, this measure is used to rank samples by a weighted k-nearest-neighbor (KNN) approach. The proposed method has claimed to work well for the PPI prediction in yeast.

11.3 Bioinformatic Applications of Random Forest and Variants

In the past decade, RFs have been successfully applied to various problems in computational biology. The popularity of RFs in this field arises from the fact that RF can be applied to a wide range of data types, even if the problems are nonlinear or involve complex high-order interaction effects. RF and its variants have been applied on a variety of bioinformatic problems, such as gene expression classification, mass spectrum protein expression analysis, biomarker discovery, sequence annotation, PPI prediction, or statistical genetics. The following survey tries to cover some representative applications.

11.3.1 Analysis of Microarray Gene Expression Data

The advent of DNA microarray technology [37] has enabled researchers to measure the expression levels of large numbers of genes simultaneously. The resultant large-scale data sets have stimulated a large body of research in bioinformatics which also created great challenges for computational techniques. Most microarray gene expression data sets suffer from the commonly known "curse-of-dimensionality" issue where the dimensionality is huge (up to several tens of thousands of genes), and the sample size is small (normally up to hundreds). Moreover, high ratio of noise and variability from microarray experiments raise even more challenges. As shown in Fig. 11.2, computational methods normally treat the microarray data as an $N \times M$ matrix, where N is large, M is small, and $N \ll M$.

One important task in biomedical research is to distinguish disease samples from nondisease samples as well as to classify different disease subtypes [39]. The sample could be a patient, a tissue, or even tissue parts whose features are expressed values of a set of genes or proteins, i.e., the so-called "molecular signature or profile." For using gene expression data to classify disease versus nondisease samples, Lee et al. [26] carried over an extensive study to compare the KNN approach, various versions of linear discriminant analysis (LDA), bagging trees, boosting, and RFs under the same experimental settings. They found that RF was the most successful technique used on the seven microarray data sets they tested.

A closely related popular topic tries to identify a set of biomarkers (normally genes) from gene expression datasets that could maintain high classification accuracy of samples when used alone. Fast and efficient feature selection techniques have attracted lots of attentions since the related data sets are high-dimensional and small. Gene–gene interactions are importance factors to consider when selecting features for disease classification; however, popular univariate selection methods could not take them into account. Thus, researchers have proposed a number of techniques to capture the correlations between genes using RFs based variable importance [2, 15, 40, 46]. Several related methods have been covered in Subsect. 11.2.2. These

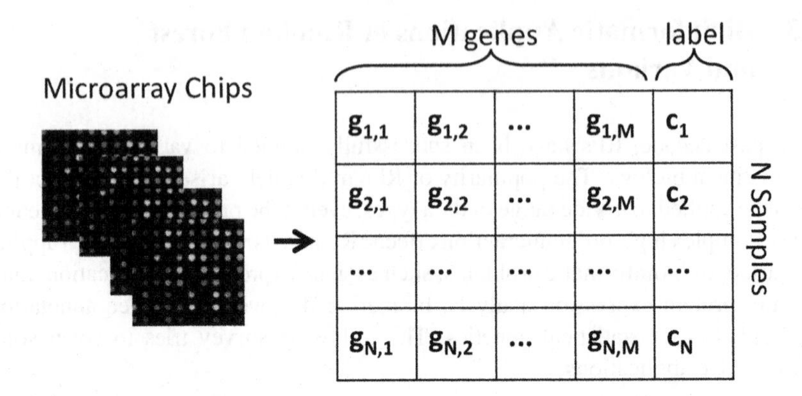

Fig. 11.2 Schematic illustration for gene expression of microarray data. Figure modified from [47]. From the computational perspective, the microarray data is described as an $N \times M$ matrix. Each row describes a sample and each column represents a gene except the last column which means the class label of each sample. $g_{i,j}$ is a numeric value representing the gene expression level of gene j in the ith sample. c_i is the class label of the ith sample [47]

importance measures could be used to filter the original feature set and then the classification model could be retrained which might be a better fit. For instance, the "enriched random forest" method, proposed by Amaratunga et al. [2], claims to improve the RF performance on ten real gene expression data sets by selecting top-ranked features using a weighted random sampling scheme for biomedical sample classification. Diaz-Uriarte et al. [15] showed that RF is able to preserve predictive accuracy while yielding smaller gene sets selected for the analysis of microarray data when compared to LDA, KNN, and SVM.

In summary as an important subfield in bioinformatics, using gene expression microarray has emerged as popular tools to identify common genetic factors that influence health and disease. Random forest methods and its feature importance measures provide the state-of-art performance for analyzing and identifying patients' molecular profiles from gene expression data sets.

11.3.2 Analysis of Mass Spectrometry-Based Proteomics Data

Modern mass spectrometry technologies allow the determination of proteomic fingerprints (e.g., expression levels of many proteins) of body fluids like serum or urine. Differently from DNA microarrays which only relate to genetic (static) factors of diseases, mass spectrum measurements can be used to diagnose the dynamic status or to predict the evolution of a disease. In modern biology, mass spectrometry technology grows to be an attractive framework for cancer diagnosis and protein-based biomarker detection [5].

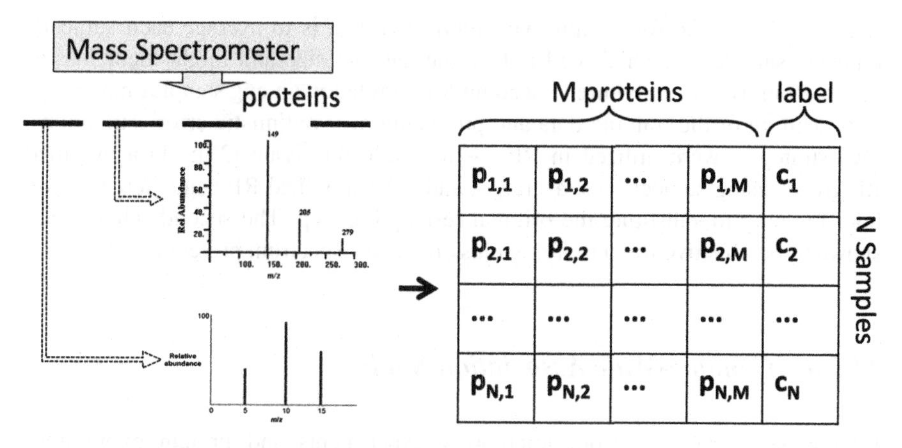

Fig. 11.3 Schematic illustration of mass spectrometry-based proteomics data sets. Figure modified from [47]. The proteomics data generated by mass spectrometer are very similar to gene microarray data in terms of the computational analysis. Differently from microarray data describes the abundance of a protein or peptide in the sample

Figure 11.3 provides a schematic description of mass spectrometry-based proteomics data sets. A typical mass spectrum sample is characterized by thousands of different mass/charge (m/z) ratios on x-axis, and their corresponding signal intensity values are on y-axis. A set of samples' mass spectrum features are treated as a data matrix by computational mining methods. Such mass spectrum data sets are also characterized by a small number of samples and a very high-dimensional feature space. Like DNA microarray data, this "curse-of-dimensionality" issue requires the computational algorithm to select the most relevant features and to make the most use of the limited data samples [47].

Random forest holds a unique position in analyzing mass spectrometry-based proteomics data for clinical classifications [18, 20, 22–24], since it considers feature interactions in learning and is well suited for high-dimensional data samples. For instance, RF has been demonstrated by Izmirlian et al. [22] in classifying SELDI–TOF (surface-enhanced laser desorption/ionization time of flight) proteomic data well with the advantages of robustness to noise and less dependence on tuning parameters. Later, Geurts et al. [18] presented a related tree ensemble approach named "extra trees" [17] which selects at each node the best among K randomly generated splits. Unlike RFs which are grown with multiple sample subsets, the base trees of extra trees are grown from the complete training set and by explicitly randomizing the splits. The approach was successfully validated on two SELDI-TOF data sets for the diagnosis of rheumatoid arthritis and inflammatory bowel diseases.

Recently, Kirchner et al. [24] showed that a RF-based approach is feasible to achieve real-time classification of fractional mass in mass spectrometry experiments. Similarly, Karpievitch et al. [23] proposed a modified RF, named as "RF++" to deal with cluster-correlated data. Many mass spectrometry-based studies produce cluster-correlated data where there exist replicated samples for the same subject.

A common practice for dealing with replicated data is to average each subject's replicate sample set, which will reduce the data set size and might incur loss of information. However, failure to account for correlation among samples may result in overfitting of the training data and producing over optimistic error estimations. Two strategies were utilized in RF++ to tackle this issue [23]: (1) a modified RF grown using subject-level averages, and (2) a modified RF using subject-level bootstrapping to substitute the original resampling step. The second scheme was shown to be effective for classifying clustered mass-spectrum proteomics data.

11.3.3 Genome-Wide Association Study

Like gene expressions from microarray experiments and protein expressions from mass-spectrum based technologies, comparing the genomes (whole DNA sequences) of different samples can also give critical information of different diseases [47]. More importantly, such studies, termed as "genome-wide association study" (GWAS), can help to determine the susceptibility of each different individual to complex diseases, as well as the response to different drugs based on individuals' genetic variations [45].

With the revolutionary advancements of next-generation sequencing technologies, huge volumes of high-throughput sequence data have become easily obtained and extremely cheap. This information has largely enhanced biologists' knowledge of many organisms and also expanded the impact of the genomes on biomedical research. Genomewide association study is becoming increasingly important for clinical decision support with respect to the diagnosis of complex diseases [45].

GWAS computational task involves scanning markers across the complete sets of DNA sequences, or genomes, from many people to find genetic variations associated with a particular disease or a biological symptom. One important concept in GWAS is the so-called "SNPs" (single nucleotide polymorphisms), which is generated from the following procedure. GWAS studies normally compare two groups of samples, (people with or without the disease) by extracting DNA from each person's sample of cells. DNA is then spread on gene chips which could read millions of DNA sequences. Rather than reading the entire DNA sequence, GWAS usually reads the SNPs which are markers indicating the DNA sequence variation at a single nucleotide position. It is estimated that the human genome has approximately seven million of SNPs [25].

To fully understand the basis of complex disease, it is critical to identify the important genetic factors involved, and the complex relationships between these factors. Many complex diseases such as diabetes, asthma, or cancer arise from a combination of multiple genes which often regulate and interact with each other to produce the disease. Therefore, the goal of studying GWASs for these diseases is to identify the complex interactions among multiple SNPs and together with environmental factors which may substantially increase the risk of developing these diseases [45]. This difficult task is commonly formulated into simpler tasks which

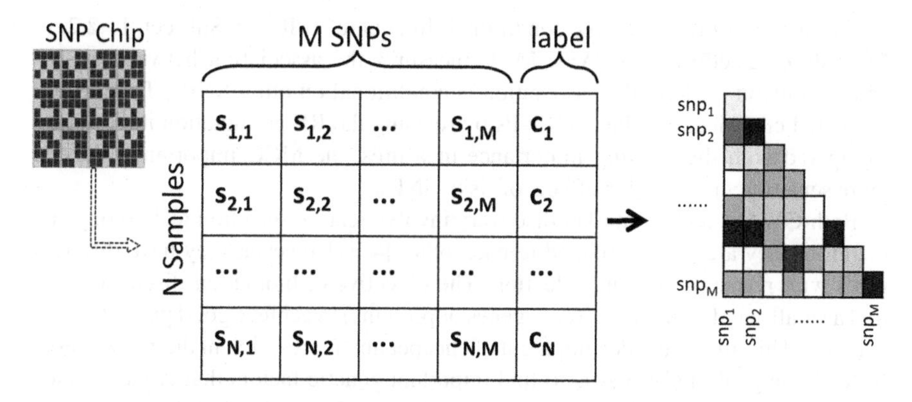

Fig. 11.4 Schematic illustration of pairwise SNP–SNP interaction effects on sample classification. The data matrix obtained from the SNP chip is similar to DNA microarray studies except that each column describes a SNP variable. The pairwise SNP–SNP interactions are schematically illustrated as the gray boxes in the right heat map where darker colors indicating stronger interactions and associations with the disease of interest. Figure modified from [47]

try to identify pairwise SNP–SNP interactions or SNP-environment interactions. Figure 11.4 provides a schematic illustration of pairwise interaction relationship between multiple SNPs. Again, the set of samples (N) and their SNP features (M) could be treated as a data matrix from computational perspective (see Fig. 11.4).

Owing to the intrinsic ability to consider multiple SNPs jointly in a nonlinear fashion [32], RF [6] has become a popular choice of many recent GWAS studies for SNP–SNP interaction identification [3, 4, 9, 30, 45]. Using the feature importance estimated from RF, it is possible to identify important SNP subsets that are associated with the outcome of the disease.

RF is especially useful to identify features that show small marginal contributions individually, but gives a larger effect when combined together. For example, the initial attempt from [28] utilized RF permutation importance (Subsect. 11.2.2.2) as a screening procedure to identify small numbers of risk-associated SNPs among large numbers of unassociated SNPs using 16 complex disease models. RF was concluded to outperform Fisher's exact test when interactions between SNPs exist. Later, a similar study from Bureau et al. [7] used a similar RF importance measure and extended the concept on pairs of predictors, in order to capture joint effects. These early studies normally limited the number of SNPs under analysis to a relatively small range (30).

Recent studies developed feature importance variants from RF to a much larger dimensional range, e.g., several hundred thousands of candidate SNPs. Besides, the issue of correlated variables are also taken into account which commonly exist in GWAS data. Cheng et al. [9] investigated the power of random forests in identifying SNP interaction pairs by proposing the "depth importance" measure (Subsect. 11.2.2.3) from RF trees. It was applied to analyze the complex disease of age-related macular degeneration. Later, Wang et al. [44] proposed an alternative importance measure, "maximal conditional chi-square" (MCC in Subsect. 11.2.2.3), for feature selection in GWASs. MCC measures the association between a SNP and the outcome where the association is conditional on other SNPs. The method estimated empirical P-values of SNPs by revising the RF permutation importance. Compared with the existing importance measures, the MCC importance showed more sensitivity to complex effects of risky SNPs.

Both GWASs and biomarker discovery involve feature selection technology and therefore they are closely related to each other [47]. However, they have different goals with respect to feature selection. The objective of biomarker discovery is to find a small set of biomarkers (e.g., genes or proteins) to achieve good prediction accuracies. This allows the development of cheaper and more efficient diagnostic tests. Instead, the goal in GWASs is to find important genetic factors that are associated with the outcome symptoms and to estimate the significance level of the association.

11.3.4 Protein–Protein Interaction Prediction

Protein–protein interactions are critical for virtually every biological function in the cell. However, experimental determination of pairwise PPIs is a labor-intensive and expensive process. Therefore, predicting PPIs from indirect information is an active field in computational biology. Recently, researchers suggested supervised learning for the task of classifying pairs of proteins as interacting or not. Three independent studies [10, 27, 33] compared the performance of multiple classifiers in predicting protein interactions. In all three studies, RF achieved the best performance on this task when integrating various biological features such as gene expression, gene ontology features, and sequence data. Figure 11.5 shows a schematic illustration of how a RF performs information integrations for the task of classifying pairs of proteins as interacting or not in yeast.

Most of the early studies have been carried out in yeast or in human [34], which aimed to predict protein interactions within a single organism (called "intraspecies PPI prediction"). More recently, researchers extended RF to predicting PPIs between organisms (called "interspecies PPI prediction"), especially between host and pathogens. For instance, Tastan et al. [43] applied the supervised RF classification framework to predict PPIs between HIV-1 viruses and human proteins. By integrating multiple biological information sources, RF defined the state-of-art performance for this task. Figure 11.6 shows a schematic illustration of protein interactions between HIV-1 and human proteins.

11.3.5 Biological Sequence Analysis

Computational analysis of biological sequences is a classic and still expanding subfield in bioinformatics. Biological sequence describes continuous chains of nucleotide acids (DNA) or amino acids (protein) which can be categorized based

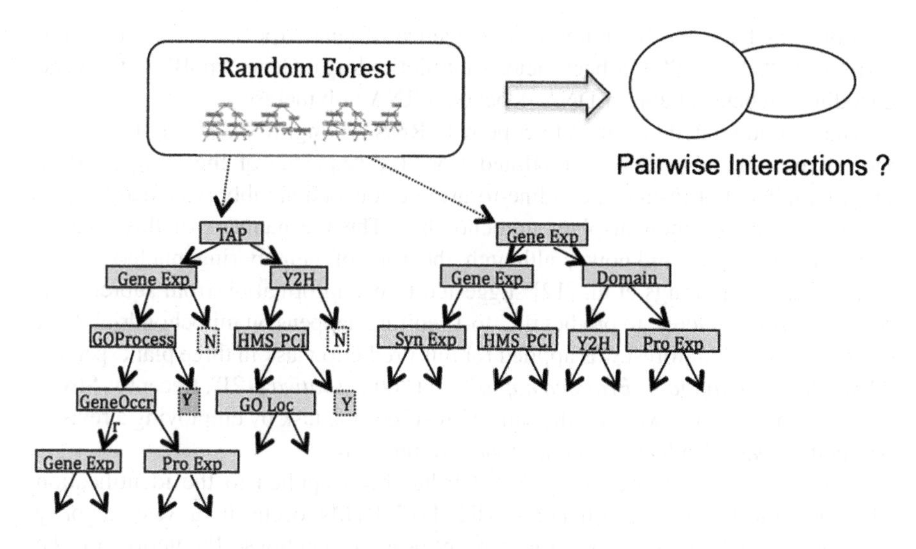

Fig. 11.5 Evidence was integrated using a random forest classifier for protein–protein interaction prediction. Figure modified from [35]

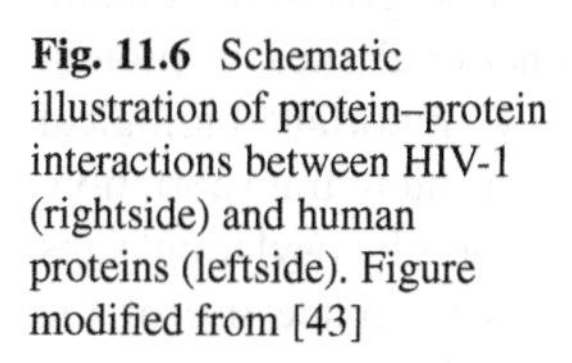

Fig. 11.6 Schematic illustration of protein–protein interactions between HIV-1 (rightside) and human proteins (leftside). Figure modified from [43]

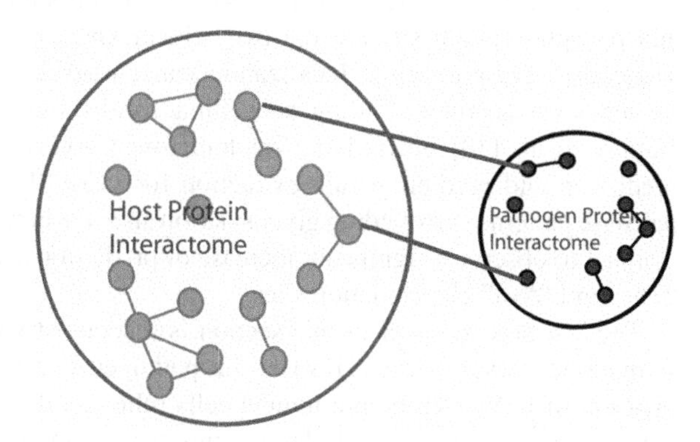

on the underlying molecule type: DNA, RNA, or protein sequence. Since more and more species genomes have been sequenced, this area remains one of the most important in bioinformatics. With biological mutations and evolution, sequence data sets are usually enormous and complex, where efficient and accurate learning models become critical factors [8].

Though there exist enormous biological sequence mining tasks, this section covers only four typical ones where RF achieved good results. All these tasks try to computationally identify the functional properties of subregions (sites) of DNA or protein sequences.

The first type of task is to predict the phenotypes (symptoms) based on protein sequence or DNA sequence. Segal et al. [38] utilized RFs to predict the replication capacity of viruses, such as HIV-1, based on amino acid sequence from reverse transcriptase and protease. Similarly, Cummings et al. [13] used RFs to model the relationships between the amino acid sequence of gene "rpoB" and the rifampin resistance ("rifampin" is a bactericidal antibiotic drug). Gene "rpoB" is the gene encoding the beta subunit of DNA-dependent RNA polymerase.

The second related task tries to cope with RNA editing. RNA editing represents the process whereby RNA is modified from the sequence of the corresponding DNA template. For instance, cytidine-to-uridine conversion (abbreviated as C-to-U conversion) is common in plant mitochondria. The mechanisms of this conversion remain largely unknown, although the role of neighboring nucleotides is emphasized. Cummings et al. [12] suggested to use information from subregions' flanking sites of interest to predict if C-to-U editing happens on mitochondrial RNA sequences. Random forest was applied for this prediction task in three plant species: "*Arabidopsis thaliana*", "*Brassica napus*", and "*Oryza sativa* [12]". Recently, Strobl et al. [41] proposed to work on the same C-to-U editing task by employing a revised RF method based on learning conditional inference trees.

The third typical biosequence task RF has been applied to the identification of "Post translational modifications (PTMs)." PTMs occur in a vast majority of proteins and are essential for certain protein functions. Prediction of the sequence location of PTMs is an important step in understanding the functional characterization of proteins [19]. Among many possible PTMs, glycosylation site and phosphorylation site are the two critical kinds of functional sites in protein sequences. Their accurate localization can elucidate many important biological process such as protein folding, subcellular localization, and protein transportation. Hamby et al. [19] utilized the random forest algorithm for glycosylation sites prediction and prediction rule extraction for yeast. Their work made use of the pairwise patterns surrounding glycosylation sites for better predictions. The authors claimed to observe a significant increase of prediction accuracy in the prediction of "Thr" and "Asn" glycosylation sites.

The last task to cover in this section is associated with HIV-1 viruses. Human Immunodeficiency Virus (HIV) is the pathogen causing the disease AIDS. The invasion of HIV-1 Virus into human cells relies on the contact of its glycoprotein "gp120" with two human cellular proteins, a receptor, and a coreceptor. The type of coreceptor is crucial for the aggressiveness of the virus and the available treatment options. Hence, Dybowski et al. [16] proposed to predict coreceptor usage based on the viral genome sequences. A random forest-based method is developed to predict coreceptor usage for new sequences using structures and sequences of "gp120." The good accuracy achieved in [16] made random forest a strong candidate for computational diagnosis of viral diseases.

11.3.6 Some Other Related Applications

Moreover, RF has been tried on many other biomedical domains. For instance, RF [14] shows to be a powerful statistical classifier in computational ecology. Cutler et al. [14] compared the accuracies of RF and four other commonly used statistical classifiers on three different ecological data sets describing: (1) invasive plant species' presence in US California, (2) the rare lichen species' presence in the US Pacific Northwest, and (3) the nest sites for cavity nesting birds in Utah. RF showed high classification accuracy in all three applications.

Another interesting application is for computational drug screening [29, 36], where panels of cell lines are used to test drug candidates for their ability to inhibit proliferation. Riddick et al. [29] built regression models using RF to predict drug response for 19 Breast Cancer and 7 Glioma cell lines. RF was used in three specific ways: (1) feature selection of drug gene expression signatures based on RF permutation importance, (2) removing outlier cell lines based on RF proximity, and (3) RF multivariate regression model for predicting continuous drug response.

More applications of RFs can be found in other different fields like quantitative structure-activity relationship modeling [42], nuclear magnetic resonance spectroscopy [31], or clinical decision supports in medicine in general [11].

11.4 Summary

With the data explosion in modern biology, machine learning algorithms are becoming increasingly popular. Since the data complexity is always rising, as a nonparametric model, RF provides a unique combination of prediction accuracy and model interpretability. This chapter mainly focused on explaining the notable extensions and applications of RF in bioinformatics. The covered references are by no means an exhaustive list, but are topics which have received much attention. We therefore sincerely apologize to related papers that are not covered in this chapter.

References

1. Altmann, A., Toloşi, L., Sander, O., Lengauer, T.: Permutation importance: a corrected feature importance measure. Bioinformatics **26**(10), 1340 (2010)
2. Amaratunga, D., Cabrera, J., Lee, Y.: Enriched random forests. Bioinformatics **24**(18), 2010 (2008)
3. Bao, L., Zhou, M., Cui, Y.: nssnpanalyzer: identifying disease-associated nonsynonymous single nucleotide polymorphisms. Nucleic Acids Research **33**(suppl 2), W480 (2005)
4. Barenboim, M., Masso, M., Vaisman, I., Jamison, D.: Statistical geometry based prediction of nonsynonymous snp functional effects using random forest and neuro-fuzzy classifiers. Proteins: Structure, Function, and Bioinformatics **71**(4), 1930–1939 (2008)
5. Barrett, J., Cairns, D.: Application of the random forest classification method to peaks detected from mass spectrometric proteomic profiles of cancer patients and controls. Statistical Applications in Genetics and Molecular Biology **7**(2), 4 (2008)
6. Breiman, L.: Random forests. Mach. Learn. **45**, 5–32 (2001). DOI 10.1023/A:1010933404324
7. Bureau, A., Dupuis, J., Falls, K., Lunetta, K.L., Hayward, B., Keith, T.P., Van Eerdewegh, P.: Identifying snps predictive of phenotype using random forests. Genet Epidemiol **28**(2), 171–82 (2005). DOI 10.1002/gepi.20041

8. Chen, X., Jeong, J.: Sequence-based prediction of protein interaction sites with an integrative method. Bioinformatics **25**(5), 585 (2009)
9. Chen, X., Liu, C.T., Zhang, M., Zhang, H.: A forest-based approach to identifying gene and gene–gene interactions. Proc Natl Acad Sci USA **104**(49), 19,199–203 (2007). DOI 10.1073/pnas.0709868104
10. Chen, X., Liu, M.: Prediction of protein–protein interactions using random decision forest framework. Bioinformatics **21**(24), 4394 (2005)
11. Chen, X., Wang, M., Zhang, H.: The use of classification trees for bioinformatics. Wiley Interdisciplinary Reviews: Data Mining and Knowledge Discovery **1**(1), 55–63 (2011)
12. Cummings, M., Myers, D.: Simple statistical models predict c-to-u edited sites in plant mitochondrial rna. BMC Bioinformatics **5**(1), 132 (2004)
13. Cummings, M., Segal, M.: Few amino acid positions in rpob are associated with most of the rifampin resistance in mycobacterium tuberculosis. BMC Bioinformatics **5**(1), 137 (2004)
14. Cutler, D., Edwards Jr, T., Beard, K., Cutler, A., Hess, K., Gibson, J., Lawler, J.: Random forests for classification in ecology. Ecology **88**(11), 2783–2792 (2007)
15. Diaz-Uriarte, R., de Andrés, S.: Variable selection from random forests: application to gene expression data. Arxiv preprint q-bio/0503025 (2005)
16. Dybowski, J.N., Heider, D., Hoffmann, D.: Prediction of co-receptor usage of hiv-1 from genotype. PLoS Comput Biol **6**(4), e1000,743 (2010). DOI 10.1371/journal.pcbi.1000743
17. Geurts, P., Ernst, D., Wehenkel, L.: Extremely randomized trees. Mach. Learn. **63**, 3–42 (2006)
18. Geurts, P., Fillet, M., De Seny, D., Meuwis, M., Malaise, M., Merville, M., Wehenkel, L.: Proteomic mass spectra classification using decision tree based ensemble methods. Bioinformatics **21**(14), 3138 (2005)
19. Hamby, S., Hirst, J.: Prediction of glycosylation sites using random forests. BMC Bioinformatics **9**(1), 500 (2008)
20. Hanselmann, M., Ko the, U., Kirchner, M., Renard, B., Amstalden, E., Glunde, K., Heeren, R., Hamprecht, F.: Toward digital staining using imaging mass spectrometry and random forests. Journal of Proteome Research **8**(7), 3558–3567 (2009)
21. Hothorn, T., Hornik, K., Zeileis, A., Wien, W., Wien, W.: Unbiased recursive partitioning: A conditional inference framework. Journal of Computational and Graphical Statistics **15**(3), 651–674 (2006)
22. Izmirlian, G.: Application of the random forest classification algorithm to a seldi-tof proteomics study in the setting of a cancer prevention trial. Annals of the New York Academy of Sciences **1020**(1), 154–174 (2004)
23. Karpievitch, Y., Hill, E., Leclerc, A., Dabney, A., Almeida, J.: An introspective comparison of random forest-based classifiers for the analysis of cluster-correlated data by way of rf++. PloS one **4**(9), e7087 (2009)
24. Kirchner, M., Timm, W., Fong, P., Wangemann, P., Steen, H.: Non-linear classification for on-the-fly fractional mass filtering and targeted precursor fragmentation in mass spectrometry experiments. Bioinformatics **26**(6), 791 (2010)
25. Kruglyak, L., Nickerson, D.A.: Variation is the spice of life. Nat Genet **27**(3), 234–6 (2001). DOI 10.1038/85776
26. Lee, J., Lee, J., Park, M., Song, S.: An extensive comparison of recent classification tools applied to microarray data. Computational Statistics & Data Analysis **48**(4), 869–885 (2005)
27. Lin, N., Wu, B., Jansen, R., Gerstein, M., Zhao, H.: Information assessment on predicting protein–protein interactions. BMC Bioinformatics **5**(1), 154 (2004)
28. Lunetta, K., Hayward, L., Segal, J., Van Eerdewegh, P.: Screening large-scale association study data: exploiting interactions using random forests. BMC Genetics **5**(1), 32 (2004)
29. Ma, Y., Ding, Z., Qian, Y., Shi, X., Castranova, V., Harner, E., Guo, L.: Predicting cancer drug response by proteomic profiling. Clinical Cancer Research **12**(15), 4583 (2006)
30. Meng, Y., Yu, Y., Cupples, L., Farrer, L., Lunetta, K.: Performance of random forest when snps are in linkage disequilibrium. BMC Bioinformatics **10**(1), 78 (2009)

31. Menze, B., Kelm, B., Masuch, R., Himmelreich, U., Bachert, P., Petrich, W., Hamprecht, F.: A comparison of random forest and its gini importance with standard chemometric methods for the feature selection and classification of spectral data. BMC Bioinformatics **10**(1), 213 (2009)
32. Moore, J., Asselbergs, F., Williams, S.: Bioinformatics challenges for genome-wide association studies. Bioinformatics **26**(4), 445 (2010)
33. Qi, Y., Bar-Joseph, Z., Klein-Seetharaman, J.: Evaluation of different biological data and computational classification methods for use in protein interaction prediction. Proteins: Structure, Function, and Bioinformatics **63**(3), 490–500 (2006)
34. Qi, Y., Dhiman, H., Bhola, N., Budyak, I., Kar, S., Man, D., Dutta, A., Tirupula, K., Carr, B., Grandis, J., et al.: Systematic prediction of human membrane receptor interactions. Proteomics **9**(23), 5243–5255 (2009)
35. Qi, Y., Klein-Seetharaman, J., Bar-Joseph, Z.: Random forest similarity for protein–protein interaction prediction from multiple sources. In: Proceedings of the Pacific Symposium on Biocomputing (2005)
36. Riddick, G., Song, H., Ahn, S., Walling, J., Borges-Rivera, D., Zhang, W., Fine, H.: Predicting in vitro drug sensitivity using random forests. Bioinformatics **27**(2), 220 (2011)
37. Saeys, Y., Inza, I., Larrañaga, P.: A review of feature selection techniques in bioinformatics. Bioinformatics **23**(19), 2507 (2007)
38. Segal, M.R.: Machine learning benchmarks and random forest regression. Technical Report, Center for Bioinformatics & Molecular Biostatistics, University of California, San Francisco (2004)
39. Statnikov, A., Wang, L., Aliferis, C.: A comprehensive comparison of random forests and support vector machines for microarray-based cancer classification. BMC Bioinformatics **9**(1), 319 (2008)
40. Strobl, C., Boulesteix, A., Kneib, T., Augustin, T., Zeileis, A.: Conditional variable importance for random forests. BMC Bioinformatics **9**(1), 307 (2008)
41. Strobl, C., Boulesteix, A., Zeileis, A., Hothorn, T.: Bias in random forest variable importance measures: Illustrations, sources and a solution. BMC Bioinformatics **8**(1), 25 (2007)
42. Svetnik, V., Liaw, A., Tong, C., Culberson, J.C., Sheridan, R.P., Feuston, B.P.: Random forest: a classification and regression tool for compound classification and qsar modeling. J Chem Inf Comput Sci **43**(6), 1947–58 (2003). DOI 10.1021/ci034160g
43. Tastan, O., Qi, Y., Carbonell, J., Klein-Seetharaman, J.: Prediction of interactions between HIV-1 and human proteins by information integration. In: Pac Symp Biocomput, vol. 516 (2009)
44. Wang, M., Chen, X., Zhang, H.: Maximal conditional chi-square importance in random forests. Bioinformatics **26**(6), 831 (2010)
45. Wang, W.Y.S., Barratt, B.J., Clayton, D.G., Todd, J.A.: Genome-wide association studies: theoretical and practical concerns. Nat Rev Genet **6**(2), 109–18 (2005). DOI 10.1038/nrg1522
46. Wu, X., Wu, Z., Li, K.: Identification of differential gene expression for microarray data using recursive random forest. Chin Med J **121**(24), 2492–2496 (2008)
47. Yang, P., Hwa Yang, Y., Zhou, B., Zomaya, Y., et al.: A review of ensemble methods in bioinformatics. Current Bioinformatics **5**(4), 296–308 (2010)
48. Zhang, H., Yu, C., Singer, B.: Cell and tumor classification using gene expression data: construction of forests. Proceedings of the National Academy of Sciences **100**(7), 4168 (2003)

Index

C. Zhang and Y. Ma (eds.), *Ensemble Machine Learning: Methods and Applications*, DOI 10.1007/978-1-4419-9326-7, © Springer Science+Business Media, LLC 2012